# Basic Concepts of
# Vegetable Science

## Second Revised and Enlarged Edition
### A Compendium for JRF, SRF, ARS, NET, SET and PhD

**Neeraj Pratap Singh**

International Panacea Limited
E-34, 2nd Floor, Connaught Circus
New Delhi 110 001 (INDIA)

**CBS**

## CBS Publishers & Distributors Pvt Ltd

New Delhi • Bengaluru • Chennai • Kochi • Kolkata • Mumbai
Bhopal • Bhubaneswar • Hyderabad • Jharkhand • Nagpur • Patna
• Pune • Uttarakhand • Dhaka (Bangladesh) • Kathmandu (Nepal)

Basic Concepts of
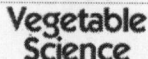

**ISBN:** 978-93-85915-21-5

Copyright © Publisher

**CBS Reprint:** 2016, 2018, 2020

First Edition: 2004

Second Revised and Enlarged Edition: 2005

Reprinted: Feb. 2007

Published by Satish Kumar Jain and produced by Varun Jain for

**CBS Publishers & Distributors** Pvt Ltd

4819/XI Prahlad Street, 24 Ansari Road, Daryaganj, New Delhi 110 002, India.
Ph: 23289259, 23266861, 23266867     Website: www.cbspd.com
Fax: 011-23243014     e-mail: delhi@cbspd.com; cbspubs@airtelmail.in.
*Corporate Office:* 204 FIE, Industrial Area, Patparganj, Delhi 110 092
Ph: 011-4934 4934     Fax: 011-4934 4935     e-mail: publishing@cbspd.com;
publicity@cbspd.com

**Branches**

- **Bengaluru:** Seema House 2975, 17th Cross, K.R. Road, Banasankari 2nd Stage, Bengaluru 560 070, Karnataka
  Ph: +91-80-26771678/79     Fax: +91-80-26771680     e-mail: bangalore@cbspd.com
- **Chennai:** 7, Subbaraya Street, Shenoy Nagar, Chennai 600 030, Tamil Nadu
  Ph: +91-44-26260666, 26208620     Fax: +91-44-42032115     e-mail: chennai@cbspd.com
- **Kochi:** 42/1325, 1326, Power House Road, Opp KSEB Power House, Ernakulam 682 018, Kochi, Kerala
  Ph: +91-484-4059061-65     Fax: +91-484-4059065     e-mail: kochi@cbspd.com
- **Kolkata:** No. 6/B, Ground Floor, Rameswar Shaw Road, Kolkata-700014 (West Bengal), India
  Ph: +91-33-2289-1126, 2289-1127, 2289-1128     e-mail: kolkata@cbspd.com
- **Mumbai:** 83-C, Dr E Moses Road, Worli, Mumbai-400018, Maharashtra
  Ph: +91-22-24902340/41     Fax: +91-22-24902342     e-mail: mumbai@cbspd.com

*Representatives*

| | | | | | |
|---|---|---|---|---|---|
| • Bhopal | 0-8319310552 | • Bhubaneswar | 0-9911037372 | • Hyderabad | 0-9885175004 |
| • Jharkhand | 0-9811541605 | • Nagpur | 0-9421945513 | • Patna | 0-9334159340 |
| • Pune | 0-9623451994 | • Uttarakhand | 0-9716462459 | • Dhaka | 01912-003485 |
| • Kathmandu | 977-9818742655 | | | (Bangladesh) | |
| (Nepal) | | | | | |

*Printed at* Swastik Packaging, Patparganj Industrial Area, Delhi, India

## *Preface to the Second Edition*

I am immensely delighted to place the second edition of this book before the students and the teachers. I am beholden to the readers for overwhelming response to the first edition of the book. It is gratifying to note that the book is serving the purpose for which it was written.

A drastic revision of the book in the light of numerous suggestions from our well wishing teachers, subject matter specialists, students, publisher and other reliable sources became a Herculean task. A thorough analysis of demands of the students had to be made, their genuine needs had to be assessed and the balance had to be struck between the information demanded and that actually required by them. It was necessary to ensure that the reader does not get an overdose and also that he does not suffer from the lack of information. It was kept in mind that he gets just what he needs — neither less nor more.

The text of the book of almost all the chapters has been thoroughly revised and upgraded. Accordingly some out-dated matter has been omitted and much up-to-date information cum chapters have been inducted.

I am sure that continued popular support, as enjoyed by previous edition of this book, would provide me the courage and motivation to undertake the challenge of another revision in near future. At this juncture, I would like to record my gratitude for all the teachers and students who patronize this book and, more particularly, those who kindly offered criticisms and suggestions for its improvement.

I am confident that with these innovations the book will prove to be very useful and will have a wide readership.

मंगलवार                                                  **Neeraj Pratap Singh**
31 August 2004

# मारुतकृपा

## तेरा तुझको अर्पण

अखिल ब्रह्माण्ड में वन्दनीय महाविराट रूप हे वीरवर हनुमान जी! आप ही परम सत्य हैं, सर्वशक्तिमान हैं, सर्वेश्वर हैं, सहज हैं, सुखद हैं, सर्वश्रेष्ठ, सर्वोच्च एवं सर्वोपरि हैं, आप की अनुकम्पा से सब कुछ अनायास ही सुलभ हो जाता है तथा असम्भव भी स्वयं संभव हो जाता है। हे सफल मनोरथ मारुतनन्दन आप हम सब पर कृपा कीजिए। आप की जय हो, जय हो, जय हो ...

The origin of vegetable culture in India could be traced in very ancient times. Around 10000 B.C. in Mesolithic Age the aboriginals Onge tribe of the Andamans as well as the tribals in Madhya Pradesh used to consume yams and moringa. The toys, pendants, earthenware, vases etc. of the Harappan civilisation were found to contain the shape of horticultural produce like melons. Evidence of use of pea is available on record in the Neolithic and Chalcolithic settlements in South India (2295 B.C. to 1300 B.C.). In Vedic Age (1500 B.C. to 1000 B.C.) the Aryans are known to have specially used cucurbits and bottlegourds. During the Mauryan Age (322 B.C. to 232 B.C.) horticultural science had quite developed as is evident from ample attention given to protect the vegetable crops from diseases, see to its nutrient requirements and apply science of irrigation. For the first time in Indian history, Ashoka encouraged the vegetable culture as a state policy (274 B.C. to 237 B.C.). The Portuguese on their arrival in India had introduced several new vegetable crops like potato, sweet potato, arrow-root, cassava, tomato, chillies, pumpkin etc.

In short, it can be summarised that concept of vegetable culture in India has undergone a great evolutionary change with the passage of time. Owing to tremendous development in horticultural science, systematic recommendations are available to make vegetable culture more profitable and enjoyable.

The present book **'Basic Concepts of Vegetable Science'** has been an honest attempt by the author to discuss various fundamental aspects of vegetable science. Written in lucid language with facts, figures and flow-charts etc., the book will provide great assistance to those students who are to compete for JRF, SRF, ARS, NET, SET and UPSC exams and also to undergraduates, post-graduates as well as Ph.D. students of Horticulture (Olericulture) seeking admission to various universities and institutions. The book is also supposed to serve as a reference book for horticultural scientists, subject matter specialists, university teachers, extension workers, progressive farmers and beginners for gaining knowledge before going in for vegetable cultivation.

Needless to say this book is a fine effort by the author to provide maximum knowledge of vegetable science and will have wide readership.

**Neeraj Pratap Singh**

# Contents

# Acronyms

| | |
|---|---|
| **APEDA** | Agricultural and Processed Foods Product Export Development Authority, New Delhi. |
| **AVRDC** | Asian Vegetable Research and Development Centre, Taiwan. |
| **CFB** | Corrugated Fibre Board. |
| **CFTRI** | Central Food Technological Research Institute, Mysore, Karnataka. |
| **CHES** | Central Horticultural Experimental Station. |
| **CIAH** | Central Institute for Arid Horticulture, Bikaner, Rajasthan. |
| **CIHNP** | Central Institute of Horticulture for Northern Plains, Lucknow, Uttar Pradesh. |
| **CIPHET** | Central Institute for Post Harvest Engineering and Technology, Ludhiana. |
| **CISH** | Central Institute for Sub-Tropical Horticulture, Lucknow, Uttar Pradesh. |
| **CITH** | Central Institute for Temperate Horticulture, Srinagar J & K. |
| **CPRI** | Central Potato Research Institute, Shimla, Himachal Pradesh. |
| **CPRS** | Central Potato Research Station, Patna, Bihar. |
| **CRL** | Central Research Laboratory IARI, New Delhi. |
| **CSC** | Central Seed Committee. |
| **CTCRI** | Central Tuber Crops Research Institute Trivandrum, Kerala. |
| **CVRC** | Central Variety Release Committee. |
| **FPTC** | Food Processing and Training Centre. |

| | |
|---|---|
| **HETC** | Horticultural Experiment and Training Centre. |
| **HOPCOMS** | Horticultural Producers Co-operative Marketing Society, Karnataka. |
| **HPMC** | Horticultural Produce Marketing Co-operation, Himachal Pradesh and J&K. |
| **IAHSCO** | Indo-American Hybrid Seeds Company. |
| **IIHR** | Indian Institute of Horticultural Research, Hessarghatta, Bangalore, Karnataka. |
| **IIVR** | Indian Institute of Vegetable Research, Varanasi, Uttar Pradesh. |
| **IPC** | International Potato Centre. |
| **IPGRI** | International Plant Genetic Resources Institute, Rome. |
| **ISHS** | International Society for Horticultural Science, Belgium. |
| **KRIBHCO** | Krishak Bharti Co-operative Ltd. |
| **MAHYCO** | Maharashtra Hybrid Seeds Company Ltd., Jalna. |
| **NBPGR** | National Bureau of Plant Genetic Resources, New Delhi. |
| **NHB** | National Horticultural Board, Gurgaon, Haryana. |
| **NOVDB** | National Oilseed and Vegetable Development Board |
| **NSC** | National Seeds Corporation. |
| **NVT** | National Varietal Trials. |
| **PBCS** | Potato Breeding and Certification Station, Kufri, Shimla, Himachal Pradesh. |
| **SSDC** | State Seed Development Corporation. |
| **SVRC** | State Variety Release Committee. |
| **UVT** | Uniform Varietal Trials. |

| | Vegetable Crops | Centres |
|---|---|---|
| | **Project Directorate on Vegetables** | |
| 1. | Amaranth | NBPGR (New Delhi) and Coimbatore |
| 2. | Brinjal | Hessarghatta, Bhubaneswar, Kalyani and Sabour |
| 3. | Capsicum | Katrain and Solan |
| 4. | Cabbage | Katrain and Kalpa (H.P.) |
| 5. | Carrot | Hissar (Tropical) and Srinagar (Temperate) |
| 6. | Cauliflower (Early) | Sabour and Faizabad |
| | Cauliflower (Mid-Season) | Pantnagar, Sabour and IARI, New Delhi |
| | Cauliflower (Late) | Katrain and Solan |
| 7. | Chillies | Coimbatore and Lam |
| 8. | Cucumber | Kalyani and Rahuri |
| 9. | Dolichos bean | NBPGR (New Delhi) Jabalpur and Kalyanpur |
| 10. | French bean | Hessarghatta, Rahuri and Solan |
| 11. | Garlic | NBPGR (New Delhi), AADF (Nasik) and Solan |
| 12. | Muskmelon | Durgapura, Faizabad and Ludhiana |
| 13. | Okra | NBPGR (New Delhi), Hessarghatta and Bhubaneswar |

| 14. | Onion | NBPGR (New Delhi), Rahuri and AADF (Nasik) |
|-----|-------|---------------------------------------------|
| 15. | Pointed gourd | Sabour, Kalyani and Faizabad |
| 16. | Peas | IARI (New Delhi), Jabalpur, Kalyanpur and Ludhiana |
| 17. | Pumpkin | Hessarghatta and Faizabad |
| 18. | Tomato | NBPGR (New Delhi), Hessarghatta, Ludhiana and Kalyanpur |
| 19. | Watermelon | Durgapura and Faizabad |

## New promising vegetable varieties

| Variety | Vegetable crops | Remarks |
|---------|-----------------|---------|
| Pusa Divya | Tomato | Hybrid |
| Phule Surekha | French bean | Photo-thermo-insensitive. Resistant to anthracnose, leaf crinkle; bean YVMV and wilt. |
| Pusa Sneha | Sponge gourd | Grown both in spring-summer and *kharif.* |
| Pusa Uday | Cucumber | High yielding improved variety. |
| MHY-3 | Muskmelon | Open pollinated variety |
| Palam Hriday | Radish | Pink fleshed and rich in Vit. C. |
| Durgapur Lal | Water melon | Good keeping quality and easy transportability. |
| Pusa Sadabahar | Tomato | Ideal for distant marketing. |

## Important Vegetable Research Stations in India

### ANDHRA PRADESH

- Agriculture Research Station, Lam., Guntoor.
- Demonstration Agricultural Station, Araku Valley, Vishakhapattam.
- Government Main Agricultural Station, Himayatsagar, Hyderabad.
- Government Model Orchard-cum-Nursery, Giddalpur Kurnool.
- Government Fruit Farm, Ramapachodavaram, Rajamundry.
- Government Fruit and Vegetable Farm, Sangareddy, Medak.

### ASSAM

- Agriculture Research Laboratory, Jorhat., Sibsagar.
- Ginger Research Station, Nayabunglow, Khasi Hills.
- Horticultural-cum-Composite Research Station, Dergaon Sibsagar.

### BENGAL

- Agriculture Research Institute, Tollygani.
- Horticultural Research Station, Krishanagar, Nadit
- Potato Research Station, Bhanjang, Darjeeling.

### BIHAR

- Agriculture Research Institute, Pusa, Darbhanga.
- Agriculture Research Institute, Patna.
- Potato Breeding and Research Station, Patna.

### MAHARASHTRA

- Agriculture Research Station, Karad, North Satara.
- Agriculture Research Station, Niphad, Nasik.

- Agriculture College Vegetable Farm, Poona.

## HARYANA

- Bassca Research Station, Gurgoan.
- Agriculture Station, Hansi, Hissar.

## HIMACHAL PRADESH

- Potato Development Station, Shilaroo, Mahasu.
- Potato Development Station, Ahla, Chamba.
- Potato Research Station Kufri, Shimla.
- Vegetable Sub-Station, Katrain, Kullu Valley.
- Agriculture Farm, Parala, Mahasu.

## JAMMU & KASHMIR

- Provincial Experimental Farm, Shalimar, Shrinagar.
- Provincial Experimental Farm, Golsamandar, Shrinagar.

## KARNATAKA

- Government Agricultural Farm, Ponnampet.
- Regional Farm, Yelavare, Harnahatti, Ariskere Taluk.
- Tungabhadra Agricultural Research Station, Siruguppa, Bellary.

## KERALA

- Agriculture Research Station, Ambalavayal, Cannanore.
- Agriculture Research Station, Taliparamba, Cannanore.
- Agriculture Research Station, Pattambi, Palghat
- Pepper Research Station, Panniyur.
- Tapioca and Allied Crops Research Station, Trivendrum.

## MADHYA PRADESH

- Government Experiment Farm, Chhindwara.
- Government Research Farm, Kuthalia, Rewa.
- Government Seed and Demonstration Farm, Waraseoni, Balaghat.

- Government Seed cum Demonstration Farm, Kaliprada, Kararia, Bhopal.
- Horticultural Research Station, Pachmadi, Hosangabad.

## MADRAS

- Agriculture Research Station, Aduturai, Tanjore.
- College Orchard Farm, Coimbatore.
- Pomological and Vegetable Station, Nilgiris.

## MANIPUR

- Government Agricultural Farm, Imphal.

## ORISSA

- Agriculture Research Station, Sambhalpur.
- Khurda Farm, Khurda, Puri.
- Khurda Farm, Khurda, Cuttack.
- State Research Station, Bhubaneswar, Puri.
- Sukinda Farm, Sukinda, Cuttack.
- Turmeric Research Station, Gudayagiri, Boudh Phubani.

## PUNJAB

- Tuber Crops Research Station, Jalandhar.
- Vegetable Research Station, Jalandhar.
- Bhupendra Agriculture Farm, Raeeni, Patiala.

## RAJASTHAN

- Plant Breeding and Vegetable Research Farm, Durgapura, Jaipur.

## UTTAR PRADESH

- Vegetable Farm Kalyanpur, Kanpur.
- Potato Research Farm, Farrukhabad.
- Horticultural Research Station, Saharanpur.

## UTTARANCHAL

- Potato Research Station, Kausambi, Almora.

## Area, Production of Vegetables (2002)

| | Area (mha) | Production (mt) | Yield (t/ha) |
|---|---|---|---|
| Asia | 33.31 | 571.7 | 17.2 |
| India | 5.73 | 78.2 | 13.6 |
| World | 46.96 | 787.4 | 16.76 |
| **Tomato** | | | |
| Asia | 2.24 | 53.3 | 23.81 |
| India | 0.52 | 7.4 | 14.27 |
| World | 3.99 | 108.5 | 27.20 |
| **Cucumber + Gherkins** | | | |
| Asia | 1.65 | 30.12 | 18.28 |
| India | 0.02 | 0.12 | 6.67 |
| World | 2.01 | 36.40 | 18.09 |
| **Okra** | | | |
| Asia | 0.41 | 3.82 | 9.23 |
| India | 0.36 | 3.50 | 9.72 |
| World | 0.79 | 4.90 | 6.20 |
| **Onion** | | | |
| Asia | 2.01 | 33.45 | 166.1 |
| India | 0.52 | 6.50 | 12.5 |
| World | 2.97 | 51.91 | 174.7 |

# ONE

## Vegetables – Surmountable Challenges

The basic challenge before India is to enhance the production of nutritious food in a sustainable manner which, besides feeding the country's large population, also increases the income of farmers, giving them economic security. Commercializing vegetable production can play a major role in meeting the shortage of food, particularly when many more mouths are required to be fed from the limited land resources. Vegetables being short duration crops, can give six to ten times more yield than any cereal crop in a year and thus, can provide better protection to the rising population.

### Vegetables for health and nutrition security

We eat food for sustenance of our body. The food we eat contains nourishing substances called *nutrients*. There are five major nutrients, namely carbohydrates, proteins, fats, minerals and vitamins. Vegetables are rich reservoir of these nutrients, particularly vitamins and minerals.

### Sources of nutrients

| Major nutrients | Main sources |
| --- | --- |
| Carbohydrates | Tapioca (Cassava), Sweet potato, Potato, Colocasia (Taro), Elephant's foot |
| Proteins | Agathi, Drumstick leaves, Colocasia leaves, Peas and Beans |
| Vitamins | |
| Vitamin A (Retinol) | Leafy vegetables, Carrot, Sweet potato, Pumpkin, Colocasia, Turnip green, Knol-khol leaves. |

| Major nutrients | Main sources |
| --- | --- |
| Vitamin $B_1$ (Thiamine) Vitamin $B_2$ or G (Riboflavin) | Lettuce, Cabbage, Carrot, Onion 'Greens' |
| Vitamin $B_5$ (Niacin) | Kale, Peas, Peppers, Potatoes, Spinach, Tomatoes |
| Vitamin C (Ascorbic acid) | Cauliflower, Cabbage, Knol-khol, Turnip, Tomato, Pepper, Drumstick leaves, Fenugreek leaves, Amaranth |
| Vitamin D (Calciferol) | 'Greens' |
| Vitamin E (Tocoferol) | Greens like Cabbage, Lettuce, Vegetable oils |
| Vitamin K (Quinone) | Green leafy vegetables |
| **Minerals** | |
| Calcium | Beans, Carrot, Cabbage, Cauliflower, Lettuce, Onions, Spinach, Peas, Tomato, Amaranth, Fenugreek |
| Phosphorus | Potato, Carrot, Tomato, Cucumber, Spinach, Cauliflower, Lettuce |
| Iron | Spinach, Lettuce, Cabbage, Peas, Beans, Tomato, Carrot, Bitter gourd, Onion |
| Iodine | Onion, Okra, Summer squash, Asparagus |

## Balanced diet

Vegetables play an important role in the balanced diet by providing not only energy but also supplying vital protective nutrients like minerals and vitamins. The scientists of the

Indian Council of Medical Research (ICMR) have recommended a balanced diet pattern for an average man's daily needs as given below :

## Daily requirement of an average man

| | | |
|---|---|---|
| Calories | - | 2800 |
| Protein | - | 55 g |
| Calcium | - | 450 mg |
| Iron | - | 20 mg |
| ß-carotene | - | 3000 µg |
| Vitamin C | - | 50 mg |
| Folic acid | - | 100 µg |
| Vitamin B$_{12}$ | - | 1.0 µg |
| Thiamine | - | 1.4 mg |
| Riboflavin | - | 1.5 mg |
| Niacin | - | 19 mg |
| Vitamin D | - | 5 µg |

According to the recommendation made by the ICMR, an average man with vegetarian or non-vegetarian food habbit should consume 125 g of green leafy vegetables, 100 g of roots and tubers and 75 g of other vegetables. The recommendation for an average woman is more or less same with the exception that roots and tubers should be consumed by them @ 75 g per day.

## Flavour compounds

Vegetables increase attractiveness and palatability by providing sensory appeal through their variety of colours and flavours. They contain some volatile flavouring compounds in the intact tissues while some are produced by the action of enzymes when the tissues are cut or crushed. The characteristic flavour of cole crops is due to the volatile sulphur compound dimethyl trisulphide. Similarly, the pungency and flavour of onion is due to a volatile sulphur compound allyl propyl disulphide. The garlic flavour is caused by diallyl disulphide.

## Toxic substances

In some vegetable crops, toxic substances are found. These toxic substances have adverse effects on human health. These anti-nutritional factors reduce bioavailability of the nutrients present in the vegetable crop.

## Toxic substances found in different vegetable crops

| Toxic substance | Vegetable crops |
| --- | --- |
| Apiin | Celery |
| Calcium oxalate crystals | Colocasia, Elephant's foot |
| CN glycocides, Linamarin | Cassava |
| Cucurbitacins | Cucurbits |
| Dioscorine | Yams |
| Haemagglutinine | French bean |
| Oxalic acid | Amaranth, *Portulaca, Celosia, Basella* |
| Saponine | Spinach, Tomato |
| Sinigrin | genus *Brassica* |
| Solanine | Potato |
| Solasodine | Brinjal |
| Tomatine | genus *Lycopersicon* and *Solanum* |
| Trypsin inhibitors, Phytic acid, Anti-vitamin E factor | Peas and Beans |

## Vegetables in disease prevention

### Antioxidants

These are substances which neutralise free radicals that are formed during metabolism of foods, or by smoking, or exposure to pollutants. Free radicals are neutralised by vitamins (A, $B_1$, $B_5$, $B_6$, Niacin, C and E), minerals like selenium and amino acids; some of which are present in fairly higher amounts in some vegetables. For example, most leafy vegetables, carrot, sweet potatoes, pumpkin and turnip green are rich in betacarotene (pro vitamin A); beans and peas are rich in essential amino acids.

### Bioflavonoids

Bioflavonoids increase the efficiency of vitamin C and protect the body from free radicals. One such bioflavonoid identified in onion and garlic is quercetin, which has been indicated to give protection against cancer and heart diseases.

### Other compounds

Onion and garlic contain several sulphur compounds such as allicin, allistalin, garlicin, diallyl disulphide, diallyl trisulphide and allyl propyl disulphide which are effective in reducing harmful blood cholesterol, thus preventing coronary thrombosis, heart attack and stroke. Cole crops contain indoles and dithiolthiones which are effective in prevention of cancers of the colon, rectum and breast. A hypoglycaemic ingredient, cheratin, has been isolated from the fruits of bitter gourd, which is effective against diabetes.

Diphenyl amine found in onion is also effective against diabetes. Leguminous vegetables reduce blood cholesterol concentration, thus preventing heart attack and stroke. Celery contains 3-n-butyl pthalide which is effective against hypertension. Yam contains a substance called diosgenin which is used in the manufacture of cortisone and contraceptive drugs.

## Production scenario

India is next only to China in area and production of vegetables. According to the latest estimates, vegetable crops in India occupy only 2.8% of the total cultivated land, producing 87.5 million tonnes of vegetables annually, including potato, from a cropped area of 6 m ha. India accounts for 13.38% of the world production of vegetables with a productivity of 14.9 tonnes/ha.

**Area, production and productivity of major vegetable crops of India (1998-99)**

| Vegetable | Area (Lakh ha) | Production (Lakh tonnes) | Productivity (Tonnes/ha) |
|---|---|---|---|
| Brinjal | 4.96 | 78.81 | 15.90 |
| Cabbag | 2.40 | 56.24 | 23.40 |
| Cauliflower | 2.55 | 46.91 | 18.40 |
| Okra | 3.26 | 33.80 | 10.40 |
| Onion | 4.81 | 54.61 | 11.40 |
| Peas | 2.82 | 27.04 | 9.60 |
| Potato | 12.80 | 222.94 | 17.60 |
| Tomato | 4.66 | 82.71 | 17.70 |
| Others | 20.44 | 270.20 | 13.20 |
| Total | 58.70 | 875.26 | — |

India occupies first position in cauliflower production, second in onion and third in cabbage in the world.

The area and production-wise the largest vegetable growing states are West Bengal, Orissa and Uttar Pradesh.

The present production is not sufficient to meet the requirement of 285 g of vegetables, on an average, per capita per day. At present our per capita availability is around 145 g per day. By the end of 2030, according to an estimate, we will need 151-193 million tonnes of vegetables to meet our requirement. Therefore, it is necessary that production of vegetables, including root and tuber crops, is increased at a much faster rate.

**Statewise area, production and productivity of vegetable crops (1998-99)**

| State/UTs | Area (lakh ha) | Production (lakh tonnes) | Productivity (tonnes/ha) |
|---|---|---|---|
| Bihar | 6.16 | 94.18 | 15.3 |
| Maharashtra | 3.41 | 44.79 | 13.1 |
| Orissa | 8.83 | 100.87 | 11.4 |
| Uttar Pradesh | 6.40 | 126.80 | 19.8 |
| West Bengal | 11.00 | 163.67 | 14.9 |

## Vegetables for export

The Agricultural and Processed Food Products Export Development Authority (APEDA) have identified traditional vegetables like onion, potato, bitter gourd and chilli and non-traditional vegetables like asparagus, celery, sweet pepper, paprika, sweet corn, baby corn, green peas, French bean, tomato and gherkin as having good export potential. Currently onion accounts for 70 per cent of the total foreign exchange earned from export of fresh vegetables.

**Exporting countries and vegetables being exported**

| Countries | Vegetables being exported |
|---|---|
| Thailand | Brinjal, Chillies, Okra, Multiplier onion, Garlic, Yellow onion |
| Holland | Onion, Capsicum, Cole crops, Tomato, Cucumber, Lettuce, Potato and Root crops |
| Spain | Onion, Garlic |
| Australia | Onion, Beans, Cole crops |
| Iran | Onion |
| Turkey | Onion |
| Egypt | Onion, Garlic |
| China | Onion, Garlic |

| Countries | Vegetables being exported |
|---|---|
| Argentina | Yellow onion |
| Indonesia | Multiplier onion, Garlic |
| Kenya | Beans, Peas, Okra |
| Guatemala | Asparagus |
| Morocco | Gherkin |
| Jordan, Lebanon and Syria | Assorted vegetables |

Among other vegetables, 60 per cent share goes to okra, 20 per cent to green chillies and 20 per cent to bitter gourd, French bean, capsicum and other mixed vegetables. In global vegetable markets, Thailand, Jordan, Lebanon, Syria, Kenya, Zimbabwe, Guatemala, China, Argentina, Indonesia, Egypt, Turkey, Iran, Cyprus, Australia, New Zealand and Holland are the main competition countries which export different vegetables to different countries.

**Vegetable export from India**

In India, vegetables are grown in the open and thus their cost of production is less as compared to those grown under protective cover. There is lot of opportunity for exporting these vegetables to the European and North American countries from India. Presently our share in world export market is very

**Export of major vegetables from India (1997-98).**

| Vegetables | Quantity (tonnes) | Value (Rs in lakh) |
|---|---|---|
| Cucumber | 10,765,91 | 1,783.76 |
| Garlic | 2,436.77 | 219.68 |
| Onion | 3,33,348.97 | 20,246.10 |
| Tomato | 862.55 | 41.36 |
| Mixed Vegetable | 17,956.51 | 1,923.00 |
| Total | 3,65,370.71 | 24,213.92 |

negligible. If we are to make our presence felt in the export market, considerable development is required in infrastructure and export promotion areas.

Fresh vegetable export from India has been on the rise. During 1997-98, total vegetable exports amounted to 3,65,370.71 tonnes valued at Rs. 24,213.92 lakh. During 1997-98, 3,33,348.97 tonnes of onion valued at Rs. 20,216.1 lakh was exported. Cucumber and gherkin exports have increased our exports significantly, touching an all-time high during 1997-98. These exports amounted to 10,765.91 tonnes valued at Rs.1,783.76 lakh. Other vegetables exported accounted 10,765,91 tonnes valued at Rs.1,783.76 lakh. Garlic and tomato are the other vegetables that are being exported.

# TWO

## Classification of Vegetables

Olericulture is a *Latin* term used to designate vegetable culture. The study of olericulture includes all aspects of their production practices. There are more than fifty individual vegetables. If the method of growing each is dealt with in detail, it is likely to make things complicated and cause a lot of repetition. To eliminate this avoidable repetition, it is desirable to classify the vegetables in some groups. This grouping or classification is mainly to show the relationship between the individual vegetables and to avoid repetition while describing their cultural operations.

### Basis of classification

**Botanical**

**MONOCOTYLEDONEAE**

*Family - Amaryllidaceae (Alliaceae)*

*Allium cepa* - Onion

*A. cepa* var. *aggregatum* - Multiplier onion

*A. cepa* var *vaviparum* - Top onion

*A. ascalonicum* - Shallot

*A. porrum* - Leek

*A. sativum* - Garlic

*A. fistulosum* - Welsh onion

*A. schoenoprasum* - Chive

**Family - Araceae**

*Colocasia esculenta* - Arvi, Taro or Dasheen

*Amorphophallus campanulatus* - Elephant's foot

**Family - Dioscoreaceae**

*Dioscorea alata* - Yam

*D. bulbifera* - Potato yam

**Family - Liliaceae**

*Asparagus officinalis* - Asparagus

**Family - Poiaceae (Gramineae)**

*Zea mays* var *rugosa* - Sweet corn

**DICOTYLEDONEAE**

**Family - Aizoaceae**

*Tetragonia expansa* - New Zealand spinach

**Family - Chenopodiaceae**

*Beta vulgaris* - Beetroot and *Palak*

*B. vulgaris* var. *cicla* - Swiss chard

*Spinacia oleracea* - Spinach

**Family - Compositae (Asteraceae)**

*Cichorium intybus* - Chicory

*Cichorium endivia* - Endive

*Cynara scolymus* - Artichoke

*C. cardunculus - Hathichuk*, Cardoon

*Lactuca sativa* - Lettuce

*L. sativa* var. *capitata* - Head lettuce

*Tragopogon porrifolius* - Salsify

## Family - Cruciferae (Brassicaceae)

*Armoracia rusticana* - Horse radish

*Brassica oleracea* var. *acephala* - Kale

*B. oleracea* var. *gemmifera* - Brussels sprouts

*B. oleracea* var. *botrytis* - Cauliflower

*B. oleracea* var. *italica* - Sprouting broccoli

*B. oleracea* var. *caulorapa* - Kohl-rabi or Knol-khol

*B. napus* var. *napobrassica* - Rutabaga

*B. campestris* var. *rapa* - Turnip

*B. campestris* var. *pekinensis* - Chinese cabbage

*B. juncea* - Mustard

*Crambe maritima* - Sea Kale.

*Raphanus sativus* - Radish

*R. sativus* var. *caudatus* - Rat-tail radish

## Family - Cucurbitaceae

*Banincasa hispida* - Wax gourd

*Citrullus lanatus* - Watermelon

*Coccinea indica* - Little gourd

*Cucumis vulgaris* var. *fistulosus* - Round melon

*Cucumis melo* var. *momordica* - Snap melon

*Cucumis melo* var. *utilissimus* - Long melon

*Cucumis sativus* - Cucumber

*Cucurbita maxima* - Winter squash

*Cucurbita moschata* - Pumpkin

*Cucurbita pepo* - Summer squash

*Cyclan thera* - Meetha Karela

*Lagenaria siceraria* - Bottle gourd

*Luffa acutangula* - Ridge gourd

*Luffa cylindrica* - Smooth gourd

*Momordica balsamina* - Mokha

*Momordica charantia* - Bitter gourd

*Momordica cochinchinensis* - Sweet gourd

*Sechium edule* - Chow-chow

*Trichosanthes anguina* - Snake gourd

*T. dioica* - Pointed gourd

**Family - Convolvulaceae**

*Ipomoea batatas* - Sweet potato

**Family - Euphorbiaceae**

*Manihot esculenta* - Tapioca or Cassava

**Family - Leguminosae (Fabaceae)**

*Cymopsis tetragonolaba* - Cluster bean

*Dolichos lablab* - Indian bean or Hyacinth bean

*Glycin max* - Soyabean

*Phaseolus acontifolius* - Moth bean

*P. calcarathus* - Rice bean

*P. lanatus* - Lima bean

*P. vulgaris* - Kidney bean or French bean

*Psophocarpus tetragonolobus* - Winged bean or Goa bean

*Trigonella foenumgraecum* - Fenugreek

*Vicia faba* - Broad bean

*Vigna sinensis* - Cowpea

*V. sinensis* var. *sesquipedalis* - Asparagus bean

**Family - Malvaceae**

*Abelmoschus esculentus* - Okra

**Family - Polygonacea**

*Rheum rhaponticum* - Rhubarb

*Rumex vesicarius* - Sorrel

**Family - Solanaceae**

*Capsicum annuum* - Sweet pepper

*C. frutescens* - Hot pepper

*Lycopersicon esculentum* - Tomato

*Solanum melongena* - Brinjal or Egg-plant.

*S. tuberosum* - Potato

**Family - Umbelliferae (Apiaceae)**

*Apium graveolens* - Celery

*Daucus carota* - Carrot

*Pastinaca sativa* - Parsnip

*Petroselinum crispum* - Parsley

**Family-Zingiberaceae**

*Zingiber offinale* - Ginger

*Curcuma domestica/longa* - Turmeric

## Hardiness

| Hardy | Semi-Hardy | Tender |
|---|---|---|
| Asparagus | Beetroot | Amaranthus |
| Broccoli | Carrot | Chillies |
| Brussel's sprouts | Cauliflower | Tomato |
| Cabbage | Celery | Cluster bean |
| Chives | Globe artichoke | Colocasia |
| Collards | Lettuce | Cowpea |
| Garlic | Palak | |
| Knol-khol | Parsnip | Cucurbits |
| Kale | Potato | Snap bean |
| Leek | | Sweet potato |
| Onion | | Tapioca |
| Parsley | | Yams |
| Peas | | |
| Radish | | |
| Turnip | | |
| Spinach | | |

## Part consumed

| Parts used | Vegetable crops |
|---|---|
| Bulb | Onion, Garlic |
| Corm | Yam, Colocasia, Cassava, Elephant's foot, Rosella |
| Immature flower parts | Broccoli, Globe artichoke, Cauliflower |
| Mature fruits | Tomato, Brinjal, Muskmelon, Water melon, Pumpkin |
| Immature fruits/pods | Peas, Beans, Okra, Capsicum, Chillies, Cucumber, Gourds, Squashes |

## Classification of Vegetables

| Parts used | Vegetable crops |
|---|---|
| Leaves | Spinach beet, Arvi, Amaranth, Spinach, Cabbage, Lettuce, Fenugreek, Leek, Parsley, Celery, Chinese cabbage, Kale, Mustard, Water cress, Chard, Sorrel, Bathua, Coriander, Zimmu, Onion leaves, Radish leaves, Turnip leaves |
| Rhizome | Arrowroots, Ginger, Turmeric |
| Root | Parsnip, Beetroot, Radish, Carrot, Turnip, Rutabaga |
| Stem | Asparagus, Potato, Khol-rabi (Knol-khol), Artichoke, Celery, Parsley |
| Tuber | Potato, Sweet potato |
| Petiole | Colocasia, Celery, Parsley, Spinach |

## Methods of culture

The vegetables grouped under 13 classes are as under:

| Group | Class | Vegetable crops |
|---|---|---|
| Group 1 | Perennial vegetable crops | Asparagus, Artichoke, Jerusalem-artichoke, Sea kale, Chow-chow, Moringa, Little gourd, Pointed gourd, Spine gourd |
| Group 2 | Greens | Spinach, Chard, New Zealand spinach, Mustard, Collards, Dendelion, Amaranth |
| Group 3 | Salad crops | Celery, Lettuce, Endive, Cress, Parsley |
| Group 4 | Cole crops | Cabbage, Cauliflower, Broccoli, Brussels sprouts, Khol-rabi or Knol-khol |

17

| Group | Class | Vegetable crops |
|-------|-------|-----------------|
| Group 5 | Root crops | Beetroot, Carrot, Parsnip, Turnip, Rutabaga, Radish, Horse-radish, Spanish salsify, Palak, Methi |
| Group 6 | Bulb crops | Onion, Garlic, Leek, Shallot, Chive, Welsh onion. |
| Group 7 | Tuber crops | Potato, Sweet potato, Cassava, Yam |
| Group 8 | Peas, Pod or leguminous vegetables | Pea, Cluster bean, Broad bean, Garden bean, Lima bean, Soyabean, Cowpea |
| Group 9 | Solanaceous vegetable crops | Tomato, Brinjal, Chillies |
| Group 10 | Cucurbits | Cucumber, Ridge gourd, Watermelon. |
| Group 11 | Fibre crop | Okra |
| Group 12 | Pot herbs or greens | Spinach, Buckweed, Pigweed, Sorrel, Kale |
| Group 13 | Other root crops | Colocasia, Yams, Tapioca, Elephant's foot |

## Season of growth

Vegetable crops can be grouped into three broad groups depending on their temperature requirement for growth and development.

| Kharif | Rabi | Zaid |
|--------|------|------|
| Beans | Garden beet | Dwarf bean |
| Brinjal | Cabbage | Dwarf tomato |
| Bittergourd | Cauliflower | Summer cucumber |
| Bottlegourd | Carrot | Summer squash |
| Cowpea | Garlic | |

## Classification of Vegetables

| Kharif | Rabi | Zaid |
|---|---|---|
| Cucumber | Knol-khol | |
| Muskmelon | Lettuce | |
| Okra | Broadbean | |
| Chillies | Methi | |
| Capsicum | Onion | |
| Ridgegourd | Spinach | |
| Summer squash | Beet leaf | |
| Sweet potato | Turnip | |
| Smoothgourd | Radish | |
| Tomato | Sprouting broccoli | |
| Watermelon | Brussel's sprout | |
| Snakegourd | | |
| Ginger | | |

## Raising methods

| Method of raising | Vegetable crops |
|---|---|
| Direct sown crops | Okra, Carrot Radish, Beans, Peas, Garlic |
| Transplanted crops | Tomato, Brinjal, Chillies, Cabbage, Cauliflower |
| Planting vines and cuttings | Sweet potato, Cassava, Pointed gourd |
| Bits of tuber and corms | Potato, Yams |

## Lime requirement

| May or may not require lime | Require lime | Require more lime | Require very high lime |
|---|---|---|---|
| Potato | Carrot | Brinjal | Celery |
| Radish | Cucumber | Cabbage | Chillies |
| Tomato | Knol-khol | Cauliflower | Leek |
| Turnip | Pea | | Lettuce |
| Watermelon | Pumpkin | | Onion |
| | | | Parsnip |
| | | | Spinach |
| | | | Turnip |

## Vegetable forcing

| Cool forcing | Warm forcing | Moderately forced |
|---|---|---|
| Asparagus | French bean | Tropical cabbages |
| Garden pea | Cucumber | Radish (Local) |
| Garden beet | Brinjal | Turnip |
| Carrot | Tomato | Cucurbits |
| Late cauliflower | Capsicum | |
| Cabbage | Chillies | |
| Celery | Watermelon | |
| Lettuce | Muskmelon | |
| Onion | | |
| Spinach | | |
| Beet leaf | | |
| Radish | | |
| Turnip | | |
| Carrot | | |

## Rate of respiration

| Very high | High | Moderate | Low | Very Low |
|---|---|---|---|---|
| Asparagus | Beans | Beet | Cabbage | Onion |
| Broccoli | Lettuce | Carrot | Sweet potato | Potato |
| Pea | | Celery | Turnip | |
| Spinach | | Cucumber | Muskmelon | |
| | | | Pepper | |
| | | | Tomato | |

## Soil reaction

| Slightly tolerant (pH 6.8 - 6.0) | Moderately tolerant (pH 6.8 - 5.5) | Highly tolerant (pH 6.8 - 5.0) |
|---|---|---|
| Asparagus | Bean | Chicory |
| Beet | Carrot | Potato |
| Cabbage | Pumpkin | Rhubarb |
| Okra | Squash | Sweet potato |
| Cauliflower | Cucumber | Watermelon |
| Celery | Tomato | |
| Spinach | Brinjal | |
| Palak | Garlic | |
| Onion | Turnip | |
| Leek | Parsley | |
| Lettuce | Pea | |
| Muskmelon | Pepper | |

## Salt tolerance

| Tolerance range (Molar Conc. of NaCl) | | | | |
|---|---|---|---|---|
| Sensitive 0.25 | Medium tolerant | | Highly tolerant | |
| | 0.50 | 0.75 | 1.00 | 1.25 |
| Tomato | Chilliés | Amaranth | French bean | Bitter gourd |
| Snake gourd | Okra | Cauliflower | Ridge gourd | Ash gourd |
| | Cabbage | Onion | | |
| | Sweet potato | Radish | | |
| | | Bottle gourd | | |

## Rooting depth

| Shallow rooted (30-40 cm) | Moderately deep (50-70 cm) | Deep (80-100 cm and above) |
|---|---|---|
| Broccoli | Beet | Artichoke |
| Brussel's sprouts | Carrot | Asparagus |
| Cabbage | Chard | Cluster bean |
| Cauliflower | Cucumber | Cowpea |
| Celery | Brinjal | French bean |
| Chinese cabbage | Muskmelon | Lima bean |
| Endive | Mustard | Parsnip |
| Garlic | Pea | Pumpkin |
| Leek | Pepper | Sweet potato |
| Lettuce | Rutabaga | Tomato |
| Onion | Summer squash | Watermelon |
| Parsley | Turnip | |
| Potato | | |
| Radish | | |
| Spinach | | |

## Light duration

| Long day | Short day | Day-neutral |
|---|---|---|
| Beans | Cowpea | Beans |
| Cowpea | Beans | Cowpea |
| Lettuce | Sweet potato | Tomato |
| Radish | Brinjal | Brinjal |
| Spinach beet | Pepper | |
| Onion | Cucumber | |
| Cabbage | Pea | |
| Beet | | |

## Photosynthesis

| $C_3$ | $C_4$ |
|---|---|
| Bean | Amaranth |
| Spinach | |
| Lettuce | |
| Carrot | |
| Potato | |
| Sweet potato | |
| Tomato | |
| Sugar beet | |

## Area of seed production

| Temperate | Sub-temperate | Tropical |
|---|---|---|
| Cabbage | Late cauliflower | Tomato |
| Knol khol | Beet leaf | Brinjal |
| Kale | Spinach | Okra |
| Brussel's sprout | Celery | Chillies |
| Sprouting broccoli | Parsley | Capsicum |
| Sugarbeet | Asparagus | All cucurbits |
| Chicory | Chayote | Colocasia |
| Potato | Subtemperate radish | Ginger |
| Amaranths | Turnip | Turmeric |
| Temperate radish | Carrot | Sweet potato |
| Turnip | | |
| Carrot | | |
| Garden beet | | |

## Nature of vegetable

| Annual | Biennial | Perennial |
|---|---|---|
| Tomato | All cole crops | Asparagus |
| Brinjal | Onion | Chayote |
| Capsicum | Leek | Ginger |
| Chillies | Radish | Turmeric |
| Garden peas | Turnip | Colocasia |
| Okra | Garden beet | Cardamom |
| Amaranths | Beet leaf | |
| All cucurbits | Spinach | |
| Celery | | |
| Parsley | | |

## Method of seed production

| A. Seed-to-seed (*in situ*) | B. Replanting method (Stecklings/ Head/Curd/ Bulb-to-seed) | Both-A. and B. |
|---|---|---|
| Garden pea | Turnip | Carrot |
| Garlic | Radish | Parsley |
| Radish | Carrot | Celery |
| Spinach | Garden beet | Chinese cabbage |
| Methi | Cabbage | Cabbage |
| Carrot | Chicory | Turnip |
| Turnip | Chinese cabbage | Radish |
| Garden beet | Sprouting broccoli | |
| Bean | Sugarbeet | |
| Okra | | |
| Tomato | | |

| A. Seed-to-seed (*in situ*) | B. Replanting method (Stecklings/ Head/Curd/ bulb-to-seed) | Both-A. and B. |
|---|---|---|
| Brinjal | | |
| Capsicum | | |
| Chillies | | |
| Cucurbits | | |
| Cauliflower | | |
| Cabbage | | |

**Nature of pollination**

| Highly cross-pollinated | Often cross-pollinated | Highly self-pollinated |
|---|---|---|
| **Wind pollinated** | | |
| Amaranths | Limabean | Asparagus bean |
| Beet leaf | Egg plant | Cluster bean |
| Spinach | Okra | Dolichos bean |
| Garden beet | Chillies | French bean |
| Sugar beet | Sweet pepper | Cowpea |
| **Insect pollinated** | | Garden pea |
| All cucurbits | | Methi |
| All cole crops | | Lettuce |
| Radish | | Tomato |
| Turnip | | |
| Carrot | | |
| Onion | | |

## Breeding systems

| | Sexual | | Asexual | |
|---|---|---|---|---|
| Predominantly self-pollinated | Often cross-pollinated | Largely cross-pollinated | Dioecious | Vegetatively propagated |
| Degree of selfing 90-100% | Adapted to self-pollination but degree of cross-pollination more than 4% | Degree of crossing can be upto 100% | Entirely cross pollinated | |
| Lettuce | Bhindi | Cole crops | Spinach | Colocasia |
| Garden pea | Chillies | Root vegetables | Asparagus | Sweet potato |
| French bean | Capsicum | Onion | Pointed gourd | Pointed gourd |
| Tomato | Brinjal | Cucurbits | | Ginger |
| Potato | Limabean | Amaranths Spinach beet | | Garlic Turmeric |

## Approximate seed size (ungraded seed number per 10g sample)

| 0-500 | 700-2400 | 2500-4500 | 5000-15000 | 20,000+ |
|---|---|---|---|---|
| Broad bean-8 | Salsify-700 | Fennel-2500 | Turnip-5000 | Celery-2500 |
| Lima bean-8 | Garden beet-750 | Onion-2700 | Parsley-5400 | Water cress-38000 |
| Pumpkin-30 | Radish-110 | Cabbage-2800 | Chicory-6000 | |
| Common bean-35 | Pepper-1400 | Knol-khol-3100 | Endive-7500 | |
| Sweet corn-50 | Brinjal-2200 | Tomato-3200 | Lettuce-8600 | |
| Garden pea-50 | Sprouting broccoli-2300 | Rutabaga-3300 | | |
| Squash-70 | Parsnip-2300 | Cauliflower-3500 | | |
| Watermelon-225 | | Leek-3600 | | |
| Okra-170 | | Cress-4000 | | |
| Artichoke-250 | | Bunching onion-4500 | | |
| Mungbean-300 | | | | |
| Asparagus-500 | | | | |
| Cucumber-350 | | | | |
| Melon-350 | | | | |

## Classification of Vegetables

# Mean temperature requirements (⁰C)

| Narrow range | | | Middle range | | | Wide range | | |
|---|---|---|---|---|---|---|---|---|
| Species | High | Low | Species | High | Low | Species | High | Low |
| Artichoke | 13-18 | -5 | Brinjal | 18-30 | -12 | Broccoli | 8-27 | -19 |
| Endive | 15-28 | -8 | Pepper | 18-30 | -12 | Broad bean | 5-25 | -20 |
| Chicory | 10-20 | -10 | Celery | 10-22 | -12 | Brussel's sprout | 10-20 | -20 |
| Parsnip | 10-12 | -10 | Celeriac | 10-20 | -12 | Cabbage | 10-30 | -20 |
| Garden pea | 15-25 | -10 | Fennel | 10-23 | -13 | Leek | 10-30 | -20 |
| | | | Spinach | 12-25 | -13 | Onion | 10-30 | -20 |
| | | | Asparagus | 15-30 | -15 | Tomato | 10-30 | -20 |
| | | | Cucumber | 20-35 | -15 | Garden beet | 8-30 | -20 |
| | | | Parsley | 15-30 | -15 | Carrot | 9-30 | -22 |
| | | | Pumpkin | 15-30 | -15 | | | |
| | | | Summer squash | 15-30 | -15 | | | |
| | | | Sweet corn | 15-30 | -15 | | | |
| | | | Radish | 14-30 | -16 | | | |
| | | | Okra | 18-35 | -17 | | | |

## Botanical description of vegetables

The botanical description includes common name, scientific name, family and somatic chromosome number.

| Common name | Scientific name | Family | 2n |
|---|---|---|---|
| Tomato | *Lycopersicon esculentum* | Solanaceae | 24 |
| Brinjal | *Solanum melongena* | Solanaceae | 24 |
| African brinjal | *S. macrocarpon* | Solanaceae | 24 |
| Pepino or Melon pear | *S. muricatum* | Solanaceae | 24 |
| Jilo | *S. gilo* | Solanaceae | 24 |
| Naranjillo | *S. quitoense* | Solanaceae | 24 |
| Garden huckleberry | *S. nigrum* | Solanaceae | 24 |
| Chilli | *Capsicum annuum* | Solanaceae | 24 |
| Capsicum | *C. annuum* | Solanaceae | |
| Hot pepper | *C. frutescens* | Solanaceae | 24 |
| Husk tomato | *Physalis pruinosa* | Solanaceae | |
| Tomatillo | *P. ixocarpa* | Solanaceae | |
| Cape gooseberry | *Physalis peruviana* | Solanaceae | |
| Tree Tomato | *Cyphomandra betacea* | Solanaceae | 24 |
| Cabbage | *Brassica oleracea* var. *capitata* | Brassicaceae | 18 |
| White cabbage | *B. oleracea* var. *alba* | Brassicaceae | |
| Cape gooseberry | *Physalis peruviana* | Solanaceæ | |
| Tree Tomato | *Cyphomandra betacea* | Solanaceae | 24 |
| Cabbage | *Brassica oleracea* var. *capitata* | Brassicaceae | 18 |
| White cabbage | *B. oleracea* var. *alba* | Brassicaceae | |
| Red cabbage | *B. oleracea* var. *rubra* | Brassicaceae | |
| Savoy cabbage | *B. oleracea* var. *sabauda* | Brassicaceæ | |
| Cauliflower | *B. oleracea* var. *botrytis* | Brassicaceae | |

## Classification of Vegetables

| Common name | Scientific name | Family | 2n |
|---|---|---|---|
| Broccoli | *B. oleracea* var. *italica* | Brassicaceae | |
| Brussels sprouts | *B. oleracea* var. *gemmifera* | Brassicaceae | |
| Kale | *B. oleracea* var. *acephala* | Brassicaceae | |
| Knol-khol | *B. oleracea* var. *caulorapa* | Brassicaceae | |
| Chinese cabbage | *B. campestris* var. *pekinensis* | Brassicaceae | 20 |
| Chinese broccoli/ Chinese kale | *B. alboglabra* | Brassicaceae | 18 |
| Rutabaga | *B. napus* var. *napobrassica.* | Brassicaceae | 38 |
| Radish | *Raphanus sativus* | Brassicaceae | 18 |
| Carrot | *Daucus carota* | Umbelliferae | 18 |
| Turnip | *Brassica compestris* var. *rapa* | Brassicaceae | 20 |
| Beet | *Beta vulgaris* | Chenopodiaceae | 18 |
| Parsnip | *Pastinaca sativa* | Umbelliferae | 22 |
| Horse radish | *Armoracia rusticana* | Brassicaceae | 32 |
| Chervil | *Anthriscum cerefolium* | Umbelliferae | 32 |
| Celeriac | *Apium graveolens* var *rapaceum* | Umbelliferae | 22 |
| Skirret | *Sium sisarum* | Umbelliferae | 12 |
| Onion | *Allium cepa* | Amaryllidaceae | 16 |
| Shallot | *A. cepa* var. *ascalonicum* | Amaryllidaceae | 16 |
| Multiplier onion | *A. cepa* var. *aggregatum* | Amaryllidaceae | 16 |
| Tree or Egyptian onion | *A. cepa* x *A. fistulosum* | Amaryllidaceae | 16 |
| Japanese onion or Welsh onion | *Allium fistulosum* | Amaryllidaceae | 16 |
| Leek | *A. porrum* | Amaryllidaceae | 32 |
| Great headed garlic | *A. umpeloprasum* | Amaryllidaceae | 48 |
| Chive | *A. schoenoprasum* | Amaryllidaceae | 16, 24, 32 |

| Common name | Scientific name | Family | 2n |
|---|---|---|---|
| Rakkyo | *A. chinensis* | Amaryllidaceae | 16, 24, 32 |
| Chinese chive | *A. tuberosum* | Amaryllidaceae | 32 |
| Garden pea | *Pisum sativum* | Leguminosae | 14 |
| French bean | *Phaseolus vulgaris* | Leguminosae | 22 |
| Cowpea | *Vigna sinensis* | Leguminosae | 22 |
| Catjung cowpea | *V. unguiculata* | Leguminosae | 22 |
| Yardlong bean | *V. sesquipedalis* | Leguminosae | 22 |
| Winged bean | *Psophocarpus tetragonolobus* | Leguminosae | 18 |
| Cluster bean | *Cyamopsis tetragonoloba* | Leguminosae | 14 |
| Hyacinth bean | *Dolichos lablab* | Leguminosae | 22 |
| Broad bean | *Vicia faba* | Leguminosae | 12 |
| Sword bean | *Canavalia gladiata* | Leguminosae | 22 |
| Jack bean | *C. ensiformis* | Leguminosae | 22 |
| Moth bean | *Phaseolus acontifolius* | Leguminosae | 22 |
| Scarlet runner bean | *P. coccineus* | Leguminosae | 22 |
| Lima bean | *P. lanatus* | Leguminosae | 22 |
| Soyabean | *Glycine max* | Leguminosae | 22 |
| Yam bean | *Pachyrrhizus erosus* | Leguminosae | 22 |
| Potato bean | *P. tuberosus* | Leguminosae | 22 |
| Okra | *Abelmoschus esculentus* | Malvaceae | 72, 108, 144 |
| Cucumber | *Cucumis sativus* | Cucurbitaceae | 14 |
| Muskmelon | *C. melo* | Cucurbitaceae | 24 |
| Long melon | *Cucumis melo* var. *utilissimus* | Cucurbitaceae | 24 |
| Snap melon | *C. melo* var. *momordica* | Cucurbitaceae | 24 |
| Mango melon | *C. melo* var. *cultus* | Cucurbitaceae | — |
| Watermelon | *Citrullus lanatus* | Cucurbitaceae | 22 |
| Bottle gourd | *Lagenaria siceraria* | Cucurbitaceae | 22 |
| Bitter gourd | *Momordica charantia* | Cucurbitaceae | 22 |

| Common name | Scientific name | Family | 2n |
|---|---|---|---|
| Kakrol or Spine gourd | *M. dioica* | Cucurbitaceae | 22 |
| Sweet gourd of Assam | *M. cochinchinensis* | Cucurbitaceae | 28 |
| Pointed gourd | *Trichosanthes dioica* | Cucurbitaceae | 24 |
| Snake gourd | *T. cucumerina / anguina* | Cucurbitaceae | 24 |
| Sponge gourd | *Luffa aegyptica / cylindrica* | Cucurbitaceae | 26 |
| Ridge gourd | *L. acutangula* | Cucurbitaceae | 26 |
| Round melon | *Praecitrullas fistulosus* | Cucurbitaceae | 24 |
| Ivy gourd | *Coccinia indica* | Cucurbitaceae | 24 |
| Chow-chow | *Sechium edule* | Cucurbitaceae | 28 |
| Pumpkin | *Cucurbita moschata* | Cucurbitaceae | 40 |
| Winter squash | *Cucurbita maxima* | Cucurbitaceae | 40 |
| Summer squash | *Cucurbita pepo* | Cucurbitaceae | 40 |
| Ash or Wax gourd | *Benincasa hispida* | Cucurbitaceae | 24 |
| Sweet potato | *Ipomea batata* | Convolvulaceae | 90 |
| Tapioca or Cassava | *Manihot esculenta* | Euphorbiaceae | 36, 72 |
| Yams | *Dioscorea* spp. | Dioscoreaceae | 40 |
| Taro | *Colocasia esculenta* | Araceae | 48, 28, 36 |
| Giant taro | *Alocasia indica* | Araceae | 28 |
| Elephant's foot | *Amorphophallus campanulatus* | Araceae | 26, 28 |
| Spinach beet | *Beta vulgaris* var. *bengalensis* | Chenopodiaceae | 18 |
| Spinach | *Spinacia oleracea* | Chenopodiaceae | 12 |
| Amaranth | *Amaranthus* spp. | Amaranthaceae | 32, 34, 64 |
| Fenugreek | *Trigonella foenum-graecum* | Leguminosae | 16 |
| Indian spinach | *Basella* spp. | Basellaceae | 48, 44, 60 |

| Common name | Scientific name | Family | 2n |
|---|---|---|---|
| Agathi | *Sesbania grandiflora* | Leguminosae | |
| Water leaf | *Ipomoea aquatica* | Convolvulaceae | |
| Curry leaf | *Murraya koenigii* | Rutaceae | 18 |
| Swiss chard | *Beta vulgaris* var. *cicla* | Chenopodiaceae | — |
| New Zealand spinach | *Tetragonia expansa* | Aizoaceae | — |
| Portulaca | *Portulaca oleracea* | Portulacaceae | 54 |
| Sorrel | *Rumex vesicarius* | Polygonaceae | — |
| Drumstick | *Moringa oleifera* | Moringaceae | 28 |
| Lettuce | *Lactuca sativa* | Compositae | 16, 18 |
| Celery | *Apium graveolens* | Umbelliferae | 22 |
| Parsley | *Petroselinum crispum* | Umbelliferae | 22 |
| Endive | *Cichorium endivia* | Compositae | |
| Chicory | *C. intybus* | Compositae | |
| Chervil | *Anthriscum cerefolium* | Umbelliferae | |
| Cress | *Lepidium sativum* | Cruciferae | |
| Water cress | *Nasturtium officinale* | Cruciferae | 16, 32 |

# THREE

## Vegetable Farming Systems

The various farming systems prevalent and followed in different regions of the country for vegetable cultivation have a significant role to play in the increased production and productivity to meet the total requirement of the alarming population.

Farming systems for growing vegetables can be grouped into different types depending on the situations, topography, agroclimatic conditions and requirement of the farming community in particular areas. These may be :

### Flat land farming

The fields/land is flat either a plateau type at the top of hill or in valley areas. The steepness is not much. The fields are large, uniform and fertile. The cultivation is very primitive, either this existed since old times or was developed later. The cultivation is mostly done with the help of bullocks or now-a-days tractors are also used where road facility exists. The farmers are rich and earn their livelihood from agriculture.

**Low land :** The situations of land are flat, located at the bottom or one side of hill. The soils are clay or clay loam, not well drained, mostly rain water or snow melted water get accumulated. In such areas underground drainage systems need to be installed for proper draining of excess water for making soils fit for the production of radish, turnip, garden peas etc.

**Upland:** The fields are large enough, flat and open valley at the side of the hill or both sides of river. The soils are fertile and well drained. The water does not accumulate in the fields. The snowfall received in such areas from November to March and water after snow melting get either drained or absorbed. These land situations are excellent for quality production of temperate vegetable varieties as rains are scanty at maturity stage.

**Top of hill :** The land situation are flat, soils are well drained, fertile, rich in humus. Snow fall during winter is a regular feature. The seeds are sown in September for e.g., garden peas and remain under snow upto March-April and after snow melting, the plants grow and harvesting is done in July-August. The other vegetables are raised from March to September.

## Terraced farming

Topographically slopy, undulating and steep situations may be available for growing vegetable crop under terraced farming. The terraces may be narrow to wide, long or short arranged on the hill side looks from far as steps in some places. The cultivation in medium and wide strips can be done with the help of small statured bullocks while in narrow strips manual digging is followed. The vegetables are grown accordingly. However, the terraced fields may have gentle slope or too steep.

## Shifting cultivation

The shifting cultivation is also called '*Jhoom Cultivation*' and mostly common in Assam, Arunachal Pradesh, etc. The piece of land found in the forest or on the side of the river is cleaned of grasses, bushes, stones, etc. and is cultivated till the fertility is exhausted. The soil mostly loamy rich in humus is quite suitable for production of vegetable crops *viz.* ginger, colocasia, turmeric, elephant's foot, chow-chow etc. profitably. After few years the cultivation is stopped and new pockets or site is exploited. This type of farming is done by a particular tribe or community and they also live there by having a temporary shelter for easy maintenance of crops.

## Subsistence farming

Growing of vegetables is done on a small scale only to meet the farmer's own demand or at the most given to the neighbourhood or relatives. The producer has the tendency to produce only those vegetables which he needs in bulk fresh or for future use. The vegetables may be produced in the kitchen garden itself. The farmers in addition have to maintain the milch cattle for fresh milk, sheep and goats for wool and meat, poultry birds for eggs and meat, growing mushroom for regular supply to the family, seasonal and permanent indoor as well outdoor flower plants for aesthetic value and beautification of the house and its surroundings. Sometimes vegetables when in flower look beautiful, provide cheap and fresh vegetables along with aesthetic value. The cultivation of vegetables is done only to meet the family requirement or the farmer has the tendency to grow only those vegetables which his family needs from time to time.

## Cluster farming

This is very common farming system followed for growing vegetable crops on the land holdings of the farmers of the same village or hamlet, the fields are located in the same area though belonging to different farmers but in a single cluster. Each cluster may be raising one, two or more vegetables. This helps in following the common cultural practices, post-harvest handling etc. More grading and marketing etc. can be conveniently done economically.

## Mixed farming

Growing of vegetable crops along with cereals, pulses, fruit etc. to meet the annual demand for those by the family. The area put under each crop will depend on the size of land holding and requirement of each component by the family. However, the scenario is changing and the area under other crops is being replaced by high value vegetables and seed crops. The vegetable production is highly remunerative as compared to these crops.

## Orchard farming

The space available between and within row of fruit plants may be old or young plantation in each zone is used properly for raising vegetable crops. The orchards, may be of citrus fruits, peach, plum, apricot, apple etc. are used for cultivation. Fruit crops and vegetable crops are harvested to have double benefit out of the land holding of the farmer. It also helps in covering the risk of the natural hazards *viz.* hail, frost, etc. If one crop fails the other crop is stand-by for meeting the loss. The cultivation of vegetable crops is easily done in young orchards when the fruit plants are small. The field operations are done conveniently and there is no competition for light, nutrition and water. The areas where well-established orchards are there, the shade loving vegetables like ginger, turmeric, colocasia, cucurbits, garden peas and beans are grown successfully particularly when the trees are not in bearing. By raising vegetables, the orchards are automatically maintained and there is no weed growth and looks neat and clean. The weed population is not allowed to grow and unable to harbour the fruit plants of food material. Moreover, the expenditure involved for the check of weeds is also saved and the same is utilized somewhere else.

## Riverbed cultivation

This is very a old practice to grow most of the vegetables in the river beds having fertile soil, and if not supplemented with organic manure, the vegetables particularly the cucurbits are planted in November-December onwards when the water in the rivers get receded. The soil during day time is warm enough as compared to other areas and moreover these are protected with *sarkanda* or any other means from the northern chilly winds. The subsequent sowings are done and regular harvests are taken and the markets are supplied with fresh vegetable produce at a time when there is no vegetable available in the market from February to May. The farmers get better prices for their produce as they get early yield, ease in irrigation, high net return per unit area and additional crop. Other cultural practices are the same as for general raising of vegetables.

## Kitchen or home gardening

Kitchen or home gardening is growing of vegetable crops in the residential houses to meet the requirements of the family all the year round. Every individual is concerned with kitchen or home gardening. Irrespective of the fact whether the individual is a villager or city dweller, kitchen gardening should be a part of his home.

Kitchen gardening is the most ancient type of gardening. It aims at an efficient and effective use of land for growing essential vegetables for daily use of a family.

In most kitchens for gardening backyard of the house is selected. Often there may not be any choice in the shape of the garden, but where possible, a rectangular garden is preferred to a square one. By close attention to succession cropping and inter cropping, five per cent of land may be made to supply adequate vegetables for an average family of husband, wife and three children. Moreover, the most economic utilisation of space can be obtained by:

● making use of the fence on three sides for cucurbits during summer and rainy season, and peas in winter, and the fourth side for beans ;

● utilizing the ridges which separate the beds for

growing root crops;

- raising single-stemmed, staked tomato plants on one side and amaranth and other leafy vegetables on the other side of footpath; and

- chillies, ginger, turmeric, coriander etc., can be grown as intercrop in perennial block where some quick growing fruit crops like guava, citrus, banana, papaya and pineapple etc; may be planted.

## Market gardening

Market gardening is a branch of vegetable farming which produces vegetables for supply to the consumers in the local market. Most of the market gardens are located within 15 to 20 km of a city. The cropping pattern in these gardens will depend on the demands of the local market. The market gardener will like to grow early varieties to catch the early market. He must be a versatile person as he will have to grow number of vegetables throughout the year. He should also be a good salesman as he may have to sell his own produce. The land being costly, intensive methods of cultivation are followed for vegetable production.

## Truck gardening

This is the type of garden which produces special crops in relatively large quantities for distant markets. Truck gardens in general follow a more extensive and less intensive method of cultivation than market gardens. The word *'truck'* has no relationship with a motor truck but is derived from French word *troquer* meaning *'to barter'*. The commodity raised are usually sold through middlemen. The location of this type of garden is determined by the soil and climatic factors suitable for raising particular crops. The truck gardener should be a specialised person. He should be proficient in large scale production and handling of some special crops. The cost of land may be cheap and cheap labour may be available. He follows the mechanised method of cultivation. His cost of cultivation is therefore less. The net income is also less as this includes the cost of transport and the charges of the middleman. With the development of a quick and easy transport system, the distinction between the market and truck gardens in continuously diminishing. Only few vegetables which i.e. potato, onion, chillies, pumpkin etc. are suitable for truck gardening.

## Processing garden

Canned, dehydrated or frozen vegetables, picklings and fermented products are prepared in processing factories. For these a particular cultivar is suitable and grown for the purpose. This type of farming which produces vegetables or their cultivar with a sole objective of supplying them to the processing factories is termed as *'vegetable gardening for processing'*. The growing of vegetables and their selection are made in such a way that there is continuous supply of raw materials to the factories. Vegetables for canning purpose are usually grown in areas of good climatic condition with relatively low cost of production. These gardens aim at growing only a few vegetables in bulk with high yield and regular supply over a longer period. Tomatoes, peas, snapbeans are some of the vegetables suited for canning. Vegetables such as broccoli, spinach, lima bean are good for freezing while white cultivars of onion and potatoes are suitable for dehydration.

## Vegetable gardening for seed production

Good seed is the base of any successful farming industry. Seed production is rather a specialised field of vegetable growing. Soil, climate and disease free conditions are factors influencing location of seed producing areas. A thorough knowledge of a proper crop, its growth habit, mode of pollination, proper isolation distance are all of prime importance. The handling of the seed crop, curing, threshing, cleaning, grading, packing and storage all need specialised knowledge. Generally, the nucleus or breeder's seed is produced by the person or organisation which gives out the variety. The foundation seed is multiplied by the Government departments or by organisations like the National Seed Corporation (NSC). The third and the fourth stages of multiplication, that is registered and certified seeds, are usually multiplied by growers. This is an expanding industry in India and has a good future. India has varied climatic conditions extending from the temperate Himalayas to the tropical south. All the vegetable seeds can be profitably grown here. There is a big potentiality of exporting vegetable seeds to foreign countries. India at present does not import any vegetable seeds. To expand foreign trade in this industry, the quality of seeds produced must be raised.

## Garden for vegetable forcing

This type of farming is adopted to grow vegetables out of normal season. This is a highly specialized type of farming involving special growing structures with temperature and humidity control etc.

Vegetable forcing requires some special structures like glass house, cold frames and hot bed. Forcing the germination of melons in winter months by warmths of cowdung manure may also be grouped under this type. Only limited number of vegetables like tomato, cucumber, lettuce, radish, beet, cauliflower etc., are preferred. In temperate region where the temperature is very low tomatoes, cucumbers, etc., cannot be grown in open during cooler months. To meet the requirement under these conditions, they are forced to grow in specialized structures where congenial condition for the crop is provided artificially. It is the most intensive type of vegetable growing as the forcing structures are not allowed to remain idle.

The total cost of cultivation is highest in vegetable forcing as compared to any other type of farming. However, vegetables produced through this system give a high return.

## Floating garden

Floating garden is seen on the Dal Lake of the Kashmir valley. Most of the summer vegetables are supplied to Srinagar from these gardens. A floating base is first made from the roots of *Typha* grass which grows wild in some parts of the lake. All the interculture operations and occasional sprinkling of water are done from boats. This type of vegetable cultivation is a specialised technique and an art in itself.

# FOUR

## Vegetable Production Techniques

Agroecological zone is defined as a land unit in terms of major climates, suitable for certain range of crops and cultivars. The Indian Council of Agricultural Research, New Delhi, has recognized eight agro-climatic zones for effective land-use planning.

### Agro-climatic zones of India

| Agro-climatic region | States |
|---|---|
| 1. Humid Western Himalayan Region | Jammu and Kashmir, Himachal Pradesh, Kumaon and Garhwal. |
| 2. Humid Bengal-Assam Region | West Bengal and Assam. |
| 3. Humid Eastern Himalayan Region | Bay Islands, Arunachal Pradesh, Nagaland, Manipur, Mizoram, Tripura, Sikkim, Meghalaya and Andaman and Nicobar Islands. |
| 4. Sub-humid Sutlej-Ganga Alluvial Plains | Punjab, Delhi, Uttar Pradesh plains and Bihar. |
| 5. Sub-humid to Humid Eastern and South Eastern Uplands | Eastern Madhya Pradesh, Orissa and Andhra Pradesh. |
| 6. Arid Western Plains | Haryana, Rajasthan, Gujarat, Dadra and Nagar Haveli, and Daman and Diu. |
| 7. Semi-arid Lava Plateaus and Central Islands | Maharashtra, Western Central Madhya Pradesh and Goa. |
| 8. Humid to Semi-arid Western Ghats | Karnataka, Tamil Nadu, Kerala, Pondicherry and Lakshadweep Islands. |

During 1985-90, the Planning Commission accepted 15 broad agro-climatic zones based on physiography and climate for effective planning. These zones are: 1. Western Himalayan Region, 2. Eastern Himalayan Region, 3. Lower Gangetic Plains Region, 4. Middle Gangetic Plains Region, 5. Upper Gangetic Plains Region, 6. Trans-Gangetic Plains Region, 7. Eastern Plateau and Hills Region, 8. Central Plateau and Hills Region, 9. Western Plateau and Hills Region, 10. Southern Plateau and Hills Region, 11. East Coast Plains and Hills Region, 12. West Coast Plains and Ghats, 13. Gujarat Plains and Hills Region, 14. Western Dry Region, and 15. The Island Region.

However, the National Bureau of Soil Survey and Land Use Planning (NBSS&LUP) Nagpur delineated the country into 21 agro-ecological regions using physiography, soil's bioclimatic types and growing periods. This approach is comprehensive and can be used for the delineation of vegetable crop's growing zone.

### Cultivation of vegetable crops in different agro-ecological zones

| Agro-ecological region | Area and choice of vegetable crop for the region | |
|---|---|---|
| | Geographical range | Vegetable crop grown |
| 1. Western Himalayas. Cold arid | This region comprises cold and agro-eco region of Western Himalayas covering Ladakh and Gilgit districts of J & K. | During summer period Tomato, Cole crops and Brinjal. |
| 2. Western Plains and Kutch peninsula. Hot arid | This eco-region consists of western parts of Rajasthan, southern parts of Haryana and Punjab, the Kutch peninsula and northern parts of Kathiawar peninsula. | Cucurbits and Cole crops. |

| Agro-ecological region | Area and choice of vegetable crop for the region | |
|---|---|---|
| | Geographical range | Vegetable crop grown |
| 3. Deccan Plateau. Hot arid | The region comprises the Deccan plateau which includes Raichur and Bellary of Karnataka and Anantapur in Andhra Pradesh. | Okra, Brinjal and Cucurbits do well with supplement irrigation. |
| 4. Northern Plains and Central Highlands. | The region consists of Northern plains, Central highlands and Gujarat plains which are characterised by hot and dry summers and cool winters. | Besides Cole crops, all vegetable crops have potential with a modification in production technology and cultivars. |
| 5. Central (Malwa) Highlands and Kathiawar Peninsula. Hot semi-arid | The region includes the western part of Madhya Pradesh, eastern part of Rajasthan and Gujarat state. This is characterised by hot and dry summer and mild winter. | All the vegetables except Cole crops can be grown. |
| 6. Deccan Plateau. Hot semi-arid eco-region | The region includes. Maharashtra; northern part of Karnataka and Andhra Pradesh. | Potato and Onion have potential in the region. |
| 7. Deccan Plateau and Eastern Ghats. Hot semi-arid | Deccan plateau and Eastern Ghats cover major part of Andhra Pradesh. | Solanaceous vegetables and Cucurbits can be grown in the region. |
| 8. Eastern Ghats (Tamil Nadu) Uplands and Deccan plateau. Hot semi-arid | The region comprises Deccan plateau, Tamil Nadu uplands and western part of Karnataka. This is characterised by hot dry summer and mild winter. | All vegetables except peas are grown in this region. |

| Agro-ecological region | Area and choice of vegetable crop for the region | |
|---|---|---|
| | Geographical range | Vegetable crop grown |
| 9. Northern Plains. Hot sub-humid | The Northern plains, hot humid eco-region with alluvium soil comprises the Northern Indo-Gangetic plains. | All vegetables depending upon the season can be grown. |
| 10. Central Highlands (Malwa and Bundelkhand). Hot sub-humid | This region is characterised by medium and deep black soil and covers a part of Central highland, including the district of Raisen, Sagar, Bhopal, Sehore,Shahjahanpur and Hoshangabad (M.P.). | All vegetables have potential depending upon the season. |
| 11. Deccan Plateau and Central Highlands | The eco-region with red and black soils is characterised by hot summer and mild winter. | Cucurbits, Cauliflower, Radish, Carrot, Cowpea, Chillies, Brinjal, Okra, Spinach beet, Onion, Fenugreek, Amaranth. |
| 12. Eastern Plateau (Chhattisgarh). Hot sub-humid | This eco-region with red and yellow soils is characterised by hot summer and cool winter. | Cucurbits, Radish, Carrot, Cowpea, Chillies, Brinjal, Okra, Spinach, Amaranth, Garlic. |
| 13. Eastern Plateau (Chhota Nagpur) and Eastern Ghats. Hot sub-humid | This eco-region with red loamy soil is characterised by hot summer and cool winter. | All potential vegetables which can be grown all round the year. |
| 14. Eastern Plains. Hot sub-humid | The eco-region has alluvial soils and is characterised by cool winter and hot summer. | The region can grow all kinds of vegetables. |
| 15. Western Himalayas. Warm sub-humid (inclusion humid) | The eco-region has brown forest and podozolic soils and is characterised by warm sub-humid to cool-humid climate. | All temperate vegetables have potential in the zone. |

| Agro-ecological region | Area and choice of vegetable crop for the region | |
| --- | --- | --- |
| | Geographical range | Vegetable crop grown |
| 16. Assam and Bengal Plains. Hot humid | This eco-region has alluvium-derived soil. | Most of the vegetable crops except peas have potential in this region. |
| 17. Eastern Himalayas. Warm perhumid | This eco-region has brown hill soils. It passes through the northern tip of West Bengal, northern most part of Arunachal Pradesh and Sikkim. | In the warm season, all kinds of vegetables can be grown. |
| 18. North-Eastern hills (Purvanchal). Warm perhumid | This eco-region has red and laterite soils. The region represents hilly states of Nagaland, Meghalaya, Manipur, Mizoram and South Tripura. | Cucurbits, Radish, Peas, Cowpea, Chillies, Brinjal, Okra, Spinach beet, Amaranth, Tapioca. |
| 19. Eastern Coastal Plains. Hot sub-humid | This region has alluvial soils. It covers the eastern coastal plains, extending from Cauvery delta to Gangetic delta. | All the vegetable crops except Cole crops are grown. |
| 20. Western Ghats and Coastal Plains. Hot humid, perhumid | The eco-region has red, lateritic and alluvial soil covers. Sahyadri and western coastal plains of Maharashtra, Karnataka and Kerala states. | A number of vegetables except Cole crops are grown in this region. |

| Agro-ecological region | Area and choice of vegetable crop for the region | |
|---|---|---|
| | Geographical range | Vegetable crop grown |
| 21. Islands of Andaman and Nicobar and Lakshadweep, Hot perhumid | This eco-region has red loamy and sandy soils and is characterised by tropical conditions with little difference in mean soil temperature in winter and summer. | Cucurbits, Chillies, Brinjal, Okra, Amaranth, Basellarubra. |

## Soil

Vegetable crops require soils as growing medium. The soils provide mechanical support, nutrients and water for plant growth. The demand of vegetable crops from soils are mainly water, nutrients and other growth-controlling elements. However, a number of factors affect roots, which absorb nutrients and moisture. To ensure development of an efficient root system, the soils must contain adequate supply of air and water and have low bulk density. Most of the vegetable crops require well-drained soils, free from hard pan and a layer of calcium carbonate needed for optimum growth and yield. The saline, saline-sodic and sodic soils should be avoided.

All vegetable crops require a particular range of soil pH. Though some of them can tolerate acidic as well as saline conditions, however, most of them grow properly only when soil pH ranges between 6.5 to 7.5.

| Soil conditions | Vegetable crops |
|---|---|
| Slightly acidic | Asparagus, Beet, Cabbage, Okra, Cauliflower, Celery, Spinach, Palak, Onion, Leek, Lettuce, Muskmelon |
| Moderately acidic | Beans, Carrot, Pumpkin, Squash, Cucumber, Tomato, Brinjal, Garlic Turnip, Parsley, Pea, Pepper |

| Soil conditions | Vegetable crops |
| --- | --- |
| Highly acidic | Chicory, Potato, Rhubarb, Sweet potato, Watermelon. |
| Low salt | Tomato, Snake gourd |
| Medium salt | Chillies, Okra, Cabbage, Sweet potato Amaranth, Cauliflower, Onion, Radish, Bottle gourd |
| High salt | French bean, Ridge gourd, Bitter gourd, Ash gourd |
| Very low lime | Potato, Radish, Tomato, Turnip, Watermelon. |
| Low lime | Carrot, Cucumber, Knol-khol, Pea, Pumpkin. |
| High lime | Brinjal, Cabbage, Cauliflower. |
| Very high lime | Celery, Chillies, Leek, Lettuce, Onion, Parsnip, Spinach, Turnip. |
| Fertile | Tomato, Potato, Brinjal, Okra, Chillies, Sweet potato, Onion, Garlic, Radish, Carrot, Beetroot, Peas, Cabbage, Cauliflower, Knol-khol, Brussel's sprouts, Sprouting Broccoli, Celery. |
| Non-fertile | Cluster bean, Cowpea, French bean, Indian bean, Broad bean, Lima bean, Peas, Palak, Methi, Amaranthus. |
| Sandy loam | Cabbage, Brinjal, Tomato, Cluster bean, French bean, Indian bean. Turnip, Beetroot, Colocasia, Asparagus, Elephants' foot, Yams, Potato, Sweet potato, Cucurbits, Leek. |

| Soil conditions | Vegetable crops |
| --- | --- |
| Loam | Cauliflower, Brussel's sprouts, Sprouting Broccoli, Chilli, Okra, Broad bean, Onion, Garlic, Celery. |
| Loamy | Cowpea, Lima bean, Pea, Winged bean, Carrot, Radish. |

## Climate

Climate is the second major factor that influences crop yield and, therefore, it is essential to know the climatic requirements of vegetable crops to get optimum production. The climate of a region is the totality of the effects of its various components *viz.* temperature, light, humidity, rainfall, wind direction, wind velocity, and such phenomena as fog, frost, thunder gale etc. Almost all components of the climate influence vegetable crops. All are closely interrelated. The effect of each is modified by others. All vegetable crops have certain natural tolerance limits towards the climatic components beyond which they do not grow normally.

**Choice of vegetable crops for various types of climate**

| Climatic conditions | Choice of the vegetable crops |
| --- | --- |
| Cool | Beetroot, Cabbage, Cauliflower, Carrot, Celery, Garlic, Kale, Knol-khol, Lettuce, Methi, Onion, Palak, Pea, Potato, Radish, Spinach, Turnip, Winter squash. |
| Hot | Beans, Brinjal, Tapioca (cassava), Cowpea, Colocasia, Cucumber, Gourds, Melon, Pumpkin, Summer squash, Okra, Pepper, Sweet potato, Tomato. |

| Climatic conditions | Choice of the vegetable crops |
| --- | --- |
| Frost-free | Asparagus, Broccoli, Brussel's sprouts, Cabbage, Chives, Collards, Garlic, Knol-khol, Kale, Leek, Onion, Parsley, Peas, Radish, Spinach, Turnip. |
| Semi-frost | Beetroot, Carrot, Cauliflower, Celery, Globe-artichoke, Lettuce, Palak, Parsnip, Potato. |
| Frosty | Amaranth, Chillies, Tomato, Cluster bean, Colocasia, Cowpea, Cucurbits, Snap bean, Sweet potato, Tapioca, Yams. |

## Vegetable varieties

### Landrace

These are primitive varieties, which had evolved without a systematic and sustained plant breeding effort. They are storehouses of genetic variability and, ordinarily, are adapted to the local soil type, climatic condition etc.

### Strain

A group of plants of common lineage, though not taxonomically distinct from others of the species of variety, are distinguished on the basis of some character or characters, such as productiveness, vigour, resistance to cold, disease etc. or other ecological or physiological characteristics. A strain is known as *variety* when it is released for commercial cultivation by a variety release committee.

### Cultivar

It is the name given to all cultivated variants of plants. It is also called *cultural variety/cultivated variety*. Cultivar is an assemblage of cultivated plants which is clearly distinguishable by any character (morphological, physiological, cytological, chemical or other) and when reproduced (sexually or asexually) retains it distinguishing characters.

**Variety**

It is subdivision or a group of plants within a species which is characterized by growth, plant fruit, seed or other characters by which it can be differentiated from other seed of the same kind.

- **Hybrid:** A hybrid variety is $F_1$ generation from a cross between two different strains.

- **Composite:** Varieties produced by open-pollination among a number of outstanding strains usually not tested for combining ability with each other.

- **Synthetic:** In cross-pollinated species; a variety obtained by mating in all possible combinations with a number of lines that combine well with each other.

- **Elite:** They are improved varieties, as result of systematic breeding (location specific) and seed multiplication done only by the well trained scientific personnel.

- **Cybrids:** Somatic hybrids obtained by fusion the complete protoplast of one species with such protoplasts of another species, which either have no nucleus or have an inactive nucleus.

> Cybrid = Protoplast of species A + Cytoplast of species B

- **Clone:** A group of individuals of common ancestry which have been propagated asexually (vegetatively), usually by cutting or natural multiplication, *viz.* bulbs or tubers.

## Important vegetable varieties

| Vegetable | Varieties |
|---|---|
| Amaranths | Japanese yellow, Pusa Kiran, Kirti, Choti Chauli, Badi chauli, Pusa chauli, Co-1, Co-2, Co-3 |
| Asparagus | Perfection, Selection-841 |
| Beet leaf | Banerjee Giant, Long Standing Pusa, Harit, All Green, Pusa Palak, Pusa Jyoti, Jobner Green, HS-23, Palak No. 51-16. |
| Bhindi | Pusa Sawani, Harbhajan, P-7, Parbhani Kranti, P-8, Varsa Uphar, Arka Anamika, Arka Abhay, Punjab Padamini, A-4, PH-57, Hissar Unnat (HRB-55) |
| Bitter Gourd | Solan Hara, Solan Safed, Priya |
| Bottle Gourd | **(i) Long varieties:** Pusa Summer Prolific Long, Pusa Meghdoot, Hot Long Green, Pocha's Long Round<br><br>**(ii) Round varieties:** Pusa Summer Prolific Round, Pusa Alankar, Pusa Manjri, Punjab Round |
| Brinjal | **(i) Long varieties:** Pusa Purple Long, Pusa Purple Cluster, ARU-10, ARU-20, Pusa Anupam (Kt-4), Shalimar Improved, Arka Nidhi, Arka Keshav, Arka Neelkanth, Surya<br><br>**(ii) Round varieites:** Pusa Purple Round, Pant Rituraj, Hissar Syamal (H-8), Dilruba, Pusa Kranti, Black Beauty, Pant Samrat, Punjab Bahar, ARU-2, Pusa Bhairav |
| Broad Bean | Large Podded, Small Podded, Pusa Sumeet |
| Brussel's Sprout | **(i) Dwarf:** Brussel's Dwarf, Catskill Early Dwarf, Dwarf Gem |

| Vegetable | Varieties |
|---|---|
| | **(ii) Intermediate:** Hild's Ideal, Long Island, Half Dwarf Rubine |
| | **(iii) Tall:** Danish Prize, Amager, Express |
| Bunching Onion | PBO-I |
| Cabbage | **Early:** Golden Acre, Pusa Mukta (Sel-9), Chaubatia Early |
| | **Mid:** Pride of India, Pusa Drum Head, Aru Glory, Green Express |
| | **Late:** Large Late Drum Head, September, Green Challenger |
| Carrot | **Temperate:** Nantes, Chanteney, Kashmir Beauty |
| | **Semi-Temperate:** Pusa Yamdagni, Danver Half long, A Plus Desi |
| | **Asiatic:** Pusa Indian, Long Red |
| Cauliflower | **Mid-season:** Improved Japanese, Dania, Pant Subra |
| | **Late:** Snowball-16, Pusa Snow Ball-1, Pusa Snowball-2, PSK-1, Hira Moti and Pusa Him Jyoti |
| Chicory | Kalpa Sel-1, K-13 |
| Chillies | Solan Yellow, Pachhad Yellow, Hot Portugal, Hungarian Wax, Sweet Banana, Pb. Lal, Shalimar Long, Pant C-1, Pusa Jwala, Pusa Sadabahar, Surajmukhi |
| Chinese cabbage | Solan Selection, Palampur Green |
| Celery | Wright Grove, Giant Standard Bearer, Utah-92-70, Golden Self Blanch |
| Colocasia | Local cutlivar |
| Coriander | Indian Type, European Types, Bulgarian |

| Vegetable | Varieties |
|-----------|-----------|
| Cucumber | Straight Eight, Japanese Long Green, Khira-75, Khira Local, Khira-90, Poinsett, Pusa Sanyog, Temperate types, Kh-1 |
| French Bean | **Dwarf:** Contender, Arka Komal, Premier, VL Boni-1, Pusa Parvati, Pant Anupum, Pant Bean-2, Jampa |
| | **Tall:** Kentucky Wonder, SVM-1, Lakshmi (P-37), Black Queen |
| Garden Beet | Crimson Globe, Detroit Dark Red |
| Garden Pea | **Early varieties:** Arkel, Matar Ageta, Early Badger, Pusa, Vipasa, Palam-Priya, Pant Uphar, Sylvia |
| | **Mid-season:** Bonvillae, Lincoln, Azad pea, VL-3, VL-7, Shalimar Matar, Solan Nirog |
| | **Late varieties:** Kinnauri, Early Giant |
| Garlic | **Small segmented:** GHC-1, Selection-1, T-56-4, T-54-4, Solan Phavari, Pusa Selection-10, ARU-52, G-50 |
| | **Large segmented:** Local cultivar |
| Ginger | Him Giri, Kandaghat Selection, Maran, Kerala Local, Local cultivars |
| Knol-khol | White Vienna, Early White Vienna, Purple Vienna, Large Green, King of Market, DPKK-1. |
| Leek | American Flag, Mammoth Colossal, Price Taker, Mussel Berry, PPL-1 |
| Lettuce | **Head Type:** Great Lake, PH-lettuce, Slobolt |
| | **Open Type:** Alamo, Chinese Yellow, Simson Rubi |
| Methi | **Common:** IC074, Pusa Early Methi No. 14 & 47 EC-491 |

| Vegetable | Varieties |
|---|---|
| | **Kausri:** Bunching, Kasuri-Methi |
| Onion | Patna Red, Pusa Red, N-53, Agri Found Dark Red, Brown Spanish, Bellary Red, Early Grano, Pusa Ratnar, N-25, VL-67, VL-3, Arka Kalyan, Patna Red, Punjab Red Round, Arka Kirtiman, Arka Lalima |
| Parsley | Moss Curled, Karlina, Humburg, Curled Leaf, Doubled Curled, Champion |
| Potato | Kufri Chandermukhi, Kufri Jyoti, Kufri Navtal, Kufri Badshah |
| Pumpkin | Solan Badami, Large Red, Large Round, Red Flesh, Local cultivar |
| Radish | **Temperate:** White Icicle, French Breakfast, R.R. White Tipped, Sel-9 |
| | **Tropical:** Japanese White, Chinese Pink, Pusa Red |
| | **Semi-temperate (Asiatic):** Pusa Himani, Pusa Chetaki, Pusa Reshmi |
| Red Cabbage | Red Acre |
| Sardha Melon | Selection-1, Selection-9 |
| Spinach | Virginia. Savoy, Early Smooth leaf, Shannihon (AG-41), All Green, Prickly Seeded, Bloomsdale, Long Standing, Pusa Bharati |
| Sprouting Broccoli | **Green:** Italian Green, Green Head, DPBF-1, DPGB-2, Pusa Broccoli |
| | KTS-1 |
| | **Purple:** DPPB-1 (Palampur Samridhi) |
| Summer Squash | Pusa Alankar, Australian Green Early Yellow Prolific |

| Vegetable | Varieties |
|-----------|-----------|
| Sugar-beet | Ramonskaya-06, Erose type-E |
| Tomato | **Indeterminate varieties:** Solan Gola, Yaswant (A-2), Sioux, Marglobe, Naveen, Ptom-9301, Shalimar-1, Shalimar-2, Angurlata, Solan Bajr, Solan Shagun |
| | **Determinate varieties:** Roma (EC-13513), Rupali, MTH-15, Ptom-18, VL-1, VL-2 |
| | **Other varieties:** NT-3, Sel.-18, HS-101, Pusa Rubi, Punjab Chhuhara, Hissar Arun, Sonali, Ajanta, Swarna, Century, Sakti, Arka Alok |
| Turnip | **Temperate type:** Purple Top White Globe, Golden Ball, Snow Ball, Early Milan |
| | **Tropical type:** Pusa Chandrima, Pusa Swarnima, Pusa Sweti, Pusa Kanchan |
| Vegetable | Pusa Sweta, Red Prince, Yardd Long Bean |
| Cow pea | Check Barasati, Pusa Dofasli, Improved Pusa Dofasli |

## Important hybrids of vegetable available in public and private sector

| Variety | Hybrid in Public Sector | Hybrid in Private Sector |
|---------|-------------------------|--------------------------|
| Brinjal | Arka Navneet, Azad Hybrid, Vijay, Pusa Hybrid-5, Pusa Hybrid-6, Suphal, Pant Hybrid-1, Pusa Hybrid-9 | MHB-1, MHB-2, MHB-3, MHB-10 (Kalpataru), MHB-39 (Ravaiyya), AHB-2, AHB-4, ARBR–216, Sumerx-19, HOE-404, Nemabakar, . NDB-1, NDBH-6, (Narendra Hybrid Brinjal-2), BDBH. 1, ARBH-201 (Shyamal), Aman |

| Variety | Hybrid in Public Sector | Hybrid in Private Sector |
|---|---|---|
| Tomato | Karnataka, Vaishali, Rashmi, Pant Hybrid-2, Pant Hybrid-10, Kt-1, Kt-2, Kt-3, Kt-4, Pusa Hybrid-1, Hybrid-10, Hybrid-2, Hybrid-4, FM-1, FM-2, Arka Shreshta, Vishal Arka, Arka Vardan, Arka Abhijit | Sheetal, Rupali, Mangla, Naveen, MTH-2, MTH-6, Sonali Gulmohar, Priti, Ratna, Sadabahar, Lerica, Vijay, Vipual, SC-9, GS-12, SH-18, Sutton Gram, Sutton Gram Sweet cluster, Sweet-Heart, Sutton Gram Prolific, Sutton Gram Wonder, Masuraj, Arth-3, Arth-4, ATH-1, NARF-101 Avinash-2, BSS-71, BSS-171, BSS-175, Madhuri (BSS-39), BSS-20 (Meenakashi), BASS-99 (Manisha), BSS-98 (Megha) |
| Capsicum | Bharat, KT-1, Pusa Dipti | Early Bountry, Sutton Gram Giant, Indira, Larie, Swarup, Hira, Green Gold, Anupum |
| Chilli | Hybrid-1, Chamatkar | Agni, HOE-808, Tejasvani, Hybrid Chilli, ARCH-236, BSS-141, BSS-138 (Gayatri) |
| Muskmelon | Punjab Hybrid-1, M-4, Hybrid-3, Hybrid Swarna, MH4-3, MH-10 | MHC-5, MHC-6, MTH-6, MTH-1 |
| Water melon | Arka Jyoti, Arka Manik, Pusa Bedana | Madhu, Milan, Mohini, Santripti, MHW-15, Amrit, Honey, MHW-4, MHW-5, MHW-6, MHW-11, Nath-101, Nath-102, Suruchi |
| Bhindi | DVR-1 | GOSH-2, GOH-3, Hybrid-6, Sheela-101 |
| Bottlegourd | Pusa-Manjari, Pusa Meghdoot, PBO6-2 | Warad, MGH-1 |
| Carrot | Hybrid No-1 | HG-202, HC-203, Carrot No.1 |
| Cauliflower | Pusa Hybrid-2, Himani, Sweti | Moon, Raja, Rani, Neeta |
| Cabbage | | BSS-50 (Bajrang), BSS-32 (Suvarna), BSS-44, BSS-115 (Sudha), Sri Ganesh Gol. |
| Cucumber | Pusa Sanyog | Hybrid Priya, Liberty |
| Summer squash | Pusa Alankar | |
| Onion | Almora Hybrid, Hybrid BL-67, Arka Kirtiman, Arka Lalima | Bobby |

## Cropping systems in vegetables

Most of the vegetable crops are of short duration and fit in with a number of sequences which result in greater production per unit area and time. The success of these vegetable rotations depends upon the selection of proper varieties, adjustment of sowing time, adequate application of organic manures, fertilizers and irrigation, control of weeds, insect-pests and diseases and timely harvesting of the crops. The timeliness of these cultural operations become a highly critical factor in a successful cropping system programme in case of vegetables.

### A. Multiple cropping

Multiple cropping is the system of cropping in which more than one crop is grown on the same unit of land in a year without any gaps. The main aim of multiple cropping is to make effective use of modern farm technology and farm input with minimum cost of cultivation and without affecting soil fertility.

### Types

**Sequential cropping:** When two or more crops are grown one after the other in short succession on the same piece of land.

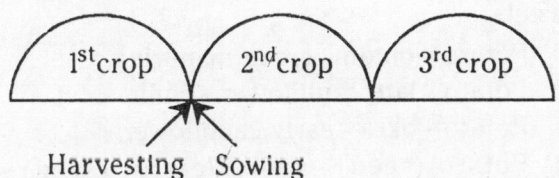

**Inter cropping:** When two or more crops are grown at the same time on a piece of land. Inter cropping intensifies farming in both space and time, while sequential in time only.

## B. Mixed cropping

It is a system of growing two or more crops on the sme piece of land with the major crop either mixed sown or in alternate line in one season. For example:

- ❖ Turmeric + Maize
- ❖ Potato + Radish
- ❖ Potato + Coriander
- ❖ Cabbage + Radish
- ❖ Cauliflower + Radish
- ❖ Cabbage + Coriander
- ❖ Cowpea + Cluster bean
- ❖ Okra + Radish
- ❖ Brinjal + Radish
- ❖ Cabbage + Lettuce + Radish

## C. Crop rotation

Crop rotation means the raising of crops one crop after the other systematically based on scientific principles so that higher yields are obtained without any depletion of soil fertility. The crop rotation may be for one year, two years, three years or so.

The following rotation may be practiced :-

**(a) Vegetable farms located away from the main markets:**

- (i) Potato - onion - green manuring.
- (ii) Potato - late cauliflower - chilli.
- (iii) Potato - okra - early cauliflower.
- (iv) Potato (seed) - radish/carrot (seed) - okra (seed).

**(b) Vegetable farms located near the main markets:**

- (i) Brinjal (long) - late cauliflower - bottle gourd.
- (ii) Cauliflower - tomato - okra.
- (iii) Potato - muskmelon - radish.
- (iv) Spinach - knol-khol - onion/green chilli.

*Vegetable Production Techniques*

## D. Succession

The succession means the growing of a second or even third crop in one year on the same piece of land after the previous crop or crops have been harvested and the soil retitled. Succession is followed in home gardens where activities are very intensive. Short duration and quick growing crops make succession remunerative.

## E. Companion/Double/Intercropping

The system involving raising of crop(s) on interspace available between row of the main crop on the same land. For example:

| Main crop | Companion crop |
| --- | --- |
| Okra | Chinese cabbage |
| Onion, garlic | Beet root, tomato, lettuce, carrot |
| Garden pea | Carrot, turnip, radish, cucumber, beans |
| Potato | Beans, cabbage |
| Radish | Garden pea, lettuce, cucumber |
| Tomato | Onion, lettuce, carrot, radish, Chinese cabbage |
| Turnip | Garden pea |
| Cole crops | Potato, tomato, garden beet, onion |
| French bean | Potato, cucumber |
| Cucumber | Pole beans, radish, bhindi, garden peas |

## F. Relay cropping

Its concept has been derived from relay race in which four runners run in the field having flag in their hand. The 1st runner passes on its flag to the succeeding partner and 2nd to 3rd and 3rd to the 4th runner. Relay cropping can be defined as growing two or more crops simultaneously during the part of the life cycle of each succeeding crops simultaneously during the part of the life cycle of each. Succeeding crops planted before harvesting the proceeding crop. Generally the 2nd crop planted after the 1st crop has reached its reproductive stage of growth but before it is ready for harvest. For example:

1.  Radish is sown before harvesting of potato.

2.  In the furrows of potato crops seeds of cucurbits are sown.

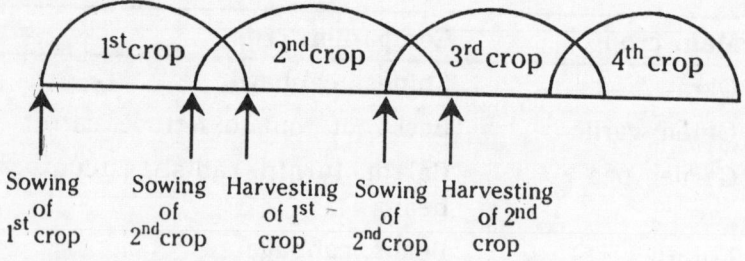

### G.   Vertical cropping

The plants are grown with support like peas and pole beans will climb naturally but tomatoes, cucumber, squash melons and gourds can be trained to climb upward. Wire fence, poles, trellis etc. can be used to support plants, tomato can be grown with staking. Peas and beans can be grown on fence. Cucumber and other vines can be trained to grow on trees and shrubs at the border of the garden.

### Nursery management

Nursery management indeed is a very important operation in successful production of vegetables especially which are raised through seedlings in properly managed nursery beds. As like cradle, nursery is the place where young seedlings are raised and nurtured before transplanting them in the main field. On the basis of nature of vegetables raised for production can be grouped as:

| Raising method | Vegetable crops |
|---|---|
| A. Direct seeding | Okra, radish, turnip, carrot, garden beet, garden peas, beans, spinach, cucurbits, chicory, beet leaf, asparagus |
| B. Nursery raising | Tomato, brinjal, capsicum, chillies, cole crops, onion, lettuce, celery, parsley, Chinese cabbage, amaranths, asparagus, potato (TPS) |
| C. Both direct seedling and nursery raising | Lettuce, amaranths, celery, parsley, Chinese cabbage |
| D. Steckling raising | Radish, turnip, carrot, garden beet, chicory |
| E. Vegetatively propagating | Asparagus (crown), ginger, turmeric (rhizome), colocasia (corms), potato (tubers) |

## Advantages

(i) It is very convenient to look after the tender seedlings.

(ii) Timely and careful plant protection measures are possible.

(iii) Most favourable growth medium is provided.

(iv) Seedlings are in a protected place and usually timed for early crops.

(v)   It ensures uniform growth and establishment of plants in the field.

(vi)  There is economy of land and seed, and more time is available for field preparation.

## Soil preparation

The soil should be fine, moist and firm to provide better germination and excellent medium for the growth of seedlings. Sufficient quantity of well decomposed farmyard manure should be mixed thoroughly in the soil. Usually the soil mixture for pots should have two parts of garden soil, one part of sand and one part of leaf mould. In case of nursery bed, the width should not be more than 120 cm and the length 150 cm. This width facilitates weeding and watering without trampling the bed. The bed is usually kept raised about 15 cm high so as to provide proper drainage of excess water and the level of the bed surface is also slightly raised in the centre with a little slope on the two sides.

## Sowing

The common practice is to broadcast seeds in the nursery bed but line sowing is preferred for proper germination and to facilitate weeding, hoeing and plant protection operations. The rows are usually kept about 5 cm apart. The size of nursery bed to raise seedlings to plant a hectare of tomato, brinjal and capsicum should be 50 m x 1.2 m, 54 m x 1.2 m and 180 m x 1.2 m, respectively. In case of cole crops like cauliflower, cabbage and knol-khol, a nursery bed of 115 m x 1.2 m would be sufficient to raise seedlings for one hectare plot. These could be divided into smaller beds instead of making a single bed. Small seeds should be still lighter in heavy soils. Covering the seeds in furrows or rows may be done gently by fingers or with the aid of a wooden strip, followed by a light watering with a sprinkler.

## After-care

Watering of the bed is done uniformly and gently so as to avoid a packing of the soil or washing away of the soil or the soil covering. It is always advisable to make the nursery beds near the source of irrigation water.

In the beginning, if the sun is too warm, the heat of the mid-day sun may be avoided by covering the seedlings with a thin layer of leaves, twigs or thatch. During summer, the seedlings may be protected against warm wind and sunshine. When the seedlings are more than 2.5 cm tall, too much of shade and water would make them yellow, succulent, lanky and susceptible to insect pests and diseases like *damping-off*, especially in humid and warm weather. The diseased plants should be removed immediately. A week before transplanting, the seedlings may be exposed to full sunshine and the number of watering reduced so that the seedlings become hardy to bear the shock of transplanting.

## Transplanting

If seedlings are allowed to grow too tall, they become weak and may start flowering. Transplanting should be done as soon as the seedlings are about 4 to 8 weeks old, 10 to 15 cm tall and have formed about 3 to 4 true leaves. The bed must be watered 24 hours before uprooting the seedlings for transplanting so that they may not suffer from desiccation. Transplanting should always be done in the evening so that plants may establish themselves in the cool atmoshpere at night and may recover from the shock of transplanting before sunrise.

Transplanting should be done as early as possible after carefully removing the plants without damaging the roots. The soil should be prepared thoroughly before transplanting. Each seedling is placed vertically in the centre of the hole made in the field and the soil near the roots is pressed down with the fingers to make the soil firm. This may be followed by immediate watering. During transplanting, care should be taken to protect the seedlings against wilting by frequently sprinkling water on them and by covering the root zone by moist soil or leaves. Regular watering is necessary after transplanting. Seedlings not doing well may be removed and replaced by new ones. Any attack of disease or insect pest must be controlled immediately.

## Category

On the basis of ability to be transplanted different vegetable crops may be classified into following categories.

| Planting response | Vegetable crops |
|---|---|
| A. Easily survive by transplanting | All Cole vegetables, Chinese cabbage, lettuce, tomato, brinjal, capsicum, chillies |
| B. Needs care in transplanting | Onion, leek, celery, parsley, amaranths |
| C. Not suited to transplanting | Garden pea, turnip, radish, carrot, garden beet, spinach, gourds, okra, methi |
| D. Suits both transplanting and direct seedling | Chinese cabbage, chillies, amaranths |
| E. Polypack planting | All cucurbit mostly, cucumber bittergourd, squashes, melons, pumpkins, tomato (hybrid) |

## Age

Seedlings should always be transplanted at their suitable age. Generally, it has been seen that over-aged seedlings have higher percentage of mortality and slow establishment rate. Bolting in onion and premature seeding in cabbage are the best examples of the negative effects of planting over-aged seedlings in the field. Thus, for transplantation age of the seedling is an important factor which determines the outcome of the crop.

**Suitable age of transplantable seedlings of some vegetable crops**

| Vegetable Crops | Age in weeks |
| --- | --- |
| Brinjal | 3–4 |
| Brussel's sprout | 6–8 |
| Cabbage | 4–6 |
| Cauliflower | 3–4 |
| Celery | 6–8 |
| Leek/Lettuce/Parsley | 4–5 |
| Onion | 6–8 |
| Sprouting broccoli | 3-4 |
| Tomato | 3-4 |

## Double transplanting

In double transplanting system, after retaining the young seedlings at a different place for some time, they are finally transplanted in the main field where they grow and give economic yield. Double transplanting is most successful in tomato, brinjal, cabbage and cauliflower. It is commonly seen that seedlings grow in little space in the primary nursery. There they face stiff competition for growth as a result of which some become weak, feeble and lanky. Besides, such seedlings, when transplanted in the main field, show poor survival and establishment and the ultimate yields are also not high. The best way is to grow seedlings by double transplanting method in which they are planted in secondary nursery they get wide space and more quantum of growth essentials without facing any competition. Seedlings obtained from secondary nursery are strong and stocky with well-developed root system which result in early and cent per cent survival and establishment in the main field. Planted seedlings pick up early and fast rate of growth as they get opportunity for normal period of growth that results in very high yield potential.

The necessity for opting double transplanting is increased when the plot is not ready for transplanting the crop, which is common in intensive cropping system. Under such circumstances, it is advisable to return seedlings to secondary nursery instead of retaining them in primary nursery.

## Propagation

Vegetable crops are propagated by seed and their vegetative parts. Some vegetable crops are raised by sowing of seeds directly in the main fields, like the crops of okra, beans, leafy vegetables and cucurbits, whereas other vegetables respond well when they are raised by planting seedlings *viz.* tomato, brinjal, chillies, cabbage, cauliflower, knol-khol. For this purpose, seeds are sown first in the nursery and seedlings are prepared for planting in the main field. Besides, there are a number of vegetable crops which are propagated by vegetative means. For instance, potato, sweet potato, elephant's foot, yams, colocasia, onion, pointed gourd, chow-chow, spine gourd and garlic are propagated by their specialised structures.

## Planting materials for vegetable crops

| Planting material | Vegetable crops |
|---|---|
| Bulb | Onion |
| Bulbil | Garlic |
| Clove | Garlic |
| Crown | Asparagus, Rhubarb, Horse radish |
| Corm | Chinese water chestnut, Tannia |
| Cormel | Giant Taro, Colocasia (Taro), Tannia |
| Limb cutting | Drumstick |
| Minisett | Yams |
| Root cutting | Pointed gourd, Breadfruit, Horse radish, Cassava |
| Rhizome | Arrowroots |
| Stem cutting | Ivy gourd, Chekurmanis, Ceylon spinach (Waterleaf), Basella |
| Seed | Almost all vegetable crops |
| Sucker/Off shoot | Globe artichoke, Giant Taro |
| Shoot tip | Giant Taro |
| Sett | Onion, Jerusalem artichoke |
| Tuber | Potato, Yams, Jerusalem artichoke, Elephant's foot |
| Vine cutting | Pointed gourd, Sweet potato, Yams, Coleus (Chinese potato) |
| Whole (entire) fruit | Chow-chow |

## Sowing time

The time of sowing seeds and plants of particular species in the open determine the success or failure of the crop to a considerable extent. Even if we use good seeds, satisfactory crop will not be obtained unless the sowing or planting is done at the optimum time. The planting time should be determined by taking into consideration the soil and weather conditions, the kind of the crop and the time when the produce is desired for vegetable purpose.

**Sowing time for raising vegetable crops**

| Sowing time | Vegetable crops |
|---|---|
| Jan-Feb | French bean, Bitter gourd, Snake gourd, Watermelon, Muskmelon |
| Feb-March | Okra, Cucumber, Brinjal, Colocasia Amaranth, Rhubarb |
| May-June | Cauliflower, Rhubarb |
| June-July | Okra, Cowpea, Bottle gourd, Bitter gourd, Snake gourd, Ridge gourd, Round melon, Pumpkin, Cucumber, Brinjal, Chilli, Sweet potato, Colocasia, Amaranth. |
| June-Sept | French bean |
| Aug-Sept | Tomato, Cauliflower, Broccoli, Brussel's sprouts |
| Aug-Jan | Radish, Carrot |
| Sept-Nov | Pea, Cabbage, Knol-khol, Lettuce, Potato, Fenugreek, Beet leaf, Garlic, Turnip, Cauliflower |
| Nov-Dec | Tomato, Brinjal, Chilli, Onion |

## Spacing

Spacing refers to distance between crop rows (inter-row spacing) and between plants within a row (intra-row spacing). Plants grow and develop with the resources of the environment (soil and atmosphere). If there are more plants per unit area, the intra-plant (within the plant and among the plant components) and inter-plant (among the plants in the area) competition for the resources (space, moisture, nutrient, sunlight, $CO_2$, etc.) will increase. If there are few plants per unit area, the environmental resources will remain underutilised. The total yield in either case will be less. There is always competition for growth factors at all levels (micro and macro) and optimum use is essential. Plant population (number) per unit area should be such that all plants get equal opportunity to utilise the resources for their optimum growth and development.

## Spacing for vegetable crops

| Spacing (cm) | Vegetable crops |
| --- | --- |
| 15 x 10 | Onion, Garlic |
| 20 x 5 | Beet leaf |
| 30 x 5-10 | Okra, French bean, Cucumber, Turnip, Carrot, Amaranth, Pea, French bean, Cowpea, Radish |
| 45 x 30 | Okra, Knol-khol, Lettuce, Colocasia |
| 45 x 45 | Chilli, Cauliflower, Sprouting broccoli |
| 60 x 15 | Potato |
| 60 x 30 | Sweet potato |
| 60 x 45 | Okra, Tomato, Cauliflower, Cabbage, Sprouting Broccoli, Brussel's sprout |
| 60 x 60 | Brinjal, Brussel's sprout |
| 75 x 75 or 90 x 90 | Cassava, Yams |
| 100 x 300 | Watermelon, Muskmelon |

| Spacing (cm) | Vegetable crops |
|---|---|
| 150 x 250 | Bitter gourd, Snake gourd, Ridge gourd, Cucumber |
| 150 x 300 | Pumpkin |
| 300-400 x 75-100 | Bottle gourd |

## Seed rate

The quantity of seed to be used depends upon its purity, viability, the time of planting, the condition of the soil, the size and vigour of the plants and the possible ravages of diseases and insects. Seeds known to possess low viability should be sown more thickly than those having high percentage of germination. Under unfavourable soil and weather conditions, heavier seed rate should be used than when the conditions are favourable for quick germination. If longer time is required for germination of any given kind of seed, the rate of planting should be heavier so as to obtain a proper stand because sometimes damage to germinating seed is directly related to seed rate. However, the seed rate depends upon the final distance of sowing/planting in the field.

## Recommended rates of seed per hectare for vegetable crops

| Seed rate (kg/ha) | Vegetable crops |
|---|---|
| 0.3-0.4 | Brussel's sprouts, Sprouting broccoli |
| 0.4-0.5 | Brinjal, Cauliflower, Knol-khol, Lettuce |
| 0.5-0.6 | Tomato, Cabbage |
| 0.75-1.0 | Amaranth |
| 1.5-2.0 | Chilli |

| Seed rate (kg/ha) | Vegetable crops |
|---|---|
| 2-3 | Bitter gourd, Snake gourd, Pumpkin, Watermelon, Cucumber, Turnip |
| 3-5 | Ridge gourd, Onion, Asparagus, Round melon, Carrot |
| 5-8 | Radish |
| 8-10 | Okra (*kharif*) |
| 15-20 | Okra (summer-spring), Beet leaf |
| 20-30 | Cowpea, Fenugreek, French bean (pole type) |
| 60-80 | Pea, French bean (bush type) |
| 800-1000 | Colocasia |
| 2000-2500 | Potato |
| **No. of vine cuttings** | |
| 2,000-2500 | Pointed gourd |
| 40,000 | Sweet potato |

## Water management

Water management is one of the most important factor for maximisation of yields. A fractional application of irrigation water is based on the knowledge of the consumptive use of water by crop and relationship between the moisture status of root zone and yield potential of the crop. Depletion of soil moisture results in reduction of yield. Increase in soil moisture stress produces moisture deficit in the plants. Therefore, assessment of the optimum need of water for the root zone is a must.

**Optimum soil moisture range at which irrigation has to be given for some vegetable crops**

| Vegetable crops | Optimum soil moisture tension or per cent available moisture depletion | Soil depth for measurement (cm) |
| --- | --- | --- |
| Beetroot | 0.25 bar | 1.8 |
| Brinjal | 20 - 40 % depletion | 0-30 |
| Cabbage | 50% depletion | 0-30 |
| Capsicum | 20-40% | 0-30 |
| Cauliflower | 0.25-0.30 bar | 15 |
| Okra | 0.50 bar | 18 |
| Onion | 0.50-0.65 bar | 8-10 |
| Potato | 0.30 bar | 15 |
| Radish | 0.25 bar | 18 |
| Sugar beet | 0.20 bar | 25 |
| Tomato | 40-50 bar | 0.30 |
| Turnip | 0.25 bar | 18 |

## Water requirement

Water requirement of a vegetable crop is the quantity needed for normal growth and yield. The water need of a vegetable is specific for a location. The amount of water varies from field to field depending on the water storage in the soil, weather situation, water supply and its quality, depth of water table, vegetable crop and its variety, stage of plant growth and management practices.

In general, however, the vegetables can be grouped under two categories according to their moisture needs. These are:

| Water loving | Stress sensitive |
|---|---|
| Cabbage | French beans |
| Cauliflower | Pea |
| Radish | Cowpea |
| Carrot | Garlic |
| Onion | Tomato |
| Potato | Brinjal |
| Spinach | Okra |
| . | Squash |
| | Muskmelon |

Water is needed mainly to meet the demand of evaporation (E), transpiration (T) and metabolic needs of the plants, all together known as consumptive use (CU).

$$CU = E + T + \text{Water needed for metabolic purpose}$$

Water used in the metabolic activities of vegetable crops in negligible and is often less than 1% of the quantity of water passing through the plant. Evapotranspiration is, therefore considered as equal to consumptive use. Different losses like percolation, seepage, runoff etc; occur during transport and application of irrigation water. Water is needed for special operations such as land preparation, transpiration, leaching etc; Water requirement (WR) of a vegetable crop, therefore, includes evapotranspiration, application losses and water needed for special purposes.

$$WR = ET + \text{Application losses} + \text{Water for special purpose}$$

Water requirement is demanded whereas the supply consists of contribution from irrigation water, effective rainfall (ER) and soil profile contribution including that from shallow water table. Range of water requirement of different vegetable crops are presented in table:

## Water requirement (mm) of vegetable crops

| Vegetable crops | Water requirement |
|---|---|
| Bean | 300-500 |
| Pea | 350-500 |
| Onion | 350-550 |
| Cabbage | 380-500 |
| Potato | 500-700 |
| Garden beet | 550-750 |
| Tomato | 600-800 |

### Critical periods of water use

Vegetable needs a fairly constant supply of soil moisture throughout the growing season. Normally water stress in early stages delays maturity and reduces yields but later in the growing season, during bolting, flowering, setting and maturity affects quality yield. Usually, vegetables grown for their foliage i.e. vegetative phase, needs uniform moisture throughout the vegetative phase while in reproductive phase as bolting, flowering, fruit/seed set and maturation, maximum water is required. However, water stress during anthesis, dehiscence and stigma receptivity can result in a poor pollination directly affecting the fruit/ pod yield and quality. Moreover, moisture stress during anthesis causes flowers to abort resulting in incomplete or no filling on pods and thus malformed pods resulting yields. Excess irrigation and at the time of when not required are detrimental for proper growth and development and affects quality and yields. Too much soil moisture, combined with high nitrogen level produced more vegetative growth in solanaceous vegetables and cucurbits, delays fruit/pod maturity and produces poor quality fruit yields.

## Moisture sensitive stages of vegetable crops

| Vegetable | Critical stage |
| --- | --- |
| Asparagus | Fern growth |
| Beans | Flowering and pod formation |
| Broccoli | Head formation and enlargement |
| Cabbage | Head formation and enlargement |
| Carrot | Root enlargement |
| Cauliflower | Frequent irrigation from planting to harvest |
| Celery | During establishment and rapid growth |
| Cucumber | Flowering and fruit enlargement |
| Egg plant | Flowering and fruit enlargement |
| Lettuce | Head development |
| Muskmelon | Flowering and fruit development |
| Onion | Bulbing and enlargement |
| Pepper | Transplanting, fruit setting and development |
| Potato | Tuber initiation to tuber maturity |
| Pumpkin | Flowering and fruit development |
| Summer squash | Flowering and fruit development |
| Sweet corn | Tasseling, silking and ear filling |
| Tomato | Flowering, fruit setting and enlargement |
| Watermelon | Blossoming to harvesting |

The total required quantities of water should be distributed through several light irrigations instead of flooding the field unevenly once or twice. To facilitate even distribution of water, the field should be well levelled in advance. Where levelling is not possible, sprinkler irrigation may be adopted, though this method requires high initial investment and all growers may not be able to afford it. However, an efficient and economical method of irrigation should be within the reach of cultivators for obtaining the desired yields.

## Drainage

The excess water in the field whether it comes from over-irrigation or from natural source or rain/snow water accumulation needs to be drained from the field to ensure normal crop growth. As poorly drained soils not only harm the vegetable crop seed directly but creates various problems in scheduling the mechanical farm operation, invites and promotes the development of diseases and pests.

## Nutrient management

In most of the vegetable producing areas, the soil has been cultivated traditionally for many years and ultimately soil fertility gets lost. Manures and fertilizers are applied to replenish the losses and to maintain the soil fertility.

Organic matter forms a very important source of various plant nutrients, has binding effect saving the soil from erosion, improves the texture and structure, water holding capacity, increases the soil temperature and finally improves the productivity which helps in raising better vegetable crops.

The decaying organic matter serves the useful functions in the soils, supports organisms, like bacteria, fungi and animal forms, improves texture and structure of soils, increases the retention power of soil for moisture and nutrients, improves aeration, helps in decomposing, the process of decompositon generates some heat which increases the temperature of the cool soils to enable in better quick seed germination and proper growth of the vegetable plants.

## Sources of nutrients

**Farm Yard Manure (FYM):** FYM is decomposed mixture of dung and urine of farm animals along with litter and left-over material from roughages or fodder fed to the cattle. On an average well-decomposed FYM contains 0.5 per cent N, 0.2 per cent $P_2O_5$ and 0.5 per cent $K_2O$. The entire amount of nutrients present in FYM is not available immediately. About 30 per cent nitrogen, 60 to 70 per cent of phosphorous and 70 per cent of potassium are available to first crops of vegetables. Researches have shown that certain vegetables have better response to FYM while for other is comparatively less. Now-a-days, since FYM is available in limited quantities, it would be profitable to apply the organic manure only to responsive vegetable crops.

### Response of vegetable crops to FYM (t/ha)

| Very high (25-30) | High (20-25) | Moderate (15-20) | Low (10-15) |
|---|---|---|---|
| Ginger | Tomato | French bean | Bhindi |
| Colocasia | Brinjal | Garden pea | Bitter gourd |
| Cauliflower | Capsicum | Garden beet | Pumpking |
| Cabbage | Chillies | Beet leaf | Summer squash |
| Onion | Cucumber | Spinach | Bottle gourd |
| Garlic | Turmeric | Lettuce | Melon |
| Sprouting broccoli | Chinese cabbage | Celery | Methi |
| | Turnip | Parsley | Chicory |
| | Radish | Broad bean | |
| | Carrot | | |
| | Brussel's sprout | | |
| | Asparagus | | |
| | Leek | | |

**Fertilizers:** Fertilizers, usually referred to as *chemical fertilizers* or *commercial fertilizers*, are the synthetic or naturally occuring chemical compounds applied to the soil to supply plant nutrients for crop growth. The high nutrient content in easily available form has led to the worldwide adoption of fertilizer use for increased and sustained crop production. That the proper use of fertilizer greatly increases crop production is well established. The use of fertilizers makes it possible to introduced extra supplies of nutrients into cycles of growth and decay, thereby improving soil fertility. It is a mistake to think that fertilizers are the substitutes for manures which have their role to play. It is advisable to use recommended doses of commercial fertilizers along with manure to enhance fertilizer use efficiency ecofriendly.

Even though every essential element is added singly or in mixture, nitrogen, phosphorous and potassium are generally added all over the world.

Urea is white, crystalline, hygroscopic and freely soluble in water. It is the most widely used nitrogenous fertilizer in vegetable crops. It contains highest percentage of nitrogen (46%) among solid fertilizers. It is an ideal fertilizer for foliar spray of nitrogen in vegetable crops.

Diammonium phosphate (DAP) is most widely accepted phosphatic fertilizers among vegetable growers. It is a completely water soluble and granular fertilizer. It is available in two grades: 16-48-0 and 18-16-0.

Potassium chloride or muriate of potash (MOP) is the most common and cheap fertilizer among potassic fertilizers. It contains 58-60% $K_2O$. It is suitable for most of the vegetable crops except potato.

## Requirement of manure and fertilizers by vegetable crops

| Vegetable crops | Farmyard manure (t/ha) | Fertilizer N:P:K (kg/ha) |
| --- | --- | --- |
| Amaranth | 10-15 | 80 : 40 : 40 |
| Beet leaf | 10-15 | 80 : 40 : 40 |
| Bitter gourd | 15-20 | 60 : 80 : 60 |
| Bottle gourd | 30-35 | 60 : 80 : 60 |
| Brinjal | 20-25 | 100 : 60 : 40 |
| Cabbage | 25-30 | 120 : 80 : 40 |
| Carrot | 10-15 | 60 : 40 : 40 |
| Cauliflower | 20-30 | 120 : 80 : 40 |
| Chilli | 20-25 | 90 : 60 : 40 |
| Colocasia | 10-15 | 100 : 80 : 80 |
| Cowpea | 15-20 | 40 : 50 : 60 |
| Cucumber | 10-13 | 50 : 50 : 70 |
| French bean | 15-20 | 50 : 80 : 60 |
| Knol-khol | 25-30 | 100 : 50 : 40 |
| Lettuce | 10-15 | 80 : 40 : 40 |
| Muskmelon | 30-35 | 80 : 50 : 50 |
| Okra | 12-15 | 60 : 40 : 40 |
| Onion | 20-25 | 100 : 60 : 60 |
| Pea | 15-20 | 30 : 70 : 50 |
| Potato | 25-30 | 120 : 80 : 80 |
| Pumpkin | 15-20 | 75 : 8 0 : 80 |
| Radish | 15-20 | 50 : 40 : 40 |

| Vegetable crops | Farmyard manure (t/ha) | Fertilizer N:P:K (kg/ha) |
| --- | --- | --- |
| Ridge gourd | 10-15 | 60 : 80 : 60 |
| Round melon | 15-20 | 60 : 80 : 60 |
| Snake gourd | 15-20 | 60 : 80 : 60 |
| Sweet potato | 15-20 | 80 : 60 : 60 |
| Tomato | 20-25 | 100 : 60 : 60 |
| Turnip | 10-15 | 50 : 50 : 60 |
| Watermelon | 30 - 35 | 80 : 50 : 50 |

**Green manure:** Green, undercomposed plant material used as manure is called *green manure*. Green manuring is growing in the field plants, usually belonging to leguminous family and incorporated into the soil after sufficient growth. The most important green manure crops suitable for vegetable growing field are sunhemp, dhaincha, *Pillipesara,* clusterbean, *Sesbania rostrata.*

Several advantages accrue due to addition of green manures. Organic matter and nitrogen are added to the soil. Growing deep-rooted green manure crops and their incorporation facilitates in bringing nutrients to the top layer from deep layer. Green manuring improves soil structure, increases water holding capacity and decreases soil loss by erosion. growing of green manure crops in off-season reduces weed proliferation and weed growth. Green manuring helps in reclaiming alkaline soils. Root-knot nematodes can be controlled very effectively in vegetable crops by green manuring.

**Biofertilizer:** The term biofertilizer or which can be more appropriatly called '*microbial inoculants*' can be generally defined as preparations containing live or latent cells of efficient strains of nitrogen fixing, phosphate solubilizing or cellulolytic micro-organisms used for application to seed, soil or composting areas with the objective to increase microbial processes to augment the extent of the availability of nutrients in a form which can be easily assimilated by plants.

Recently, application of biofertilizers in vegetable crops has been found very effective. These harper atmospheric nitrogen with the help fo specialized soil micro-organisms. Such nitrogen fixing micro-organisms are either free living in the soil and symbiotic with plants.

The nitrogen is fixed symbiotically by the bacteria, *Rhizobium* inoculated leguminous vegetable crops efficiently. Each legume vegetable should be inoculated for that particular leguminous vegetable crop.

**Cross-inoculation groups of *Rhizobium***

| *Rhizobium* spp. | Cross inoculation grouping | Vegetable types |
|---|---|---|
| *R. leguminosarum* | Pea group | *Pisum* (Pea) |
| *R. phaseoli* | Bean group | *Phaseolus* (Kidney bean) |
| *R. meliloti* | Alfalfa group | *Trigonella* (Methi) |
| *R.* sp. | Cowpea group | *Vigna* (Cowpea) *Cymopsis* (Cluster bean) |

The beneficial effect of *Rhizobium* inoculation under the field conditions is noticed by increase in the yield of the succeeding vegetable crops.

Blue Green Algae (BGA) is mostly grown on wet lands. This condition is highly suitable for its growth. They are free living and can fix atmospheric nitrogen. The most important species of BGA are *Anabaena* and *Nostoc*. The amount of nitrogen fixed by blue-green algae ranges from 15 to 45 kg N/ha.

*Azolla* is a floating fresh water fern, an important organic manure and comprises an algal symbiont (*Anabaena azollae*). *Azolla pinnata* is the most common species occuring in India. A thick mat of *Azolla* supplies 30 to 40 kg N/ha in subsequent vegetable fields.

*Azotobacter* is capable of fixing 20 to 30 kg N/ha. It can applied by seed inoculation, seedling dip or by soil application. *Azotobacter* can be used for onion, brinjal, tomato, cabbage, potato and sugarbeet very effectively under different agroclimatic conditions.

*Azospirillum lipeferum* and *A. brasilense* are two important species of *Azospirillum*. It is a common soil inhabitant in tropical conditions. The seed inoculation with this bacteria has been found useful in amaranths.

Bacteria, especially those of genera *Pseudomonas* and *Bacillus* as well as fungi genera *Penicillium* and *Aspergillus* play an important part in solubilizing the insoluble form of soil phosphorus. *Phosphobacterin* as a biofertilizer in potato has given good results.

Roots of most of the vegetable crops (perhaps 97%) become infected with fungi present in the native soils and form mycorrhizal, which spells out a 'symbiotic' (that is intimate) and 'mutualistic' (that is mutually beneficial) association between a non-pathogenic fungi or a weakly pathogenic fungus and the living root cells, primarily cortical and epidermal cells. In the process of exchange (mutually beneficial), the fungus concerned receives organic nutrients from the plant, and, in return, augments the plant capacity to absorb more mineral nutrients and water. The greatest advantage of *mycorrhizae* is probably to increase the absorption of essential ions that normaly diffuse slowly though needed in large quantities, particularly phosphate, $NH_4^+$, $K^+$ and $NO_3^-$. In onion, inoculation with *endomycorrhizae*, *Vesicular Arbuscular Mycorrhizae* (VAM) at nursery stage and in the main field recorded and additional yield of 22.93 t/ha and saved 25% recommended dose of NPK fertilizers.

## Mulching

Mulches are covering material to the soil surface and mulching is a created/acceptable age-old cultural practice around crop plants to improve growing conditions with natural or synthetic materials. They are mainly used to reduce or increase the temperature. High temperature effects seed germination, vegetative growth, flowering, fruit set and yield. Mulch also improves the quality of vegetables by avoiding their direct contact with the soil. The role of mulching has become of great significance and is used extensively in vegetable crop production.

### Principles

❖ Changing and modifying the micro-climate all round the plant(s).

❖ Maintain soil temperature and moisture, avoiding weed competition, soil structure and biological activity.

❖ Check evaporation, soil erosion and conserve soil moisture.

❖ Check weeds which cause shade effect and act as hibernation place for insect (aphids) and alternate host for disease pathogens.

❖ Add organic matter in due course of time (organic mulch) and improve soil fertility.

❖ Silver-coloured plastic, reflects lights acts as effective repellents for insects/birds etc.

❖ In hills where mostly unfavourable weather conditions prevail, the greatest role of mulching is wanting.

❖ Most of the vegetables applied with mulch early in season, brings improvement in quality.

## Objectives

❖ To conserve moisture.

❖ To reduce infiltration rate.

❖ To reduce run-off and soil erosion.

❖ To maintain soil temperature.

❖ To increase fertilizer use efficiency.

❖ To save young germinating and tender plants from direct sun.

❖ To project vegetables/seed crop from heavy rains, snow, hails and cold freezing injury.

❖ To increase plant growth and yields.

❖ To check weed growth and maintain cleanliness.

❖ To promote earlier harvest.

❖ To add organic matter in soil.

❖ To stimulate microflora in soil.

## Methods

i)    The mulch material, both organic and inorganic are applied on the soil surface in the form of continuous layer having a specified thickness depending upon the type of material used.

ii)   By doing cultural operations i.e. stirring uppermost crust of soil surface with the help of any farm implement which helps in breaking the soil surface crust down to few centimeters depth. This operation helps in breaking the water capillaries through which evaporation takes place and the soil water evaporation is checked.

The main objective of both of these methods is to create barrier to the moisture loss and to alter temperature by a few degrees, and to effect certain attributes pertaining to micro-climate of seed plant especially near root zone.

## Mulch materials

### Natural

❖    Plant residues - hay, straw, leaves, stubble, dry orchard leaves.

❖    Animal manure.

❖    Peat moss.

❖    Wood products - bark, wood chips, saw-dust, tree droppings etc.

❖    Paper.

These are of organic in nature and get degraded in one or two growing seasons. These are cheap, easily and locally available. These are applied at the rate of 4-6 tonnes/ha dry form, 10-12 tonnes/ha, green leaves and cattle manure in 4-7 cm thick layer.

## Synthetic

❖ Polythene film-white, black and transparent (300-600 gauge).

❖ Metal foils.

❖ Asphalt mulch.

These are made up of plastic or polythene films mainly of white reflective, white, opaque black, black, blue and transparent (clear) coloured, most extensively used for commercial vegetable production. In hilly areas, in some pockets, the use of polythene mulch is picking up in vegetable/seed crop growing.

## Others

❖ Polyvinyl chloride sheet

❖ Photo-degradable plastic mulch

❖ Aluminium foil

❖ Polythene tunnels

❖ Silver-coloured strips

❖ Black fertene

❖ Alkathene

❖ Bitumen paints

## Interculture operations

The main aim of interculture is to keep the soil loose and destroy weeds in between the rows and plants of the vegetables. The hoeing is done to pulverise the soil, break the crust and capillaries to check water losses through evaporation. During the interculture operations, weeds are also pulled out and thrown. The interculture is done till the vegetable crops attain growth and the field is clean. However, the shallow frequent cultivation should be given without harming the roots/root hairs of the vegetable plants and provide soil mulch. Avoid deep cultivation, as it may destroy the vegetable plant root located in 30-60 cm in the soil depth. Weeding should be started as soon as vegetable plants are set in the field. To check the weed growth and for making the soil loose around the plants; the field may be tilled 2-3 times after irrigation when the soil comes into working conditions. Hoeing in sandy, silt and loam soils is easy but in clay loam and loamy soil it is not so easy.

## Weed management

Weeds are found everywhere from crop lands to wastelands and even at the bottom of the sea. Most of them are harmful to crops. Some of them are useful to mankind too, as they have medicinal values. However, for the successful production of vegetable crops, weed management is of utmost importance. In fact, none of the vegetables is known to be free from weeds. The incidence of weeds varies with climate, soil, management practices etc. No doubt, weeds do harm vegetable production (loss ranges from 30% to 45%) but at high incidence only. Though much damage is caused by weeds during the early stages of crop's growth, later when crop canopy develops, most weeds are suppressed. The extent of suppression depends on the spread of the crop. Upright growing crop plants do not have the ability to suppress weed. In this case, dense planting of crop proves helpful in suppression of many types of weeds and thereby economises vegetable production. Control of weeds at the right time and in the right way increases production potential of vegetable crops, helps in adopting production practices effectively and efficiently, and makes vegetable cultivation profitable.

## Association of weeds with vegetable crops

| Vegetable crops | Associated weeds |
|---|---|
| Tomato | *Carthamus oxycantha, Chenopodium album, Cirsium arvense, Cynodon dactylon, Cyperus rotundus, Echinochloa crusgalli, Orobanche cernua, Spergula arvensis, Vicia hirsuta* |
| Brinjal | *Blumea spp., Chenopodium album, Cynodon dactylon, Cyperus rotundus, Orobanche aegyptica* |
| Chillies | *Ageratum conyzoides, Amaranthus spinosus, Argemon mexicana, Blumea spp., Cannabis sativa, Chenopodium album, Cirsium arvense, Convolvulus arvensis, Cynodon dactylon, Cyperus rotundus, Euphorbia hirta, Leucas asperta, Orobanche cernua, Vicia hirsuta, Trianthema monogyna* |
| Cabbage | *Chenopodium album, Cynodon dactylon, Cyperus rotundus, Digitaria sanguinallis* |
| Cauliflower | *Chenopodium album, Cynodon dactylon, Cyperus rotundus, Digitaria sanguinallis* |
| Garlic | *Amaranthus viridis, Cynodon dactylon, Cyperus rotundus* |
| Onion | *Amaranthus viridis, Coronopus didima, Cynodon dactylon, Cyperous rotundus, Echinochloa colonum, Tribulus terres* |
| Potato | *Alternanthera sessilis, Argemone maxicana, Asphodelus tenuifolius, Blumea spp, Cannabis sativa, Chenopodium album, Cirsium arvense, Convolvulus arvensis, Cynodon dactylon, Cyperus rotundus, Euphorbia hirta, Melilotus alba, Orobanche ramosa, Wedelia calendulacea* |

| Vegetable crops | Associated weeds |
| --- | --- |
| Peas | *Anagalis arvensis, Asphodelus tenuifolius, Blumea laura, Carthamus oxycantha, Chenopodium album, Convolvulus arvensis, Cynodon dactylon, Cyperus rotundus, Digera arvensis, Fumaria parviflora, Polygonum bistorta, Viola spp.* |
| Radish | *Chenopodium album, Convolvulus arvensis, Cynodon dactylon, Cyperus rotundus, Poa annual* |
| Carrot | *Chrysanthemum segetum, Cyperus rotundus, Orobanche ramosa, Oxalis corymbosa, Portulaca oleracea, Trianthama monogyna* |
| Beetroot | *Chenopodium album, Polygonum plebijum, Cynodon dactylon* |
| Ginger, Turmeric | *Cyperus rotundus, Cynodon dactylon, Amaranthus viridis* |
| Asparagus | *Chenopodium album, Convolvulus arvensis, Cynodon dactylon, Cyperus rotundus, Ranuculus repens* |

## Chemical weed control

Use of herbicides can control weeds very effectively in vegetable crops resulting in efficient use of irrigation, fertilizers and plant protection measures. Weed-free crops give higher yield. Weed control by application of chemical herbicides is easy to adopt, requires very short time and costs less than other methods. Besides, it is extremely useful in minimum tillage operation.

# Chemical weed control in vegetable crops

| Vegetable crops | Recommended weedicide | Dose (kg/ha) | Time of spray |
|---|---|---|---|
| Cabbage | Oxyfluorfen | 1.00 | Just after |
| | or Fluchloralin | 2.00 | transplanting |
| Cauliflower | Oxyfluorfen | 1.25 | 24 hrs after |
| | or Pendimethalin | 2.00 | transplanting |
| Radish | Butachlor | 3.00 | 24 hrs after |
| | or Alachlor | | sowing |
| Knol-khol | Butachlor | 3.00 | 24 hrs after |
| | or Fluchloralin | 2.00 | sowing |
| Turnip | Alachlor | 1.87 | After sowing |
| | or Fluchloralin | 0.90 | |
| Pea | Fluchloralin | 0.90 | After sowing |
| | or Alachlor | 1.00 | |
| French bean | Alachlor | 3.00 | After sowing |
| Cowpea | Alachlor | 3.00 | After sowing |
| Cluster bean | Alachlor | 3.00 | After sowing |
| Okra | Fluchloralin | 3.00 | After sowing |
| Tomato | Alachlor | 3.00 | One day after |
| | or Fluchloralin | 3.00 | transplanting |
| Brinjal | Fluchloralin | 1.50 | One day after transplanting |
| Chilli | Alachlor | 2.00 | 24 hrs after |
| | or Butachlor | 2.00 | transplanting |
| Carrot | Fluchloralin | 3.00 | After sowing |
| | or Linuron | 2.00 | |
| Beetroot | Alachlor | 3.00 | After sowing |
| | or Linuron | 2.00 | |
| Onion | Fluchloralin | 2.00 | After transplanting |
| Garlic | Oxyfluorfen | 2.00 | After transplanting |
| | or Fluchloralin | 2.00 | |
| Cucumber, Bottle gourd, Ridge gourd, Ash gourd, Bitter gourd, Pumpkin and Squash | Alachlor | 3.00 | 24 hrs after seed sowing |
| | or Butachlor | 3.00 | |
| | or Fluchloralin | 3.00 | |
| Lettuce | Fluazifopbutyl | 2.00 | After sowing |
| Amaranth | Fluazifopbutyl | 2.00 | After sowing |
| Spinach | Fluazifopbutyl | 3.00 | After sowing |
| Fenugreek | Pendimethalin | 1.20 | After sowing |

## Staking

Some of the weak vegetable plants especially cucurbits, pole varieties of garden pea, French bean, tomato etc. need support to get themselves exposed to open environment for getting more sunlight. This exposure to sunlight helps for more physiological and photosynthetic activities leading to more growth, flowering, pod/fruits and ultimately higher yields of better quality.

### Objectives

❖ Vegetable plants exposed to more diffused light.

❖ Proper utilisation of inputs like manures, fertilizers, irrigation water etc.

❖ Elimination of most pests (disease, insect-pests and weeds).

❖ Makes harvesting/picking of ripe fruits/pods easy.

❖ Helps in maintaining clean produce from soiling etc, and improves quality of vegetable yields.

❖ If stakes are used in dwarf and semi-dwarf vegetables/ varieties, helps in increasing quality yields.

### Material used

Normally the staking/support used are the house roofs, trees, bushes, big rocks, particularly for cucurbits because in these the vine length varies from few meters to several meters depending on the soil type, organic manure applied, irrigation water used and type of stake used. The staking material in use are:

❖ Tree branches (cheap and best for vegetable crops)

❖ Bamboo sticks

❖ Sticks rope

❖ Sticks + wire netting + wire trellis supported by strings

❖ Sticks + nylon threads

On a commercial scale the tree branches used in case of tomato in some areas, after tomato crop is over in July, pole-type bean seed sowing is done in between rows of tomato and the same stakes are used for providing support to the bean crop. Similarly during winter, tall medium varieties of garden pea seeds mixed with wheat seeds are sown as mixed crop, the wheat plants act as support to the weak plants of garden peas. In areas where bamboo is growing, is used for staking for bitter gourd crops.

## Pruning and training

Pruning away some unwanted branches within that tangled mass of vegetation lets in light and air. Some vegetables crops are either grown as such or with pruning and training (staking). In training, the weak plants may either be provided with stakes only without doing any pruning or both pruning and staking may be done. The plants are trained in such a way in the initial stages of the crop and thus, the vegetables plants are still young. Two to three leaders are left by removing the side laterals. The stakes should be driven soon after transplanting to minimise root damage.

### Advantages

* Increases the yields per unit area.

* Earlier maturity is ensured primarily due to a large number of first clusters/flush.

* Quality gets improved.

* Labour cost of picking or harvesting is reduced.

* Hoeing, spraying, dusting and irrigation operations can be performed conveniently.

* Elimination of pests and soil-borne diseases.

* More exposure to sunlight results increased yield.

**Disadvantages**

❖ High cost of training and pruning.

❖ Over exposure to direct sun in hot weather may have adverse effect eg. sun-scald in tomato.

❖ More chances of spreading the virus disease without adequate precaution.

## Pinching

Removing or cutting the growing point to permit more branching in the plant. This helps in producing more number of branches/plant resulting in increased yield. Pinching for the least pruning possible on a stem, push out the growing point of a shoot with thumbnail or sharp knife carefully causing minimum damage to the plant.

In latter case, frequent cleaning or dipping of knife in alcohol (sterilizing) will help to avoid transmission of virus and inoculum of disease and insects. This pinch causes growth to falter briefly. But also has another effect, the tip of the stem releases a hormone called auxin, that moves down that stem, inhibiting the growth of the lower buds. Removal of that stem tip prevents auxin flow, so lateral buds and existing side-shoots now grow more vigorously. Pinching generally is useful for slowing stem growth, to direct the energies in a tomato plant.

## Growth regulators in vegetable production

The role of plant growth regulators in various physiological processes in plants is well known. Growth regulators are known to affect seed germination, vegetative growth, nodulation tuberization, flowering, fruit and seed development, and fruit ripening and yield. These can also be used for producing polyploidy and male sterility in order to overcome inter-species incompatibilities and producing hybrid seeds. They also change the soil microflora. Further, the physico-chemical quality of the crop is also influenced by their use.

## List of growth regulators along with their important use in vegetable crops

| S. No. | Growth regulators | Vegetable crops | Attributes affected |
|---|---|---|---|
| 1. | **Auxin** | | |
| | **(i) Natural** | | |
| | IAA | Spinach, Fenugreek, Okra, Tomato, Brinjal, Cowpea, Onion, Cabbage, Cauliflower | Seed germination, fruit set and yield. |
| | **(ii) Synthetic** | | |
| | 2, 4-D, PCPA IBA | Tomato Cabbage, Cauliflower | Fruit set, fruiting and yield. |
| | 2, 4, 5-T | Potato | Seedling set, growth and yield Improving shelf life. |
| | NAA | Cauliflower, Cabbage Onion, Brinjal, Tomato | Flower drop, fruit set and yield. |
| 2. | **Gibberellins** | | |
| | Gibberellic acid (GA) | Potato | Breaking dormancy |
| | | Cucurbits, Tomato | Sex expression, fruiting and yield. |
| 3. | **Cytokinin** | | |
| | **(i) Natural** | | |
| | Zeatin | French bean, Pumpkin, Jerusalem artichoke | Cell enlargement. |

| S. No. | Growth regulators | Vegetable crops | Attributes affected |
|---|---|---|---|
| | **(ii) Synthetic** | | |
| | B A | Lettuce | Break seed dormancy and tissue culture. |
| 4. | **Ethylene** | | |
| | Ethephon (CEPA) | Cucurbits | Flowering, fruiting and sex-expression. |
| | | Okra | Vegetative growth and apical dominance. |
| | | Tomato, Chillies | Earliness, fruit ripening and yield. |
| | | Pea | Triple response. |
| 5. | **Inhibitor** | | |
| | Abscisic acid | | To increase seed dormancy To check germination. |
| 6. | **Retardant** | | |
| | Maleic hydrazide (MH) | Pea | Growth and yield |
| | | Cucurbits | Flowering, sex-expression and yield. |
| | | Onion, Garlic | Sprout inhibition and reducing storage losses. |

Auxins like IAA, IBA, 2, 4-D and also gibberellins as seed treatment have been reported give increased percentage of germination, quick growth and better yield in tomato, pepper, brinjal and radish. $GA_3$ induces majority of long day and/or chilling, requiring plants to flower. Application of auxins, gibberellins, retardant and ethylene is known to regulate the sex in majority of cucurbit. In cucumber, gynoecious lines are maintained by spraying $GA_3$ @ 1000 ppm at 2-true leaf stage which induced staminate flowers enabling selfing and maintenance. Ethrel spray @ 50-100 at 2-3 true leaf stage increases pistillate and hermaphrodite flowers in muskmelon, cucumber, summer squash, pumpkin and thus inceases the yield greatly. Gibberellic acid spray @ 25 ppm at 2-4 true leaf stage in water melon, spray of maleic hydrazide (MH) @ 50-100 ppm at 2-4 true leaf stage in bottle gourd, round melon and summer squash, increase the pistillate flowers and subsequently yield. During rainy season, when there is flower drop and during winter season when there is problem of fruit set, the spray of PCPA @ 50-100 ppm is very effective. In tomato, two foliar sprays of CCC @ 500 ppm one week before transplanting and 4 weeks after transplanting increase the Tomato Leaf Curl Virus (TLCV) resistance in the plants. Spray of MH @ 2500 ppm at 2-3 weeks before harvesting helps in sprouting inhibition in onion and potato. Beneficial effect of CCC in induction of salt tolerance has been reported in okra.

## Thinning

Thinning is an important operation in thick sowing vegetables especially in sugarbeet, chicory, radish, turnip, carrot etc. and needs to be done to permit enought space for proper development. These roots do not develop properly and attain the optimum size according to the variety if the crop is not thinned in its early stages. In case of seed ball in sugarbeet containing 5-8 seeds in each seed ball and when sown, more than one seedling germiantes from each seed/seed ball and thinning is an essential. In cucurbit, seed is sown in hills, the plants thinned at 4-true leaf stage to maintain 2-3 plants/hill and when sown along the furrows the plants at each place are thinned from one to two. Thinning is common in commercial practice but high cost of labour has forced measures to eliminate this tedious task.

Even if thinning is required, sowing needs to be done carefully to avoid too thick a stand which greatly increases the labour. Timely thinning gives better results and delays disturbs those that remain.

## Advantages

❖ Sowing slightly more thickly than necessary is usually desirable with vine vegetables to ensure a good stand.

❖ Thinning offers an opportunity to weed out weak plants, leaving the vigorous ones.

## Earthing-up

It is the important operation needed to be done in most vegetables. The soil in between rows and plants are loosened and shifted to near around the plant particularly at bolting stage *viz.* cabbage, turnip, carrot, radish, cauliflower and in potato at tuberization and second earthing-up may done 20-25 days later to cover the developing tubers. The main aim of earthing-up is to make the plant base strong/stable to avoid lodging of the plants even if there happen to be strong winds.

## Wind breaks

In some areas, vegetables are damaged by strong winds at any stage *viz.* bolting, flowering, pod/fruit setting and at maturity and sometimes causes damage to the vegetable plants or dropping of immature fruits/pods causes losses in yield and deterioration of quality. For protecting the vegetables from such hazards, the use of wind breaks is of utmost importance. The winds or gales which at times uproot and blow the growing vegetable plants from the field. Also the harvested lot is blown or shattered and drying of fruits/pods takes place resulting in loss of yield. The loss may be some times up to 80-85 per cent.

## Wind break types

### Natural

❖ Hill top

❖ Forests

❖ Group of trees on one side of vegetative field/village.

❖ Erection of stone wall all around the field which also protects the vegetables from stray and wild animals.

## *Artificial*

❖ Planting of tall species/varieties alternately with main vegetable crop.

❖ Planting of tall crops all around the vegetable crop.

❖ Growing fast-growing species act as wind break.

❖ Installation of screen structures.

## Harvesting

Harvesting is the final agricultural operation. It is the removal of whole or part(s) of the plant at a time or at different intervals, depending on the kind of vegetable and demand. In order to achieve high quality and a good price in the market it is utmost important that harvesting is done at right condition. Early or delayed harvesting may not fetch remunerative prices in the market, and, therefore, the grower should be vigilant of the crop's maturity as well as market demand period once the crop is ready for harvesting.

## Physiological changes in vegetables due to delayed harvesting

| Vegetables | Deterioration change |
|---|---|
| Leafy vegetables | Yellowing |
| Cauliflower and broccoli | Opening of florets and softening |
| Sweet corn and garden pea | Loss of sweetness |
| Okra | Softening |
| Cucumber | Yellowing |

| Vegetables | Deterioration change |
|---|---|
| Beans | Toughening, yellowing |
| Chayote, tomatoes and peppers | Seed germination |
| Asparagus | Elongation and feathering of spears |
| Onion, garlic, sweet potato, ginger | Sprouting and rooting |
| Colocasia, carrot, radish, turnip and garden beet | Softening |

## Harvesting methods

**Picking:** Hand picking of fruits *viz.*, tomato, capsicum chillies, garden peas, beans and okra.

**Cutting:** Cutting with sharp knife, sickle or any other cutting implement *viz.* spinach, beet leaf, cabbage, Chinese cabbage, cauliflower, etc.

**Separation:** The consumable parts are separated with hand from main stem *viz.* fruits are separated by giving a jerk or by twisting and separated from the vines/stems without causing any injury/damage to the stems e.g. cucumber, squashes, bitter gourd, etc.

**Up-rooting:** The underground consumable parts are uprooted either with the help of khilana/spade etc. or ploughing and then collected e.g. potato, sweet potato, colocasia, ginger, root vegetables-radish, turnip, carrot, garden beet, etc.

**Hand pulling:** Hand pulling of onion and garlic may be done in the field and allowed to dry for 4-5 days under shade and followed by immediate topping.

## Period and stages of vegetable harvesting

| Name of vegetable | Period (days) | Stage of harvesting |
|---|---|---|
| **Cole crops** | | |
| Cabbage | 100-120 | Developed, firm compact heads |
| Cauliflower | 100-130 | Curds are compact |
| Knol-khol | 120 | When knobs are tender |
| Brussel's sprouts | 120 | Young and immature sprouts |
| Sprouting broccoli | 120-130 | Cluster of green flower buds are cut along with a few leaves and stem |
| **Fruit vegetables** | | |
| Brinjal | 70-80 | Tender fruits with stalks |
| Chillies | 60-90 | 1. Green and mature for vegetable purpose |
| | | 2. Green or ripe fruits for pickle making |
| | | 3. Full ripened fruits for drying and powder making |
| Tomato | 120-150 | 1. Mature green stage for distant market |
| | | 2. Pink stage or turning pink stage for local market |
| | | 3. Ripe fruit for home consumption |
| | | 4. Full ripe stage for immediate use in canning and pickling |
| Okra | | |
| Rainy | 70-90 | Tender green pods for vegetable |
| Spring—summer | 40-45 | " |

| Name of vegetable | Period (days) | Stage of harvesting |
|---|---|---|
| **Legume vegetables** | | |
| Cluster bean | 50-60 | Tender stage of pods |
| Cowpea | 40-50 | Tender and half grown |
| French bean | 40-50 | Young, tender and delicate |
| Indian bean | 60-70 | Tender |
| Broad bean | 90-120 | Very young pods are preferred for harvesting |
| Peas | 100-130 | When pods are well filled up with tender seeds having high sugar content |
| **Root vegetables** | | |
| Carrot | 60-70 | Fully developed |
| Radish | | |
| Tropical | 25-30 | Tender |
| Temperate | 45-60 | Tender |
| Turnip | | |
| Early | 60-80 | Fully developed roots |
| Late | 90-120 | Fully developed roots |
| Beet root | 60-70 | Fully developed roots |
| **Tuber vegetables** | | |
| Colocasia | | |
| Partial harvesting | 60-70 | Immature corms |
| Final harvesting | 45-60 | Tender leaves |
| Elephant's foot | 1 to 4 years | Developed corm |
| Yams | 220-240 | Developed corm |
| Potato | | |
| Early | 60-70 | Fully developed |
| Mid | 70-190 | Fully developed |
| Late | 90-120 | Fully developed |
| Sweet potato | 80-90 | Immature half grown |
| | 120-150 | Fully developed mature tubers |

| Name of vegetable | Period (days) | Stage of harvesting |
|---|---|---|
| **Bulbous vegetables** | | |
| Onion | 60-70 | Onion greens |
| | 70-80 | Immature bulbs |
| | 80-120 | Fully mature bulbs |
| Garlic | 130-150 | Fully mature bulbs |
| **Cucurbitaceous vegetables** | | |
| Cucumber | 60-70 | Tender |
| Gourds | 60-90 | Tender |
| Melons | 60-90 | Fully developed and mature |
| Pumpkin | 60-90 | Tender |
| Squashes | 90-120 | Fully developed and mature |
| Pointed gourd | 90-120 | Tender, developed fruits |
| Chayote | 90-120 | Tender, developed fruits |
| Kakrol | 90-120 | Tender, developed fruits |
| Little gourd | 120-150 | Tender, developed fruits |
| **Leafy vegetables** | | |
| Amaranth | 25-30 | Tender, succulent |
| Palak | 25-30 | Tender, succulent |
| Spinach | 25-30 | Tender, succulent |
| Fenugreek | 25-30 | Tender, succulent |
| Basella | | |
| Seed sown | 60-75 | Tender, succulent |
| Raised from cutting | 45-50 | Tender, succulent |
| Lettuce | 70-90 | Tender, developed leaves |
| Leek | 60-70 | Tender plants are pulled out |
| Celery | 120-140 | Soft tender leaves |
| Asparagus | 250-300 | Tender 'Spears' are cut just below the surface of the soil |

**Merits of harvesting vegetable crops at right stage**

○ It helps grower to fetch high price in the market.

○ Yield is more.

○ Good quality and flavour.

○ Less damage by insect, pests and diseases.

○ Crop may be prevented from high and/or low temperature effects.

## Yields of vegetable crops

Yield refers to the harvested produce obtained from a crop grown in a unit area of land, usually expressed as quintals or tonnes per hectare. The total dry matter produced by a vegetable crop is known as **biological yield** and a fraction of the biological yield which is useful for man is known as **economic yield**.

**Harvest Index** (H.I.) is the ratio of economic yield to biological yield. Harvest index can also be expressed as percentage.

$$\text{Harvest Index} = \frac{\text{Biological Yield}}{\text{Economic Yield}}$$

**Approximate yield of vegetable crops**

| Vegetable crops | Approximate yield (tonnes/ha) |
| --- | --- |
| Amaranth | 12.5 |
| Beet leaf | 8-10 |
| Bitter gourd | 12-15 |
| Bottle gourd | 15 |
| Brinjal | 30 |
| Cabbage | 27.5 |

| Vegetable crops | Approximate yield (tonnes/ha) |
| --- | --- |
| Carrot | 22.5 |
| Cauliflower | 27-28 |
| Chilli | 10 |
| Colocasia | 15 |
| Cowpea | 8 |
| Cucumber | 10-12 |
| French bean | 3-5 |
| Knol-khol | 17.5 |
| Lettuce | 11 |
| Muskmelon | 12.5 |
| Okra | 5-10 |
| Onion | 17.5 |
| Pea | 5-8 |
| Potato | 20-35 |
| Pumpkin | 27.5 |
| Radish | 17.5 |
| Ridge gourd | 12-27 |
| Round melon | 10 |
| Snake gourd | 12-15 |
| Sweet potato | 12.5 |
| Tomato | 22.5 |
| Turnip | 25 |
| Watermelon | 32.5 |

## Physiological disorders or non-parasitic or inanimate diseases

There are some disorders in which no primary parasite is involved. They are brought about by abnormal environmental conditions like nutrition, weather, maturity etc. Sometimes more than one factor may be responsible for the cause and hence it is termed as '*syndrome*'.

## Classification

### Due to deficiency of nutrients

#### *Boron*

Brown heart of beetroot, radish and turnip, splitting of carrot, browning in cauliflower, cracked stem of celery.

#### *Calcium*

Cavity spot in carrot, black heart of celery, hypocotyl necrosis in French bean, tip-burn of lettuce, blossom-end rot of watermelon and tomato, metsubre of taro.

#### *Potassium*

Blotchy ripening in tomato.

#### *Nitrogen*

Buttoning.

#### *Molybdenum*

Whip-tail of cole crops and turnip.

#### *Magnesium*

Chlorosis in cauliflower.

### Due to excessive nitrogen

Splitting of carrot, ricyness and hollow stem of cauliflower, bulb sprouting in garlic, hollow heart of potato.

### Due to excessive phosphorous

Pencil strip of celery.

### Due to excessive moisture

Cotyledon cracking in French bean, bulb sprouting in garlic, delayed cooking in colocasia and elephant's foot, Growth crack in sweet potato and yam bean.

### Due to excessive exposure to sun rays

Sun scald in tomato, greening of potato.

### Due to high temperature

Leafy curd in cauliflower, blossom drop and ovule abortion in French bean, blossom-end rot of watermelon and tomato, flower drop, poor fruit setting and puffiness in tomato.

### Due to low temperature

Blindness, ricyness, fuzziness, buttoning in cauliflower, delayed flowering and pod development in French bean, flower drop, poor fruit setting and puffiness in tomato, chilling and freezing injury of potato.

### Due to water deficiency

Cracking and puffiness of tomato, internal brown spot of potato.

### Due to improper pollination and fertilization

Puffiness or pockets and poor fruit setting in tomato.

### Due to poor ventilation

Black heart of potato.

### Due to deleterious effect of ethylene

Bitterness in carrot, russet spotting of lettuce.

## Major disorders

| Vegetable crops | Physiological disorders |
| --- | --- |
| Beetroot | Brown heart or Crown heart or Heart rot |
| Broccoli | Browning, Whip-tail |
| Carrot | Splitting, Cavity spot, Bitterness |
| Cauliflower | Ricyness, Fuzziness, Leafyness, Blindness, Buttoning, Browning, Hollow stem, Whip-tail |
| Celery | Black heart, Cracked stem, Pencil strip, Freezing injury |
| French bean | Delayed flowering and pod development, Blossom drop and ovule abortion, Cotyledon cracking, Hypocotyl necrosis |
| Garlic | Bulb sprouting, Splitting |
| Lettuce | Tip-burn, Russet spotting |
| Radish | Brown heart |
| Turnip | Brown heart, Whip-tail |
| Tomato | Cracking, Blotchy ripening, Puffiness Blossom-end rot, Sun scald, Flower drop and poor fruit setting, Cat face, Hail injury |
| Potato | Internal brown spot, Greening, Black heart, Hollow heart, Chilling injury, Freezing injury |
| Colocasia, Elephant's foot | Delayed cooking |
| Sweet potato, Yam bean | Growth crack |
| Taro | Metsubre |

# Description of important physiological disorders

## Beetroot, Radish and Turnip

### *Brown heart or Crown heart or Heart rot*

A physiological disorder due to deficiency of boron in which the young unfolding leaves fail to develop normally and the plants eventually turn brown or black with rough, unhealthy and greyish-coloured roots. Borax if applied @ 11.2 kg/ha in soil checks the boron deficiency symptoms.

## Carrot

### *Splitting*

A major physiological disorder of carrot in which the roots crack. Though it is usually triggered by genetic factors, but a number of other factors like heavy side dressing with nitrogen fertilizer in early stages and boron deficiency may be involved. The splitting is reduced by low application of nitrogenous fertilizers.

### *Cavity spot*

A physiological disorder due to calcium deficiency in which a cavity appears in the cortex and in most cases the subtending epidermis collapses to form a pitted lesion. It can be corrected by application of calcium.

### *Bitterness*

A storage disorder due to deleterious effect of ethylene on carrot roots which causes an increase in the total phenol content of roots and induces the formation of new compounds, including isocoumarin and engenin, which are mainly associated with the bitter flavour in carrots.

# Cauliflower

## Ricyness

A disorder characterised by premature initiation of floral buds on curd giving a velvety appearance which may result from temperatures that are higher or lower to the optimum temperature required for a particular cultivar during curd development. Selection of proper varieties for a particular time of cultivation, optimum application of nitrogenous fertilizer and planting of resistant and tolerant varieties help in minimising this condition.

## Fuzziness

Fuzziness appears as flower pedicles of velvety curds elongate. It may be hereditary or non-hereditary. Cultivation in abnormal times encourages fuzziness. Sowing at normal time minimises fuzziness.

## Blindness

The term is applied to cauliflower plants without terminal buds which do not produce marketable curds. During the early stage of plant growth, damage to growing point by insects, low temperature or frost causes blindness. Damage of growing point by insect may be avoided by proper spraying of insecticides.

## Leafyness

A physiological disorder of cauliflower in which green bracts grow out of the curd due to exposure of the curd to temperature higher than the optimum required for its development. Selection of proper varieties may help to reduce it.

## Buttoning

Formation of small curds or buttons due to deficiency of nitrogen, late planting of seedlings of early varieties or viceversa. Identifying exact cause helps overcome this malady.

## *Hollow stem*

In heavy fertilized soils, particularly with nitrogen, rapidly growing plants of cauliflower develop hollow stem and curd. It may be corrected by close spacing and optimum use of nitrogenous fertilizers.

## *Browning or Brown rot or Red rot*

A nutritional disorder of cauliflower and broccoli due to deficiency of boron resulting in the development of pinkish or rusty-brown areas on the surface of the curd. It is corrected by the application of Borax.

## *Whip-tail*

A nutritional disorder of cole crops and turnip due to deficiency of molybdenum resulting in severe reduction of the lamina, leaving the large bare mid-rib in acute condition. Whip-tail can be corrected by liming the soil and application of molybdenum.

## Celery

## *Black heart*

A physiological disorder due to calcium deficiency resulting in tip burn of the young leaves followed by drying, blackening and in severe case killing of the entire heart of the plant. It can be corrected by the application of calcium.

## *Cracked stem*

Basically a boron deficiency disorder in which affected tissues collapse, become light yellow and later form a corky layer. It can be corrected by the application of Borax.

## *Pencil strip*

A disorder associated with high soil phosphorus showing narrow brown lines on petioles. Optimum application of phosphatic fertilizers helps to minimise this condition.

## French bean

### Delayed flowering and pod development

Flower initiation and pod development are greatly delayed under subtropical temperatures, especially below 10°C where fertilization may not occur, producing small and misshaped pods.

### Blossom drop and ovule abortion

Blossoms drop and ovule abortion are common problems at high temperature especially above 35°C. Therefore, planting crop at a suitable time and place is of utmost importance.

### Cotyledon cracking

Transverse cotyledon cracking takes place when dry seeds of beans are sown in wet soils. Resistant varieties with hard seed coat, optimum seed moisture content and planting crop at suitable time are essential to avoid this disorder.

### Hypocotyl necrosis

The necrosis of hypocotyl is associated with low calcium content in seed after germination. Soil rich in calcium and magnesium offset this problem.

## Garlic

### Bulb sprouting

Sprouting of bulbs in the field occurs at maturity particularly when there is winter rain or excessive soil moisture and supply of nitrogen. This disorder is, however, not of permanent nature. It varies from variety to variety. Early planting causes sprouting. The causes and possible remedies are being studied.

### Splitting

It is noticed sometimes in some varieties and is due to delayed harvesting or irrigation after a long spell of drought.

## Lettuce

### Tip burn

Tip burn is a physiological disorder in lettuce. It results in burning or scorching of lateral margins of inner leaves of mature head. Unfavourable seasonal/climatic factors and calcium deficiency are the causes. By applying calcium chloride, this malady can be rectified.

### Russet spotting

A post-harvest disorder characterised by localised, spot-like lesions that may start either in the epidermis or in the mesophyll which, in advance stage, may show discolouration of vascular tissue and collapse of mesophyll cells resulting in pit-like depression. Ethylene accumulation to the tune of 0.1 ppm concentration is sufficient to cause this disorder during a normal transit period of 5-8 days at $5^0$C.

## Watermelon

### Blossom-end rot

Smooth, leathery, firm, dark green or brown delimited areas, 2.5 to 7.5 cm in diameter, around the point of blossom attached, caused due to faulty nutrition associated with irregular moisture supply and high temperature.

## Tomato

### Cracking

When rain follows a long, dry spell, there is cracking. High day temperature followed by low temperature with high relative humidity also causes fruit cracking. Deficiency of boron causes cracking. Soil application of boron @ 15-20 kg /ha or spraying of borax (0.25%), 2-3 times at fruiting to ripening stage reduces fruit cracking. There should be proper control of moisture, especially at fruit maturity and ripening stage.

### Blotchy ripening

Greenish-yellow to whitish patch on tomato, mostly on stem-end portion of the fruits, is called blotchy ripening. Caused mainly by potassium deficiency, it occurs due to imbalance of nitrogen and potassium deficiency. To control blotchy ripening, potassic fertilizer should be applied adequately and temperature should be controlled. Short photoperiod and relatively low day temperature reduces its incidence.

### Puffiness or Pocket

A disorder due to high or low temperature and low soil moisture in which the outer wall of the fruit continues to develop normally but the remaining internal tissue growth is retarded resulting in partially filled, less firm and light weight tomato. To reduce its incidence, maintenance of normal temperature, frequent irrigation, spraying of boric acid (10-15 ppm) and some growth regulators is required.

### Blossom end rot

Small water-soaked spots usually appear at the blossom end of the half-grown fruit which in turn becomes light to dark brown in colour, sunken and leathery. Caused by calcium deficiency, irregular moisture supply and high temperature. Calcium sulphate, calcium hydroxide and calcium chloride should be applied to overcome this disorder.

### Sun scald

Appearance of blistered water-soaked areas followed by rapid desiccation and formation of sunken spots which are grey or white in green fruits or yellowish in red fruits due to sudden exposure to sunlight, which is more serious in hot weather. Dense foliage varieties with thick pericarp developed in temperate countries are less prone to low temperature injury.

## *Flower drop and poor fruit setting*

Flower drop and poor fruit-set in tomato is a common problem. It is due to imbalance in supply of nutrition, incorrect method and time of its application and abnormal weather conditions. The poor fruit setting is sometimes due to failure of pollination or fertilization and can be overcome by the use of 2, 4-D at 1-2 ppm along with urea at 1 per cent in the form of foliar spray at flowering stage.

## *Cat-face*

A disorder characterised by the distortion of the blossom end of fruit showing ridges, furrows and blotches, maybe due to abnormal growing condition during formation of the blossom, which causes the death of the cells and black discolouration at the blossom end of the ovary. Avoiding growing susceptible varieties and providing normal growing conditions are the control measures.

## Potato

### *Internal brown spot*

It is characterised by irregular, dry, brown spots scattered through the flesh of tubers. These spots are never found in vascular region. It appears particularly in light sandy soils which are not irrigated regularly. It is not transmissible. An old variety, Craigs Defiance, is prone to this.

## *Greening*

Appearance of green colour on the tuber exposed to direct sunlight due to the presence of solanin. Avoiding exposure of tubers to direct sunlight is the control measure.

## *Black heart*

A potato disorder in which dark grey to purplish or inky black discolouration occurs in the central tissues of the tuber. In advanced stages, the affected tissues may dry out and separate thus forming cavities. It is caused due to adverse enzymatic reaction resulting from sub-oxidation, poor ventilation and high temperature (above 33°C) during storage and transportation and high soil temperature during growing and maturity of tubers in the field. Therefore, avoid storage temperature above 33°C and poor ventilation.

### Hollow heart

Hollow heart consists of an irregular cavity in the centre of tubers. In tissues surrounding the cavity, there is no decay or discolouration. Hollow heart condition appears often in varieties which bulk rapidly and produce over-sized tubers. The condition can be avoided by closer spacing of plants and avoiding excessive use of fertilizers.

### Chilling injury

Chilling injury may follow prolonged storage of tubers at temperature of about 0°C. This results in discoloured blotches in the flesh of tubers which vary from light reddish-brown to dark brown, diffused brownish black patches on skin and reduced or completely inhibited sprouting of affected tubers when planted.

### Freezing injury

The injury starts at -1.5°C when tissues of slightly frozen tubers get discoloured. Entirely frozen tubers, on thawing, present a cheese-like appearance and later the tissue breaks down into slimy watery masses.

### Colocasia and Elephant's foot

### Hard to cook

Water stagnation in the field results in tubers becoming very hard to cook. This occurs both in Dasheen as well as Eddoe varieties. To maintain the cooking quality of tubers, proper drainage should be maintained in the field.

### Sweet potato and Yam bean

*Growth crack*

Some varieties in sweet potato have a tendency to crack at maturity. This is usually due to excessive soil moisture or moisture shortage. Cracking also takes place if harvesting is delayed. The application of potassium checks tendency of cracking in tubers.

### Taro

*Metsubre*

A nutritional disorder of taro due to calcium deficiency in which the defective corm have smooth or concave top, slightly brownish in colour and of varying size. Apply calcium to overcome this disorder.

## Disease and pest management

Vegetable crop protection against disease and pest plays a very important role. Without taking crop protection measures, it is impossible to get success in vegetable production. The modern high yielding varieties, being mostly susceptible to diseases and pests, require much care with respect to management of diseases and pests.

If all immunisation-prophylaxis system of disease-pests management measures are adopted for growing a vegetable crop then there will be little chance for a disease or pest to prove serious. But in cases where preventive measures could not be followed up to the desired level due to lack of facilities and little or no time is left to take such precautionary measures, as often happens in intensive or multiple cropping system, then the use of chemicals for the management of disease and pest is bound to have good results. Chemicals which are safe, economical and easily available may be procured and given place in integrated methods of disease and pest management.

# Immunisation-prophylaxis system of disease-pest management

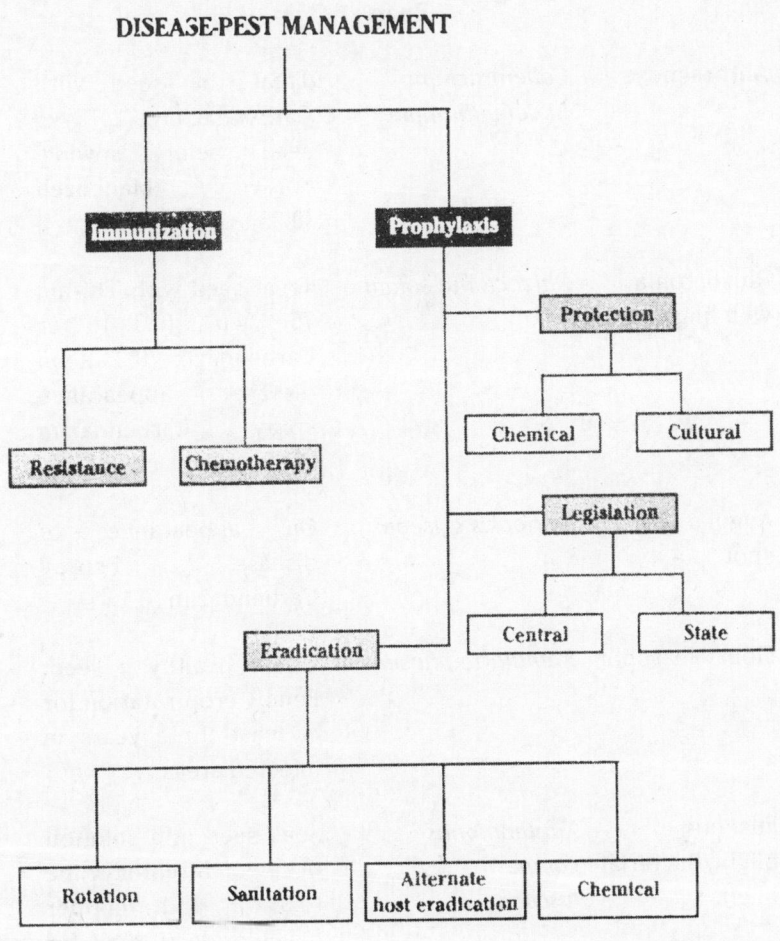

## Fungal and bacterial disease management in vegetable crops

| Disease | Scientific name | Management |
|---|---|---|
| | **Bean** | |
| Anthracnose | *Colletotrichum lindemuthianum* | Treat the seed with Carbendazim (2 g/kg seed) before sowing. Spray Mancozeb (0.25%). |
| Rhizoctonia web blight | *Rhizoctonia solani* | Treat seed with Thiram 75 wp (0.3%) or Carbendazim (0.2%). On disease appearance spray Carbendazim (0.05). |
| Angular leaf spot | *Isariopsis griseola* | On appearance of disease, spray Carbendazim (0.1%). |
| Floury leaf spot | *Ramularia phaseoli* | Use healthy seed. Follow crop rotation for at least three years in infected areas. |
| Fuscous blight/Bacterial blight | *Xanthomonas campestris.* pv. *fuscans* | Soak seed in a solution of Streptocycline (0.01%) and copper oxychloride (0.25%) for 4 hr before sowing. |

| Disease | Scientific name | Management |
|---------|-----------------|------------|
| **Cabbage, Cauliflower and Knol-khol** | | |
| Damping-off | *Rhizoctonia solani, Phytophthora* spp. and *Pythium* spp. | Treat nursery beds with formalin (5 litres/10 litres of water) at least 20 days prior to sowing. Alternatively, treat the seeds with Carbendazim (0.2%) before sowing. |
| Black rot | *Xanthomonas campestris* pv. *compestris* | Treat the seed by soaking in tap water for 30 minutes followed by hot water dip at 52°C for 30 minutes. Give a spray of Streptocycline (10/g 100 litres of water) at curd-formation stage. |
| Curd rot | *Sclerotinia sclerotiorum, Erwinia carotovora* | Give a protective spray of Mancozeb (0.25%) and Streptocycline (0.1%) to curds particularly before the likelihood of frost. |
| Stalk rot | *Sclerotinia sclerotiorum* | Spray the crop with a mixture of Carbendazim (0.05%) and Mancozeb (0.25%) from the curd initiation to pod setting stage at 10-15 day intervals |
| Downy mildew | *Peronospora parasitica* | Give hot water treatment to seed or dress with Thiram (0.3%). Spray the crop with Mancozeb (0.2%) at 10-15 days' intervals. |
| White rust or White blisters | *Albugo candida* | Spray copper oxychloride (0.3%) or Bordeaux mixture (0.8%) |

| Disease | Scientific name | Management |
|---|---|---|
| **Capsicum and Chilli** | | |
| Damping-off | *Pythium aphanidermatum, Phytophthora* spp. | Treat seed with Captan (2 g/kg seed) before sowing. Drench the nursery beds with Captan (0.2%) or Mancozeb (0.25%) and Carbendazim (0.05%) solution on appearance of symptoms. |
| Fruit-rot and leaf blight | *Phytophthora* spp. | Treat the seed with Mancozeb (0.3%). Spray Metalaxy/Mancozeb (0.25%) just before the onset of monsoon and with Bordeaux mixture (1%) or copper oxychloride (0.3%) at 8-10 days' intervals. |
| Anthracnose or die-back | *Colletotrichum capsici* | Treat seed with Thiram 75 WP (0.8%) or Mancozeb (0.2%). Spray crop with copper oxychloride (0.3%) or Mancozeb (0.25%) at fortnightly intervals |
| **Cucurbits** | | |
| Powdery mildew | *Erysiphe cichoracearum* and *Sphaerotheca fuliginea* | Spray Dinocap (0.1%) or wettable sulphur (0.2%) or Calixin (0.05%) 2-3 times at 10 days' interval after appearance of disease. |
| Anthracnose | *Colletotrichum lagenarium* | Spray Mancozeb or Hexacap (0.25%) or Carbendazim (0.1%) at 15 days' intervals. |
| Downy mildew | *Pseudoperonospora cubensis* | Spray Mancozeb at 8-10 days' intervals. |

| Disease | Scientific name | Management |
|---------|-----------------|------------|
| | | **Pea** |
| Powdery mildew | *Erysiphe polygoni* | Spray Karathane (0.05%) or wettable sulphur (0.2%) or Carbendazim or Baycor (0.05%) |
| Ascochyta foot-rot and blight | *Ascochyta pisi, A. pinodella* and *A. pinodes* | Treat seed with Carbendazim (0.25%) before sowing. Spray infected crop with Carbendazim (0.1%) or Mancozeb at flowering and afterwards at 10-15 days' intervals. |
| Wilt | *Fusarium oxysporum* f. sp. *pisi* | Treat seeds with Thiram (0.3%) or Carbendazim (0.2%) before sowing and avoid early sowing in badly infected areas. Drench the infected area with Carbendazim (0.5%). Destroy the weed hosts such as *Lathyrus vicia* etc. |
| Rust | *Uromyces pisi* and *U. fabae* | Spray Mancozeb (0.25%) or Bayleton (0.05%) on disease appearance and repeat at 10-15 days' intervals. |
| Bacterial blight | *Pseudomonas syringae* pv. *pisi* | Give a spray of Streptocycline (0.01%) on appearance of the disease on crop. |
| | | **Tomato** |
| Damping-off | *Pythium* spp. *Phytophthora* spp. | Treat seed-beds with formalin @ 5 litres/1000 litres of water, 15-20 days before sowing. Drench/irrigate nursery beds with Mancozeb (0.25%) and Carbendazim (0.05%). |
| Buck-eye rot | *Phytophthora nicotiana* var. *parasitica* | Give a spray of Metalazyl/ Mancozeb. |
| Alternaria blight | *Alternaria solani* | Treat seed with Thiram 75 WP 2.53 g/kg seed. Spray Hexacap (0.25%) or copper oxychloride (0.3%) or Mancozeb at 8-10 days' intervals. |

| Disease | Scientific name | Management |
|---|---|---|
| Septoria leaf blight | *Septoria lycopersici* | Treat seed with Thiram 75 WP 2.53 g/kg seed. Spray Hexacap (0.25%) or copper oxychloride (0.3%) or Mancozeb at 8-10 days' interval. |
| Bacterial wilt | *Ralstonia (Pseudomonas) solanacearum* | Always transplant disease-free, healthy seedlings. Follow 3-year crop rotation in infested fields by including cereals and crucifers. |
| Bacterial canker | *Clavibacter michiganense* sub sp. *michiganense* | Use disease-free seed and seedlings. Follow a 3-year crop rotation in infected fields by including non-host crops. |

## Viral and mycoplasma disease management in vegetable crops

| Disease | Vector | Management |
|---|---|---|
| | | **Bean** |
| Yellow mosaic | Transmitted by aphids | Apply Carbofuran @ 1.5 kg/na at the time of sowing. Two to three foliar sprays of Dimethoate (0.05%) or Phosphamidon (0.2%) at 10 days' interval. |
| | | **Bitter gourd** |
| Mosaic | Transmitted by aphids | Spray just after germination with Dimethoate or Phosphamidon (0.05%) at 10 days' intervals to prevent aphid vectors. |
| Witche's broom | Transmitted by leafhoppers | Application of Carbofuran @ 1.5 kg/ha at the time of sowing followed by 5-6 foliar sprays of either Phosphamidon or Oxydemeton methyl (0.05%) at 10 days' interval. |
| | | **Brinjal** |
| Little leaf | Transmitted by leafhoopers | Ten to 50 ppm of tetracycline antibiotics or 50-100 ppm of Chloremphenical. Dipping seedlings in 0.2% Carbofuran (75% WP) for 24 hr reduces vector population. |

| Disease | Vector | Management |
|---------|--------|------------|
| | | **Chilli** |
| Mosaic | Transmitted by aphids | Apply Carbofuran @ 1.5 kg at the time of transplanting. Three to four foliar sprays of Dimethoate at 10-day intervals. |
| Leaf curl | Transmitted by whitefly | Apply Carbofuran @ 1.5 kg at the time of transplanting. Three to four foliar sprays of Dimethoate at 10-day intervals. |
| | | **Cowpea** |
| Mosaic | Transmitted by aphids | Spray Dimethoate (0.05%) at 10-day intervals. |
| Yellow flecks | Transmitted by whitefly | Apply granular or emulsifiable insecticides and mineral oil. |
| | | **Cucumber** |
| Mosaic | Transmitted by aphids, seeds | Spray Dimethoate (0.05%) or Oxydemeton methyl (0.02%) at weekly interval. |
| Green mottle mosaic | Transmitted by seeds | Dry heat-treatment of seeds for three days at 70°C eliminates local infection. Soil fumigation with Methyl bromide helps reduce virus' spread through soil. |
| | | **French bean** |
| Common mosaic | Transmitted by aphids | Apply Carbofuran or Disulfotan or Phorate 10G granules @ 1.5 kg/ha at the time of sowing followed by 2-3 foliar spraying of Dimethoate (0.05). |
| Golden mosaic | Transmitted by whitefly | Soil application of Carbofuran or Disulfotan @ 1.5 kg/ha at the time of sowing. |
| Phyllody | Transmitted by leafhopper | Foliar application of Oxytetracycline hydrochloride solution (500 ppm) at weekly interval. |
| | | **Okra** |
| Enation * leafcurl | Transmitted by whitefly | Foliar spray of Dimethoate (0.05%) at 10-day intervals. |

| Disease | Vector | Management |
|---------|--------|------------|
| Yellow-vein mosaic | Transmitted by whitefly | Four to five foliar sprayings of Dimethoate (0.05%) or Oxydemeton methyl (0.02%) at 10-day intervals. Apply Carbofuran @ 1kg/ha at the time of sowing. Use resistant varieties Arka Anamika and Arka Abhay. |

**Pumpkin**

| Disease | Vector | Management |
|---------|--------|------------|
| Mosaic | Transmitted by aphids | Apply Carbofuran (1.5 kg/ha) at the time of sowing. Spray Phosphamidon (0.5%) at 10-day intervals. |
| Yellow-vein mosaic | Transmitted by whitefly | Soil application of Carbofuran @ 1.5 kg/ha at the time of seed sowing followed by foliar application of Dimethoate or Phosphamidon (0.2%) at 10-day intervals. |

**Radish**

| Disease | Vector | Management |
|---------|--------|------------|
| Mosaic I and Mosaic II | Transmitted by aphids | The disease spread can be minimised by one soil application of Carbofuran @ 1.5 kg/ha at the time of sowing, followed by 2-3 foliar sprays of either Phosphamidon (0.05%) at 10-day intervals. |
| Phyllody | Transmitted by leafhopper | Soil application of Thimet 10G, Disulfotan or Carbofuran @ 1.5 kg/ha at the time of sowing. |

**Tomato**

| Disease | Vector | Management |
|---------|--------|------------|
| Leaf curl | Transmitted by whitefly | Apply Carbofuran @ 1.5 kg/ha at the time of transplanting. Foliar application of Dimethoate (3-4 time) at 10-days' interval. |
| Tomato mosaic | Transmitted by contact and seed | Treat seed in trisodium phosphate solution followed by sodium hypochlorite for 95 and 30 minutes respectively. Sterilize all the tools. |
| Spotted wilt | Transmitted by thrips | Soil application of Carbofuran @ 1.5 kg/ha at the time of sowing. Foliar application of Dimethoate (0.05%) at 10-day intervals. |
| Fern leaf | Transmitted by aphids | Spraying of systemic insecticide like Phosphamidon (0.05%) or Oxydemeton methyl (0.02%) at 10-day intervals controls aphid vectors. |

| Disease | Vector | Management |
|---------|--------|------------|
| | | **Watermelon** |
| Mosaic | Transmitted by aphids | Rogue infected plants. Control weed hosts. Control aphid vectors. |
| Bud necrosis | Transmitted by thrips | Soil application of Carbofuran @ 1.5 kg/ha at the time of sowing is helpful. Foliar application of Dimethoate (0.05%) at 10-day intervals. |

## Insect-Pest management in vegetable crops

| Insect-Pest | Scientific name | Management |
|-------------|-----------------|------------|
| | | **Amaranth and Spinach** |
| Leaf-eating caterpillar | *Hymenia recurvalis* | Remove infested plants and parts. Spray Dichlorvos (0.5%), if required. |
| Leaf-webber | *Lamprosema indicata* | Remove infested plant parts. Spray Dichlorvos (0.05%), if required. |
| Aphids | *Aphis croccivora, Lipaphis erysimi* | Remove infested plant parts. If required, apply Monocrotophos or Phosphamidon or Dimethoate or Oxymethyl demeton (0.05%). |
| Scale | *Coccus hesperidum* | If absolutely essential ˙then only spray Monocrotophos (0.05%). |
| Stem weevil | *Hypolixus truncatules* | Spray Dichlorvos (500 g/ha), if required. |
| Blue beetle | *Altica caerulesces* | Neem seed-kernel extract (4%) may be sprayed, if required. |
| | | **Beetroot** |
| Cut-worm | *Agrotis ipsilon* | As and when observed soil drenching may be done with Chlorpyriphos (0.1%). |
| Leaf-eating caterpillar | *Spodoptera litura* | Spray neem seed-kernel extract (4%) in early stage of the larvae and for controlling large larvae spray Cypermethrin (0.0125%). |

| Insect-Pest | Scientific name | Management |
|---|---|---|
| | | **Brinjal** |
| Shoot and fruit-borer | *Leucinodes orbonalis* | Cypermethrin (0.0125%) spray at 2% flower bud damage threshold is very effective. Alternately, spray Fenvalerate (0.01%) or Cypermethrin (0.0125%) once 15-20 days after flowering. Punjab Barsati, a moderately resistant variety, may be cultivated. |
| Stem-borer | *Euzophera perticella* | Apply Carbofuran (1 kg/ha) or neem cake (500 kg/ha) while planting around the plant base. |
| Ash-weevil | *Myllocerus subfasciatus* | Apply neem cake (500-1000 kg/ha). Drench the soil with Chlorpyriphos (0.05%). |
| Midge | *Asphondylia* sp. | Spray neem seed-kernel extract (4%) or Phosphamidon (0.05%) once in 2-3 weeks after flowering. |
| Leaf-roller | *Antoba olivacea* | Spray Dichlorvos (0.1%). |
| Horn moth | *Acherontia styx* | Spray Quinalphos/ Chlorpyriphos (0.05%) or Carbaryl (0.2%). |
| Bud-worm | *Phthorimoea operculella P. blspigona* | Spray Quinalphos/Chlorpyriphos (0.05%) or Carbaryl (0.2%). |
| Mealy bug | *Coccidohystrix insolitus* | Spray fish oil rosin soap. |
| Aphids | *Aphis gossypii Myzus persicae* | Spray Monocrotophos, Phosphamidon, Dimethoate, or Oxymethyl demeton (0.05%). |
| Leafhopper | *Amrasca biguttula biguttula* | Apply Monocrotophos, Phosphamidon, Dimethoate, Oxymethyl demeton (0.05%). |

| Insect-Pest | Scientific name | Management |
|---|---|---|
| Lace-wing bug | *Urentus hystricellus U. sentis* | Spray Monocrotophos, Phosphamidon, Dimethoate, Oxymethyl demeton (0.05%). |
| Red bug | *Caridius janus* | Monocrotophos, Phosphamidon, Dimethoate, Methyl demeton may - be applied @ 0.05%. |
| Thrips | *Thrips* palmi, *Thrips tabaci* | Apply Monocrotophos, Phosphamidon, Dimethoate, Methyl demeton (0.05%) or spray neem formulations (2-3 ml/litre of water). |
| Epilachna beetle (Hadda beetle) | *Epilachna vigintioctopunctata Epilachna dodecastigma* | Spray contact insecticides like Carbaryl (0.1%)/ Quinalphos (0.05%) if serious damage is observed. |
| Termites | *Microtermes* sp., *Trinervitermes biformis* | Drench soil with Chlorpyriphos (0.1%). |
| Mite | *Tetranychus cinnabarinus* | Spray Ethion (0.05%) or neem oil (1%). |

### Cabbage and Cauliflower

| | | |
|---|---|---|
| Diamond back moth | *Plutella xylostella* | Sow two rows of mustard for every 25 rows of. cabbage. Mustard attracts 80-93% of diamond back moths, stemborer and bugs. |
| Leaf-webber | *Crocidolomia binotalis* | Apply 4% neem seed kernel extract at head-initiation between 17 and 28 days after planting. Spray contact insecticides like Quinalphos/ Chlorpyriphos. |
| Stem-borer | *Hellula undalis* | Mustard is thickly sown to have 50-60 plants/m. |
| Aphids | *Brevicoryne brassicae, Lipaphis erysimi* | Mustard as trap crop attracts aphids. If desired, spray neem seed kernel extract (4%) or Oxydemeton methyl (0.02%). |

| Insect-Pest | Scientific name | Management |
|---|---|---|
| Gram-caterpillar | *Helicoverpa armigera* | Spraying Endosulfan (0.07%) or Dichlorvos (0.1%). Large larvae should be controlled by hand picking (use a sharp thick iron needle to pick the larvae boring the head) and killing them. |
| Tobacco caterpillar | *Spodoptera litura* | Spray neem seed kernel extract (4%) in early stage of the larvae. Large larvae have to be controlled by picking (use a sharp thick iron needle to pierce the hole in head) and killing them. |
| Mustard saw-fly | *Athalia lugens proxima* | Spray contact insecticides like Quinalphos or Chlorpyriphos (0.05%). |
| Striped-flea beetle | *Phyllotreta striolata* | Spray contact insecticides like Quinalphos or Chlorpyriphos (0.05%). |
| Planted bug | *Bagrada cruciferarum* | Single spot application of Monocrotophos, Phosphamidon, Dimethoate and Oxymethyl demeton (0.05%). |
| **Capsicum and Chilli** | | |
| Thrips | *Scirtothrips dorsalis, Thrips palmi* | Spray Monocrotophos, Phosphamidon, Dimethoate, Oxymethyl demeton (0.05%) or neem formulations (2-3 ml/litre). |
| Mites | *Polyphagotarsonemus latus, Tarsonemus translucens, Tetranychus cinnabarinus* | Spray Dimethoate (0.05%) or wettable sulphur (2kg/ha). |
| Aphids | *Aphis gossypii* | Apply Monocrotophos, Phosphamidon, Dimethoate or Methyl demeton (0.05%). |
| Fruit-borer | *Helicoverpa armigera* | Spray Dichlorvos (0.1%). |

| Insect-Pest | Scientific name | Management |
|---|---|---|
| Fruit-borer | *Spodoptera litura* | Neem formulations (5 ml/litre) may be applied to young larvae. Apply Cypermethrin (75 g/ha) once the larvae get dispersed in field. |
| Cut-worm | *Agrotis ipsilon* | Drench soil with Chlorpyriphos (0.1%) |
| White grub | *Anomala bengalensis, Holotrichia consanguinea, H. reynaudi* | Apply neem cake (1.00 kg/ha), if observed every year. Drench soil with Chlorypriphos (0.1%). |

### Carrot

| | | |
|---|---|---|
| Leafhopper | *Empoasca punjabensis* | Apply Monocrotophos or Phosphamidon or Dimethoate or Oxymethyl demeton (0.05%). |
| Cut-worm | *Agrotis ipsilon* | Drench soil with Chlorpyriphos (0.1%). |

### Cluster bean

| | | |
|---|---|---|
| Aphid | *Aphis croccivora* | Spray Monocrotophos, Phosphamidon, Dimethoate, Methyl demeton (0.05%). |
| Pod-borer | *Adisura atkinsoni* | Spray Quinalphos or Chlorpyriphos (0.05%) or Carbaryl (0.5%) or Cypermethrin (0.0125%) at pod formation stage. |
| Bugs | *Megacopta (Captosoma) cribraria* | Dusting with Malathion/ Quinalphos (2% dust) reduces damage. |

### Cowpea

| | | |
|---|---|---|
| Stem fly | *Ophiomyia phaseoli* | Spray Phosphamidon or Oxymethyl demeton (0.05%) or neem seed kernel extract (4%). |
| Pod-borer | *Muruca testulalis, Lampides boeticus* | Spray Quinalphos or Chlorpyriphos (0.05%) or Carbaryl (0.15) or Cypermethrin (0.0125%) at flower-bud stage. |

| Insect-Pest | Scientific name | Management |
|---|---|---|
| Bruchid | *Callasobruchus chinensis* | Dry seeds thoroughly before storage. Add 2% edible oil (volume/weight). It prevents incidence for 6 months. |
| Aphid | *Aphis croccivora* | Apply Monocrotophos, Phosphamidon, Dimethoate, Oxymethyl demeton (0.05%). |
| Bugs | *Nezara viridula Anoplocnemis phasiana, Megacopta cribraria* | Dusting with Malathion / Quinalphos (2% dust) reduces damage. |
| Leaf-eating caterpillar | *Spilasoma obliqua, Euproctis fraterna* | Spray Quinalphos or Chlorpyriphos (0.05%) or Carbaryl (0.2%) or Cypermethrin (0.0125%) if serious. |

### Cucumber, Gourds and Pumpkin

| Insect-Pest | Scientific name | Management |
|---|---|---|
| Red pumpkin beetle | *Aulacophora foveicollis, A. lewisii* | Spray Carbaryl (0.2%) or Quinalphos/Chlorpyriphos (0.05%). |
| Serpentine leaf-miner | *Liriomyza trifolii* | Neem seed kernel extract (4%) or Triazophos (0.05%) once in 3 weeks. |
| Fruit fly | *Bactrocera cucurbitae* | Resistant pumpkin Arka Suryamukhi should be grown. Foliar spraying of Fenthion (0.05%) with 5% jaggery at fruit formation/ ripening may be given. |
| Blister beetle | *Mylabris pustulata* | Collecting and destroying flying adults is economical and best. |
| Stink bug | *Caridius brunneus C. janus, C. observus* | Spray Monocrotophos, Phosphamidon, Dimethoate, and Oxymethyl demeton (0.05%). |

| Insect-Pest | Scientific name | Management |
|---|---|---|
| Aphids | *Aphis gossypii Aphis malvae* | Spray Monocrotophos, Phosphamidon, Dimethoate or Oxymethyl demeton (0.05%). |
| Mite | *Tetranychus neocaledonicus* | Spray Dimethoate/Ethion (0.05%) or wettable sulphur (0.2%). |
| Thrips | *Thrips tabaci, Frankliniella sulphurea* | Spray Phosphamidon, Dimethoate, Oxymethyl demeton (0.05%) or apply neem cake (500 kg/ha) while sowing. |
| Plume moth | *Sphenarches caffer* | Hand-picking is the best method of control. |
| Leaf-eating caterpillar | *Diaphania (Margaronia) indica* | Give one spray of Quinalphos or Chlorpyriphos (0.05%). |
| Stem gall fly | *Lasioptera falcata* | Spray Monocrotophos, Phosphamidon, Dimethoate, Oxymethyl demeton (0.05%). |

### French bean

| Insect-Pest | Scientific name | Management |
|---|---|---|
| Stem-fly | *Ophiomyia phaseoli* | Apply Monocrotophos, Phosphamidon, Dimethoate or Oxymethyl demeton (0.05%) or neem seed kernel extract (5%). |
| Leafhopper | *Empoasca kerri* | Spray Monocrotophos, Phosphamidon, Dimethoate, Oxymethyl demeton (all at 0.05%). |
| Serpentine leaf-miner | *Liriomyza trifolii* | Spray neem seed kernel extract (4%) or neem formulations (3-5 ml/litre) or Triazophos (0.05%) 10 and 20 days after sowing. |
| Thrips | *Calliothrips indicus Megaleus distalis* | One spray of Phosphamidon, Dimethoate or Oxymethyl demeton (0.05%) or neem formulations (2-3 ml/litre). |
| Red spider mite | *Tetranychus cinnaberinus* | Spray Dicofol or Ethion (0.05%). |

| Insect-Pest | Scientific name | Management |
|---|---|---|
| **Garlic and Onion** | | |
| Thrips | *Thrip tabaci* | Spray Monocrotophos, Phosphamidon, Dimethoate, Methyl demeton (0.05%) or neem formulations (2-3 ml/litre). |
| Mite | *Aceria tulipae* | Apply Dimethoate/Ethion (0.05%) at initial stage of infestation. |
| Onion fly | *Delia antiqua* | Cultivars of *Allium fistulosum* are resistant. Apply Carbofuran/Phorate (0.5 kg/ha) or neem cake (500 kg/ha) to soil while sowing only. |
| Cut-worm | *Agrotis ipsilon* | Soil application with Chlorpyriphos (0.1%). |
| Groundnut earwig | *Euborellia annulipes* | Soil application with Chlorpyriphos (0.1%). |
| **Lablab Bean** | | |
| Pod-borer | *Helicoverpa armigera, Etiella zinckenella, Adisura atkinsoni* | Spray Quinalphos or Chlorpyriphos (0.05%) or Cypermethrin (0.0125%) at the time of flower-bud formation and again at small pod stage. |
| Aphid | *Aphis croccivora* | Apply Monocrotophos, Phosphamidon, Dimethoate or Oxymethyl demeton (0.05%). |
| Leaf-eating caterpillars | *Spilarcitia (Spilosoma) obliqua, Euproctis fraterna* | Spray Quinalphos or Chlorpyriphos (0.05%) or Carbaryl (0.2%). |
| Bugs | *Nezara viridula, Anoplocnemis phasiana, Megacopta (Captosoma) cribraria* | Dusting with Quinalphos or Malathion (2% dust) reduces damage. |

| Insect-Pest | Scientific name | Management |
|---|---|---|
| Weevils | *Alcidodes pictus A. bubo* | Apply Quinalphos or Chlorpyriphos (0.05%) or Carbaryl (0.15%). |
| Red spider mite | *Tetrancychus cinnabarinus* | Spray Dicofol or Ethion (0.05%). |
| Leafhopper | *Empoasca kerri* | Apply Phosphamidon, Dimethoate or Oxymethyl demeton (0.05%). |

## Okra

| | | |
|---|---|---|
| Leafhopper | *Amrasca biguttula biguttula* | Apply Monocrotophos, Phosphamidon, Dimethoate or Oxymethyl demeton (0.05%). |
| Shoot-and fruit-borer | *Earias vittella* | Spray Carbaryl (0.15%) or Quinalphos/Chlorpyriphos (0.05%) or Fenvalerate (0.01%). |
| Whitefly | *Bemisia tabaci* | Spray Triazophos (0.05%) or fish oil rosin soap (2%). |
| Aphids | *Aphis gossypii* | Apply Monocrotophos, Phosphamidon, Dimethoate, Methyl demeton (0.05%). |
| Mite | *Tetranychus cinnabarinus* | Spray Dimethoate/Dicofol/ Ethion (0.05%) or wettable sulphur (2 kg/ha). |
| Blister beetle | *Mylabris pustulata* | Occasional pest. Collect and destroy adults. |
| Petiole maggot | *Melanagromyza hibisci* | Occasionally becomes serious causing plant mortality. Apply Carbofuran (0.5 kg/ha) to soil while sowing or Phosphamidon (0.05%) 7 and 10 days after sowing. |
| Cotton leaf-roller | *Syllepte derogata* | Spray Dichlorvos (0.1%). |

| Insect-Pest | Scientific name | Management |
|---|---|---|
| **Radish** | | |
| Mustard saw-fly | *Athalia lugnes proxima* | One spray of Qunialphos or Chlorpyriphos (0.05%). |
| Painted bug | *Bargroda cruciferarum* | Spot application of Phosphamidon and Oxymethyl demeton (0.05%). |
| Aphids | *Lipaphis erysimi Myzus persicae* | One Application of phosphamidon, Dimethoate and Oxymethyl demeton (0.05%). |
| Striped flea beetle | *Phyllotreta striolata* | Spray Quinalphos or Chlorpyriphos (0.05%). |
| **Tomato** | | |
| Fruit-borer | *Helicoverpa armigera* | Spray with Endosulfan (0.07%) or Dichlorovos (0.1%). |
| Serpentine leaf-miner | *Liriomyza trifolii* | Apply neem seed-kernel extract (4%) Triazophos. |
| White flies | *Bemisia tabaci* | Spray Triazophos (0.05%) or fish oil rosin soap. |
| Mite | *Tetrancychus cinnabarinus* | Spray Dicofol/Ethion (0.05%) or neem oil (1%). |
| Mealy bugs | *Ferrisia virgata* | Spray fish oil rosin soap. |
| Fruit-sucking moth | *Othereis fullonica, O. materna, O. acnilla* | Spray Qunalphos/Chlorpyriphos (0.05%) or Monocrotophos (0.05%) |
| Leaf-eating caterpillars | *Spodoptera litura S. exigua* | Spray Cypermethrin (0.0125%) or neem seed-kernel extact (4%). |
| Bugs | *Nezara viridula* | Spray Monocrotophos/Phosphamidon/Dimethoate/Oxymethyl demeton (0.05%). |
| Thrips | *Thrips tabaci Calliothrips indius* | Spray neem formulations (2-3 ml/litre). |

## Disease management in tuber crops

| Disease | Scientific name | Management |
|---|---|---|
| | | **Potato** |
| Late blight | *Phytophthora infestans* | Spray Metalayxyl-based fungicides (0.25%) and Mancozeb (0.21%). Grow late blight-resistant varieties—Kufri Sutlej, Kufri Badshah and Kufri Jawahar—in plains, and Kufri Jyoti, Kufri Giriraj, Kufri Kanchan and Kufri Megha in hilly regions. |
| Leaf-spot | *Alternaria* spp | Spray Mancozeb (0.2%) + urea (2%). |
| Wart | *Synchytrium endobioticum* | Use healthy seed of immune varieties—Kufri Jyoti, Kufri Sherpa and Kufri Kanchan. |
| Black scurf and stem canker | *Rhizoctonia solani* | Treat seed tubers with 3% boric acid for 30 minutes. |
| Dry-rot | *Fusarium* spp | Treat seed tubers with 3% boric acid for 30 minutes. |
| Bacterial wilt | *Ralstonia solanacearum* | Use whole tubers for planting and sterilize cutting knife with Methanol every time while cutting seeds to avoid spread of the disease from tuber-to-tuber. Use bleaching powder @ 12 kg/ha mixed with fertilizer in furrows at planting. |

| Disease | Scientific name | Management |
|---------|-----------------|------------|
| Soft-rot | *Erwinia* spp *Pseudomonas* spp *Bacillus* spp | Treat seeds with boric acid (3%) or Benomyl (500 ppm) for 30 minutes; Mancozeb (0.2%), Captafol (0.25%) for 10 minutes. |
| Common scab | *Streptomyces* spp | Treat seed tubers with 3% boric acid for 30 minutes. |
| Latent or faint mosaic PVX and PVS | Transmitted by aphids | Use certified seed. Blind earthing-up. Rogue-infected plants. |
| Severe mosaic PVY | Transmitted by aphids | Use certified seed. Blind earthing-up and rogue infected plants. Change tuber seed every 4-5 years in north-western plains. |
| Rugose mosaic PVX+PVY | Transmitted by aphids | Change seed at every 4-5 years in north-western plains. |
| Crinkle mosaic PVX+PVA | Transmitted by aphids | Change seed at every 4-5 years in north-western plains. |
| Leaf roll (PLRV) | Transmitted by aphids | Apply 10 kg/ha Phorate at planting or spray either Methyl demeton 25 EC (1.2 litres/ha), Dimethoate (1.2 litres/ha), or Monocrotophos 40 EC (1.2 litres/ha). |

### Amorphophallus

| Disease | Scientific name | Management |
|---------|-----------------|------------|
| Collar-rot | *Sclerotium rolfsii* | Drench soil around the plants with 0.2% Captan or 0.5% Benzimidazole. |
| Mosaic | Amorphophallus mosaic virus | Use healthy corms for planting. Rogue-out infected plants from field. |

| Disease | Scientific name | Management |
|---------|-----------------|------------|
| **Cassava** | | |
| Cassava mosaic disease | Indian cassava mosaic geminivirus | Grow field-tolerant varieties H97m, H165, Sree Visakham and Sree Sahya. |
| Brown leaf-spot | *Cercospora henningsii* | Grow field-tolerant varieties H97 and Sree Visakham. Spray Bavistin (0.1%). |
| Anthracnose/die back | *Colletotrichum manihotis* | Spray Mancozeb @ 0.3% or Bavistin @ 0.1% once or twice. |
| Tuber rot | *Phytophthora drechsleri* | Improve drainage. Remove infected tubers from field and incorporate *Trichoderma viridae* into soil. |
| **Colocasia** | | |
| Leaf blight | *Phytophthora colocasiae* | Grow resistant/tolerant varieties — Muktakeshi, Jankhri, Topi-I, and Nadia Local. Spray Ridomil (0.05%) or Mancozeb (0.25%). |
| Mosaic Dasheen mosaic virus | Transmitted by aphids | Use healthy planting material. Rogue out infected plants. |
| **Sweet Potato** | | |
| Chlorotic leaf distortion | *Fusarium lateritium* | Spray Bavistin (0.001%) at monthly interval. Spray Dipotassium hydrogen phosphate or Disodium hydrogen phosphate (0.001%). |
| Feathery mottle disease Sweet potato feathery mottle potyvirus | Transmitted by aphids | Use virus-free planting material. Rogue out infected plants. |

| Disease | Scientific name | Management |
|---------|-----------------|------------|
| **Yams** | | |
| Leaf blight/ anthracnose | *Colletotrichum gloeosporiodes (Glomerella cingulata,* | Spray Mancozeb (0.2%) |
| Leaf spots | *Cercospora carbonaceae C. dioscoreae* | Spray Mancozeb or Captan (0.25%) at fortnighthy interval. |
| Mosaic Yam mosaic virus | Transmitted by aphids | Use disease-free planting material. |

## Insect-Pest management in tuber crops

| Insect-Pest | Scientific name | Management |
|-------------|-----------------|------------|
| **Potato** | | |
| Aphids | *Myzus persicae Aphis gosspyii* | Apply 10 kg/ha Phorate at planting or spray either Methyl demeton 25 EC (1.2 litres/ha), Dimethoate (1.2 litres/ha) or Monocrotophos 40 EC (1.2 litres/ha). |
| Mites | *Hemitarsonemus latus* | Spray Kelthane (2.0 litres/ha). |
| Tuber moth | *Pthorimaea operculella* | Spray Carbaryl 50 WP (2kg/ha) or Monocrotophos (1.5 litres/ha). Apply 250 g/q Quinalphos dust on stored seed. |
| Cut-worms | *Agrotis segetum, Agrotis ipsilon* | Drench soil around the plants and ridges with Carbaryl 50 WP (2 kg/ha), Endosulphan 25 EC (1.25 litres/ha) or Chlorpyriphos 20 EC (2.5 litres/ha). |
| Golden cyst nematode | *Globodera pallida, G. rostochinensis* | Grow resistant variety Kufri Swarna. Apply Carbofuran (60-75 kg/ha) in split dose, half at planting time and half at earthing-up. |

| Insect-Pest | Scientific name | Management |
|---|---|---|
| **Amorphophallus** | | |
| Aphids | *Aphis gossypii* | Spraying of Dimethoate, Quinalphos or Fenthion (0.05%). |
| Thrips | *Caliothrips indicus* | -do- |
| Leaf-eating caterpillars | *Spodoptera litura* | -do- |
| Scale insect | *Aspidiella hartii* | -do- |
| Mealybugs | *Pseudococcus citriculus* *Rhizoecus* sp. | Spray with Dimethoate (0.05%). |
| **Cassava** | | |
| Scale insect | *Aonidomytilus albus* | Spray with Dimethoate (0.05%). |
| Termite | *Odontotermes obesus* | Soil application of Carbaryl (10%) dust or spray of Chlorpyriphos (0.05%) is recommended. |
| White grub | *Leucopholis coneophora* | Soil application of Carbaryl (10%) dust are useful. |
| Spider mites | *Tetranychus cinnabarinus* *T. neocaledonicus* | Spray Methyl demeton (0.05%) in severe infestation (January-April). Spraying water at run-off level is also effective. Foliar application of urea followed by spraying of Dimethoate (0.05%). |
| Thrips | *Retithrips syriacus* | Spray Dimethoate (0.05%). |
| Spiral white fly | *Aleurodicus dispersus* | Spray neem-based products (Azadirachtin). |
| Common white fly | *Bemisia tabaci* | Spray Dimethoate (0.05%). |

| Insect-Pest | Scientific Name | Management |
|---|---|---|
| Chips borers | *Araecerus fasciculatus Lasioderma serricorne, Rhyzopertha dominica, Dinoderus minutes, Sitophilus oryzae* | Drying cassava chips to a very low moisture content (below 10%). Polythene impregnated jute bag, woven plastic bags, metal bins etc. can be used for long-time storage. |
| Flour beetle | *Sitotroga cerealella, Tribolium castaneum* | Methyl bromide and aluminum phosphide are the fumigants most widely accepted for effective and safe disinfestation. Impregnating the bags using Malathion (0.5%), Fenvalerate (0.1%) or Azadirachtin (Nimbicidin 2%) before storing the chips also reduces insect infestation. |

## Nematode management in vegetable crops

| Nematode | Scientific name | Management |
|---|---|---|
| **Ash Gourd, Cucumber, Muskmelon, Pumpkin, Ridge gourd, Watermelon** | | |
| Root-knot nematodes | *Meloidogyne* spp. | Muskmelon Hara Madhu is moderately resistant. Cucumber Bikaner is resistant. Watermelon Shahjanpuri is resistant. Ridge gourd Meerut Special and Panipati are resistant. Pumpkin Jaipuri and Dasna are resistant. Ash gourd Agra and Jaipuri are resistant. |

| Nematode | Scientific name | Management |
|---|---|---|
| **Beetroot and Carrot** | | |
| Root-knot nematodes *Meloidogyne* spp. | Stunting of plants and galls on roots. | Apply neem, cake (1 tonne/ha) or Carbofuran (1 kg/ha). |
| **Brinjal, Bell Pepper, Potato and Tomato** | | |
| Root-knot nematodes | *Meloidogyne* spp. | Apply neem/subabool leaves (0.5 kg/m$^2$) in nursery. Tomato Hissar Lalit, NTDR 1, PNR 7, Arka Varadan and Mangala are resistant. Brinjal Giant of Banaras, Gola, Gulla, Black Beauty and Manjari Gota are resistant. |
| Reniform nematode | *Rotylenchulus reniformis* | Tomato Kalyanpur I, Kalyanpur III and LA 121 are resistant. Integration of non-host crop (*Capsicum annuum*) and application of a nematicide (Carbofuran/ Phorate 1 kg/ha) 15 days after planting tomato arc helpful. |
| Cyst nematodes | *Globodera pallida, G. rostochinensis* | Crop rotation with French bean and Peas. Potato Kufri Swarna is resistant. |
| **Amaranth** | | |
| Root-knot nematode | , *Meloidogyne* spp. | Amaranth Bhubaneswar Local and Kantei Khoda are resistant. |

| Nematode | Scientific name | Management |
|---|---|---|
| **Okra** | | |
| Root-knot nematodes | *Meloidogyne* spp. | Seed treatment with Carbofuran (3%). Soil treatment with Carbofuran (2 kg/ha) at sowing. |
| **Onion** | | |
| Stem and bulb nematode | *Ditylenchus dipsaci* | Seed/bulb treatment in hot water at $46^0$C for 1 hr. |

# FIVE

## Postharvest Management of Vegetables

Application of modern technologies in vegetable culture have significantly contributed to improve vegetable production. However, efforts to prevent post-harvest losses of vegetable are meagre. The losses increase cumulatively as the produce moves down the pipe line from harvesting to its consumption by the consumer. It is estimated that the total losses of vegetables in India due to inadequate post-harvest handling, transportation and storage are at least 20-30% which is terms of money value are worth 10 billion rupees annually. Vegetables are characterized by high metabolic activities and known to possess short shelf-life. About 10-15% fresh vegetables shrivel and stale, lowering their market value and consumer acceptability. Minimizing these losses can increase their supply without bringing additional land under cultivation. It will also help to keep pollution under control. Improper handling and storage cause physical damages due to tissue breakdown. Mechanical losses include bruising, cracking, cuts, microbial spoilage by fungi and bacteria, whereas physiological losses include changes in respiration, transpiration, pigments, organic acids and flavour. About 36% of vegetables decay due to soft-rot bacteria. The losses can be minimized by proper pre and postharvest treatments.

More so, prior to preserving and processing there is need to produce vegetables that are suitable for the purposes. Some vegetables that can be processed and a few are processed are listed in table :

## Vegetables that are processed and can be processed

| Product | Are processed | Can be processed |
|---|---|---|
| Canned | Tomato, potato, Chinese cabbage, sweet potato | Lima bean, okra, asparagus, carrot, snap bean, garden pea |
| Pickled | Cauliflower, radish, turnip, ginger and sweet pepper | Cucumber, bitter gourd, onion |
| Dehydrated | Tomato, brinjal, cabbage, cauliflower, potato, sweet potato, ginger, Chinese cabbage. | Onion, carrot, garlic, bellpepper |
| Fermented | Radish | Cabagge, Chinese cabbage |
| Chips | Potato, colocasia | |
| Flour | Potato | |
| Jam & Sweet meat | Carrot, *petha* | |
| Halwa | Pumpkin | |

# Preharvest factors

## Selection of varieties

Varieties with higher yield, better keeping quality. slower ripening and longer shelf-life under ambient condition and better processing quality should be developed and commercially grown.

## Cultural operations

Cultural operations help in prolonging shelf-life of vegetables. For root crops, preparation of the soil to a fine tilth of porous nature is necessary to avoid root forking. Irregular irrigation causes cracking of carrot and radish and splitting of outer scales of onions. In onion and garlic, irrigation should be stopped 3 weeks before harvesting to ensure better keeping quality. Heavy application of nitrogenous fertilizers causes faster tissue deterioration, while essential supply of K improves keeping quality of vegetables. molybdenum deficiency in cabbage causes heart-rot, manganese deficiency in peas leads to marsh spots and excessive irrigation and fertilisation cause hollow- heart in potato.

## Preharvest treatments

Postharvest shelf-life of vegetables is improved by preharvest application of chemicals. Preharvest application of maleic hydrazide (MH) reduces sprouting of onions and potatoes during storage. In *rabi* and *kharif* onions, application of 1,500-2,000ppm maleic hydrazide, 75-90 days after transplanting reduces sprouting during 4-5 months of storage in ventilated structures. Postharvest diseases of tomato and onion can be controlled by 3 preharvest sprays of 0.2% Difolatan at 10 day interval. Similarly, preharvest application of growth promoters such as N-benzyladenine (10-20ppm) prolongs shelf-life of leafy vegetables.

## Maturity

Vegetables are harvested as and when they attain maximum size and yet are tender. Over maturity in root crops causes sponginess and pithiness. Their harvesting should not be delayed. Delay in harvesting of onion and garlic reduces their storage quality. The maturity indices of vegetables are given in table:

## Maturity indices of vegetables

| Vegetable | Maturity indices |
|---|---|
| Radish, turnip, carrot and garden beet | Large enough, desired size, mild, tender and crisp (over-mature when pithy or fibrous) |
| Potato | Tops beginning to dry out and topple down. Skin slipping from the tuber, starch content and leaf senescence are harvesting indices of potato |
| Ginger, yam, French bean, turmeric and colocasia | Large enough, pseudo-stem dries and topple down |
| Green onion | Leaves at their broadest and longest size |

| Vegetable | Maturity indices |
|---|---|
| Cow pea and snap bean | Well-filled pods that snap readily |
| Lima bean and Bean | Well-filled pods that begin to lose their greenness |
| Bhindi | Desirable fruit size reached and the tips of which can be snapped readily |
| Snake gourd | Desirable size reached and thumbnail can penetrate flesh readily (over mature if thumbnail cannot penetrate flesh readily) |
| Bitter gourd, chayote and cucumber | Desirable size reached but still tender (over mature if colour dulls or changes and seeds are tough) slicing cucumber must be medium sized dark green immature with small seeds |
| Sweet corn | Exudes milky sap when thumbnail penetrates kernel |
| Broccoli | Bud cluster compact (over-mature if loose) |
| Lettuce | Big enough before flowering unless flowers are desired |
| Cabbage | Head firm compact (over mature if head cracks) |
| Celery | Big enough before it becomes pithy |
| Garden peas | Well-filled still pods, young, firm and tender that snap easily, attractive appearance, change in colour from dark to light green. Harvest before webbing start |
| Brinjal | Desirable size reached but still tender before fruit harden or show streaks of unusual colour. The skin should be bright and glossy |
| Onion and Garlic | Tops begin to dry out and topple down. Neck tissues begin to soften. Development of red pigment and the characteristics pungency of the variety. |
| Asparagus | Spear attain full size ( 15-20 cm long) but still tender |

| Vegetable | Maturity indices |
|---|---|
| Beet leaf. spinach and methi | Leaves attain full size according to variety, tender succulent and dark green in colour |
| Colocasia leaves | Leaves and petiole attain full size, tender, succulent and or dark green colour |
| Chinese cabbage | Leaves and petiole attained full size, tender, succulent and dark to light green colour according to variety |
| Khol-khol | The knobs (bulb) attain full size but tender, green in colour (over-mature becomes fibrous or spongy) |

## Harvesting

Harvesting should be done during cooler part of the day preferably in the morning. The produce should be shifted to shade as early as possible. Harvesting during hot periods raises field heat of the produce, causing wilting and shrivelling. Harvesting during or immediately after rains should not be carried out since it creates most favourable conditions for multiplication of micro-organisms. Care in harvesting is necessary as any bruises or injuries during harvesting may later manifest as black or brown patches making them unattractive. Injury to peel may become an entry point for microorganisms, causing rotting.

Many vegetables are harvested unripe for their safe handling, transportation and marketing but they must be matured when harvested so that they can ripen later on normally and develop good eating quality.

## Postharvest factors

### Curing

Curing is conducted immediately after harvesting. It strengthens the skin. The process is induced at relatively higher temperature and humidity, involving suberization of outer tissues followed by the development of wound periderm which acts as an effective barrier against infection and water loss. It is favoured by high temperature and high humidity. Potato, sweet potato, colocasia, onion and garlic are cured prior to storage or marketing. In sweet potato, this condition is most rapid at 33°C and relative humidity of 95%. Potato tubers are held at 18°C for 2 days and then at 7°-10°C for 10-12 days at 90% relative humidity. Curing also reduces the moisture content especially in onion and garlic. Drying of superficial leaves of onion bulbs protects them from microbial infection in storage. Maximum safe temperature for onion curing at field is 37.8°C for 3-5 days. Artificial curing of onions created at 40°C for 16 hr reduces rot losses in storage.

## Degreening

Degreening is the process of decomposing green pigments in fruits usually by applying ethylene or other similar metabolic inducers to give a fruit its characteristic colour as preferred by consumers. It is applicable to tomato. The time required to degreen a fruit depends upon the degree of natural colour break and maturity. The higher the green colour and more mature a fruit is, the less time is required to reduce the chlorophyll to a desired level.

Degreening is carried out in special treating rooms with controlled temperature and humidity in which low concentration of ethylene (20 ppm) is applied. The ethylene should be supplied from a gas cylinder. These rooms are thoroughly ventilated to keep the $CO_2$ level below 1% which does not allow higher colouring. If kerosene fumes are placed outside the degreening room, they enter the room through ducts by forced ventilation. Despite the fire hazard involved, kerosene fumes produce better coloured fruits than pure ethylene. It is due to good ventilation. The best degreening temperature is 27°C. Higher temperatures delay degreening. The relative humidity should be 85-90%. Higher humidity levels cause condensation during degreening and are associated with slow degreening and increase in decay. Low humidity though, checks decay, causes excessive shrinkage, shrivelling and peel breakdown.

In another method, fruits in containers are sealed by 2 sheets of plastic film and water. The PVC (0.2 mm thick) can also be used. The ethylene is introduced from a can of 4.2 litres capacity in the film which can cover 1.2 tonnes of fruits (20 kg x 60 containers), the ethylene concentration becomes nearly 1000 ppm resulting in satisfactory colouring. After 15 hr, the film is removed enforcing the fruit to air. Degreening takes 3 days. Ethylene accelerates decomposition of chlorophyll without significantly affecting the synthesis of carotenoid pigments.

## Pre-cooling

High temperatures are detrimental to keeping quality of vegetables, especially when harvesting is done during hot day's. Precooling is a means of removing the field heat. It slows down the respiration of the produce, minimizes susceptibility to attack of microorganism, reduces water loss and eases the load on cooling system of storage or transport. Peas and okra which deteriorate fast, need prompt cooling. Sometimes stage of ripening and level of field heat of produce also determine the need for precooling. For example, unless tomatoes are above 26.7°C and ripening is to be delayed, there is no need for precooling.

In air cooling, cool air can be obtained from cold storage. Temperature should not be less than -1°C to avoid freezing. Where night temperatures are low, doors of the store rooms can be opened for cooling in the night.

In water cooling (hydrocooling), field heat is removed quickly. It is used for leafy vegetables to retain their texture and freshness. Ice can be added to bring down the temperature. However, temperature should be controlled to avoid chilling injury in cold sensitive vegetables.

## Washing and drying

Most of the vegetables are washed after harvesting to improve their appearance, prevent wilting and remove primary inoculum load of micro-organisms. Hence a fungicide/bactericide should be used in washing water. After washing, excess of water should be removed which would otherwise encourage microbial spoilage. Root and tuber crops are often washed to remove the soil adhering to these.

## Sorting and grading

Immature, diseased and badly bruised vegetables are sorted out. Most of the countries have their own set of standards of domestic trade and for international trade, standards have also been defined. Grades are based on size, weight, colour and shape. Grading is done manually or mechanically.

## Disinfestation

Vegetables are susceptible to fruit fly attacks. Disinfestation is done either by vapour heat treatment at 43°C with air saturated with water vapour for 6-8 hr; by ethylene dibromide fumigation (18-22 g of EDB/cubic meter for 2-4 hr. Residues of inorganic bromide must not exceed 10 vg/g) or by cold treatment (exposure of fruits to near freezing temperature for a specified period).

## Postharvest treatments

A complete inhibition of sprouting of cool chamber (evaporatively cooled) stored potatoes for 4 months and 5 months is achieved by spraying them with an aqueous emulsion of CIPC @ 50 mg and 100 mg/kg of tubers respectively, before completion of dormancy period.

## Waxing

Vegetables have a natural waxy layer on their outer surface which is partly removed by washing. An extra discontinuous layer of wax applied artificially with sufficient thickness and consistency to prevent anaerobic condition within the vegetables provides necessary protection against decay organisms. Waxing is especially important if tiny injuries and scratches on their surface are present. These can be sealed by wax. Waxing also enhances the gloss of vegetables. Therefore appearance is improved, making them more acceptable.

If refrigerated storage facilities are not available, protective skin coating with wax increases the storage life of fresh vegetables at ambient temperatures.

There are 2 types of wax emulsion-wax "W' and "O". The wax composition "W' does not impart any gloss to vegetables and hence where gloss is required for improving marketability of the produce, composition wax "O" is recommended. Both these emulsions contain 12% total solids.

The application of wax emulsion to freshly harvested healthy produce protects them against excessive moisture loss, higher rate of respiration, heat build-up or thermal decomposition. Texture and quality of the fresh produce is maintained as near the fresh conditions as possible for a long time.

The wax emulsion without fungicide does not protect vegetables against microbial spoilage. Therefore, suitable fungicides are added to the wax emulsion. Storage life of some vegetables with and without wax emulsion treatment is presented in table:

# Storage life of some vegetables with and without wax treatment at ambient temperature

| Commodity | Condition of commodity | Type of wax emulsion and concentration (%) | Storage life (days) Treated | Untreated | Quantity of wax emulsion at 12% concentration required for one tonne commodity (litre) |
|---|---|---|---|---|---|
| Potato | Mature, transplanted | W/12 | 80 | 40 | |
| Tomato | Green, firm | 0/9 | 14 | 7 | 5.0 |
| Tomato | Pink, firm | 0/9 | 12 | 4 | |
| Brinjal | | W/12 | 8 | 5 | |
| Carrot | Mature, topped | W/12 | 9 | 6 | |
| Capsicum | Green, firm | W/12 | 9 | 4 | |
| Cucumber | Mature | W/12 | 7 | 4 | |
| Parwal | Firm, mature | W/12 | 6 | 3 | |

## Control of ripening process

Ripening in vegetables can be retarded by using proper packing, low temperature, ethylene absorbants, skin coating of waxol, growth retardants and using fungicides for controlling their spoilage. Use of Cycocel (500mg/litre), Alar (500mg/litre), GA (250mg/litre) and menadione bisulphite (500mg/litre) significantly retards ripening.

## Ripening of fruits

Ripening transforms a physically mature but inedible plant organ into a visually-attractive taste and smell sensation. It marks the completion of development and commencement of senescence with life of a fruit and is normally an irreversible event. Ripening can be achieved by the application of ethylene.

Accurate quantity of ethylene should be used in the ripening room at regular intervals. A concentration of $CO_2$

above 1% delays ripening. Hence, thorough ventilation is essential. By use of ethephon commercially known as ethrel or CEPA (7 fluid ounces release 1 cft of (ethylene), making it alkaline using caustic soda (3 g of caustic soda for 20ml of ethephon). Calcium carbide can also be used for ripening (100 g for 100 kg fruits).

### Pre-packaging in plastic films

Pre-packaging increases the shelf-life by creating a modified atmosphere with an increase in concentration of $CO_2$ in the package. The packaging material used should provide reasonable access to oxygen. For this, breathing films like polystyrene and cellulose acetate are used. But tougher LDPE films which have high $O_2$ and $CO_2$ transmission rates are more durable. The pouches must have perforations to transmit oxygen and carbon dioxide rapidly enough for the respiration of fresh produce. The pouch used reduces bruising, facilitates inspection, reduces moisture loss (weight loss) and prevents dehydration. It also creates modified atmosphere.

In pre-packaging, leaves, stalk, stem etc. are trimmed, washed, cleaned and weighed quantities are put in pouches. Ethylene absorbants may be added to the package wherever required to retard the ripening process. Hydrated lime (calcium hydroxide) inserts may also be beneficial in controlling $CO_2$ concentration within the film package.

### Packaging

A wide range of packages-gunny (hessian) bags, woven, bamboo, weed and grass stem baskets, palmyra mats, wooden cases, earthern pots, corrugated fibre board cartons and rigid plastic crates are used. More than one type with different sizes and shapes of packaging is used for individual commodity either due to availability of the particular packing material in the local market or due to nature and cost of the material to be packed. Certain commodities which can withstand rigours of journey such as potato, onion, carrots, radish etc. are packed in gunny bags.

Wheat and paddy straw, banana leaves, dry grass etc., which are easily available locally and less costly, are used mostly as cushioning material. These are unhygienic and do not allow respiratory heat to escape from the packing box. Use of moulded pulp tray, honey comb partition, cell pack etc. have replaced the use of cushioning material in costly commodities and for export purposes. Wooden boxes are used for packing important varieties of tomato and capsicum.

## Pelletization

Loading and unloading are done manually in India. Due to low unit load, there is a tendency to throw, drop or mishandle the package, damaging the commodity. This loss can be considerably reduced by using pellet system. However, this requires the standardization of box dimensions. For each commodity it should be worked out. Once this is accomplished, mechanical loading and unloading become very easy with the fork-lift system.

## Transportation

For selecting the mode of transport, the distance to reach the destination as well as the perishability of the commodity should be considered. For highly perishables ones, there should be minimum temperature rise during transit. The commodity should be pre-cooled before packing to remove the field heat. Road transport should be preferred for perishable commodities than rail transport.

For local market, the produce is brought by bullock carts or tractor trollies. Carts, trailers and trucks used in the field should have good suspension and low tyre pressure to avoid excessive jolting of produce. They should be driven slowly. Lining of trailer with straw or leaves can also help prevent damage.

## Storage

In India, the production patterns, dietary habits and economic considerations warrant long period storage in large quantities of onion and potato. With other vegetables, the main need of storage is mostly for short periods and in many cases for a few days or weeks only. To store vegetables, low temperature and high humidity (90-95%) are required except in onion and garlic which require relatively lower relative humidity (70%). Different structures which are used for storage of mostly potatoes and onions can be used for other commodities also.

### Storage in cool dry rooms

It is mostly used for storage of seed potatoes kept on the floor or on bamboo racks with proper ventilation. The store is built of unbaked bricks. It has a thatched roof covered with tiles. Windows are provided on the side walls of the store which are kept open during night and closed during the day.

### Storage in pits

The pit is dug under the shade of a tree or a roof 75-90 cm wide and 45 cm deep. It is soaked with water and allowed to dry for 5 days. The bottom and sides of the pit are covered with 'neem' leaves. The selected potatoes are heaped up in the pit to a height of about 1m and covered with a thick layer of straw or grass. At about 30-60 cm distance from the pit, a ditch is dug and filled with water occasionally. The condition of the tubers is examined by opening the pit from the top. The rotted tubers are removed.

## Onion storage

In Panipat district of Haryana and Jalalabad in Uttar Pradesh, structures are made of bamboo/*sarkanda* nets and thatched roof with 'sirki' which is covered on top with jute cloth. The size of 3.0m x 1.2m x 1.2m has the capacity to store 40 quintals. The bottom net is fixed at about 15-20 cm height from the ground level to have aeration.

In Nasik and Pune areas, the structures have side walls made of bamboo or locally available wood, spaced 1-2cm apart. Roof is made of either thatch, asbestos sheet or tiles/ tin sheet. There is, however, no bottom ventilation but flooring is done with the help of soil, stone particles and sand and it is raised from ground level. Before loading onion, sticks are spread on the floor. The size of the structure varies from area-to-area. Normally width is kept as 1.5m and height is about 1.5m. The length may vary from 13.5 to 30m. Its capacity is 200-450 quintals.

In Bihar and Gujarat, these structures have 3-4 tiers and the floors are made of bamboo pieces spaced 1-2 cm apart. The depth of loading is generally 30-60 cm. Ventilation in structure is provided by raised platform and windows on the side walls. Cold stores are rarely used for storage of onion in India.

## Cool stores

Perishables can be effectively stored over different periods under low temperatures and high humidity condition as obtained in commercial cool stores. In India, the total installed capacity of about 2,522 cool stores is over 5 millions tonnes. Ninety per cent of this capacity is being utilized for the storage of potatoes including seed potato.

During storage, chilling injury occurs in some commodities and hence, these should be kept at optimum temperature of their storage. Physiological disorders and major post-harvest diseases of vegetables also deteriorate their quality.

## Zero energy cool chamber

The zero energy cool chamber based on evaporative cooling system is saturated with water. In summer, when outside temperature is 44°C, the maximum temperature inside the chamber never goes beyond more than 28°C, the relative humidity being 90%.

## Storage life of fresh vegetables

| Produce | Time at optimum temperature (weeks) | | |
|---|---|---|---|
| | -1 to 4°C | 5-9°C | 10°C |
| *Very perishable (0-4 weeks)* | | | |
| Asparagus | 2-4 | | |
| Bean | 1-3 | | |
| Broccoli | 1-2 | | |
| Brussel's sprout | 2-4 | | |
| Cauliflower | 2-4 | | |
| Cucumber | | 2-4 | |
| Lettuce | 1-3 | | |
| Pea | 1-3 | | |
| Rhubarb | 2-3 | | |
| Spinach | 1-2 | | |
| Sweet corn | 1-2 | | |
| Tomato (coloured) | | 1-3 | |
| *Perishable (4-8 weeks)* | | | |
| Cabbage | 4-8 | | |
| Tomato (green) | | | 3-6 |
| *Semi-perishable (6-12 weeks)* | | | |
| Celery | 6-10 | | |
| Leek | 8-12 | | |
| Marrow | | | 6-10 |
| *Non-perishable (> 12 weeks)* | | | |
| Beet root | 12-20 | | |
| Carrot | 12-20 | | |
| Onion | 12-28 | | |
| Parsnip | 12-20 | | |
| Pumpkin | | | 12-24 |
| Potato | | 16-24 | |
| Sweet potato | | | 16-24 |
| Swede turnip | 16-24 | | |

## CPRI cool home for potato

The store is a brick masonry structure measuring 9.1m x 4.6m x 3.7m. The north and south sides are double-walled with 11cm gap between them. It is filled with rice husk to provide insulation. The 23cm outer walls bear the weight of the roof. The outside of the walls are whitewashed to reflect heat. Eight ventilators (76cm x 60cm) are provided on the double walls on the northern side and three ventilators on the eastern walls. The ventilators are provided with 5 cm thick wood wool which is continuously moistened by water dripping from above through small holes in the GI pipe. The ceiling is made of plywood with 5 cm diameter holes drilled in it at 10cm x 10cm spacing so that the hot air within the room can rise above the ceiling and eventually escape into the open. The roof is made of galvanized iron sheets with provision of ample ventilation.

The potatoes are stored in 80kg bags on wooden pellets. The temperature inside the store remains 6-12°C, lower than outside during May and June.

## AADF-CIP design cool home

It is a 'C' grade brick masonry structure with, 35cm thick walls in the mud with dimensions of 5.4m x 4.5m x 2.55m having a 10 tonne capacity. The foundation of the structure is simple spread footing and the walls are solid with the insulation provided by the extra width of brick wall and mud plaster. Eight ventilators (60cm x 30cm) are provided at a height of 15cm from plinth level. A water trough is provided throughout the length of store except for the passageway. The roof is made of thatch using local materials. Ventilators are provided at the top for exit of hot air from the store.

The potatoes are stored loose on the bamboo platform above the water trough at the height of ventilator (30cm from the plinth level). Water or moist sand is kept in the water trough to facilitate evaporative cooling. As the day temperature rises, it heats up the air inside the store. The air becomes lighter and escape through the ventilators at the top. Fresh air enters from the ventilators at the bottom. gets cooled through contact with water/moist sand in the water trough of the store to replace the hot air by convection current.

### Forced evaporation cool stores

These stores are cooled with industrial desert coolers which force the cool and humidified air through the heap of potatoes.

### Two-tier onion storage structure

Its capacity is 40 quintals with the size of structure being 4.8m x 1.5m x 1.2m. The side walls are made of bamboo splits with adequate support of wooden planks. The lower base is also made of bamboo splits with adequate structural support. It is raised from the ground level by about 20 cm for proper ventilation from the bottom. The second base is at a height of about 75 cm from the bottom base. The second tier facilitates better aeration, reducing the losses. The total losses in this structure are about 36.6% over a period of 4 months.

### Bottom ventilated onion storage godown

Its capacity is 450 quintals having size of 15.8m x 4.8m x 2.4m. Ventilation is provided for free circulation from all sides including bottom ventilation. Below the floor, there is air gap of 60cm for this purpose. The columns of the structure are made of brick masonry which support the roof made of Mangalore tiles on timber trusses, purlins and rafters. There are 2 onion racks in each structure of 15m x 1.5m size with a passage of 1.8m between them for loading/unloading of onions. The onion racks are of wooden battens.

## Irradiation

Application of irradiation for suppressing sprouting and hence extension of shelf-life has been allowed in India. Sprouting of onion can be checked by gamma irradiation at a dose of 0.06-0.1 KGY. In potato, gamma irradiation at 0.IKGY can inhibit sprouting completely. It also inhibits the light induced synthesis of chlorophyll and toxic alkaloid solanin. The irradiated potatoes could be stored successfully for 6 months at 15°C with 10% loss.

# SIX

## Processing of Vegetables

Vegetables are a major source of vitamins and minerals. Since they are highly perishable, they need to be preserved and processed in various value added products.

### Heat preservation and processing

Heat is widely used in preservation of food by cooking, microwave heating, blanching, frying, canning, pasteurizing, boiling or heating foods prior to consumption. The thermal processed foods (bottled and canned) are totally sterile. In these processed foods both pathogenic and toxin-producing organisms are destroyed. Higher temperatures above optimum are lethal to bacterial spores. Enzymes are also heat sensitive and all the known enzymes lose their activity if heated to 80° and above. By the application of heat, both microbial and enzymatic spoilage can be well checked. To reduce the chemical changes in foods, heating should be done to prevent microbial and enzymatic spoilage.

### Canning (sterilization)

For canning, vegetables should be absolutely fresh. The vegetables should be tender, fully ripe and firm vegetables are considered ideal. The unit operations include sorting and grading, washing, peeling, coring and pitting, blanching, can filling, brining, lidding or clinching, exhausting, sealing and processing by heating to inactivate bacteria. Vegetables except more acidic (tomato and rhubarb) which are generally non-acidic, require to be processed at high temperature 115°-121°C (high pressure of 10-15 lb/inch$^2$) in the autoclave. The temperature and time of processing vary with size of the can. The larger the can, the greater the processing time and vice-versa.

## Bottling

Bottles which can stand high temperature, can be sealed airtight. Pasteurization, canning, freezing, carbonation and drying are used for preservation. Various tomato products are heat processed.

## Pasteurization

Pasteurization of vegetable juices by overflow method by heat is most popular. The heating of juice at $85^0$-$90^0$C for 30 minutes can kill spore forming bacteria, mould spore and enzymes. They can be flash pasteurized, i.e. at a high temperature for short time only ($90^0$-$95^0$C for 1 minutes) and then filled into containers which are sealed airtight under cover of steam to sterilize the seal and then cooled.

## Low temperature preservation and processing

Freezing and cold storage are the oldest methods of food preservation. Commercial and household refrigerators are usually run at $4.4^0$-$7.2^0$C, whereas in frozen storages frozen condition is maintained at zero degree or below. Refrigerated or cool storage preserves perishables for days or weeks depending upon the commodity. Frozen storage preserves perishables for months or even years because of very low temperature. Although low temperature is not lethal to bacterial growth and multiplication, yet retards their activities. Freezing preserves the food without major changes in its physico-chemical composition. Freezing with cryogenic liquids (liquid nitrogen at $-196^0$C or liquid $CO_2$ at $-43^0$C) is gaining importance.

For frozen peas, fresh, clean, sound, whole immature seeds should be selected washed and blanched for 2-3 minutes in boiling water to ensure adequate colour and flavour. The cooking time (blanching for quick frozen peas) may vary according to variety, maturity, and size of peas. After blanching, they are cooled and packed in polyethylene or laminate pouches. Freezing should be carried out in such a way that the range of temperature of maximum crystallization is passed quickly. The quick freezing process requires $-18^0C$ to $-25^0C$. The product is maintained at low temperature ($-18$ to $20^0C$). For frozen beans, method is followed as for peas except that blanching time varies from 4-5 minutes.

Freezing is accomplished by exposing the food to very low temperature resulting in converting the water molecules of food into ice crystals. Once the food is frozen, it has to be stored under very low temperature.

## Preservation by removing moisture

Drying (removing moisture) helps in preservation of foods. Microbes cannot grow and multiply in absence of sufficient water in their environment. Many of the enzymatic reactions are hydrolytic in nature, requiring water. Chemical reactions in food materials are slowed down when the reactions are in solid state. Hence, by removing water from the commodity, it should be possible to preserve them by checking the important spoilage agents. This principle forms the basis for dehydrated foods and for osmotic dehydration where high sugar or salt acts as a preservative. Vegetables may be dried in air, super heated steam, in vacuum, in inert gases or by direct application of heat.

### Drying, dehydration and concentration

Removal of moisture by applying heat is called *drying*. Dehydration is drying by artificially heating under controlled temperature, humidity and air flow. The basic drying processes are classified as sun-drying, atmospheric dehydration (stationary or batch process and continuous process) and vacuum dehydration. Sun-drying is still in use. These methods are accompanied by removing water. Moisture may be removed from foods by any of the methods. Dehydration (below 5%) moisture helps to preserve nutritive value of the product although heat liable vitamin losses are more. Low moisture preserved material can be fortified with vitamins under proper storage conditions. The dried products have more shelf-life if properly packed and stored. Retention of nutrients and overall quality is better in modern methods of dehydration such as freeze drying, vacuum drying, osmotic or solar dehydration as there are little changes in physico-chemical composition. Dried products go a long way as snack products. In India, dehydration of vegetables is of much greater importance to reduce the considerable spoilage. Foods are concentrated to remove moisture either by dehydration or freezing.

Concentration can be a form of preservation of vegetable pulps and juices. They are concentrated in flash evaporators or concentrators. Different types of beverages are made from concentrates. Their colour and flavour retention is better than in dehydration and drying (powders). There are various types of driers also for concentration of pulps and juices into powder for drying of vegetables such as foam mat, kiln, spray drum roller, vacuum and fluidised bed, microwave driers.

Sun-drying of vegetables is practised by solar energy widely in tropical and subtropical regions. In indirect solar drying, the vegetables are spread on trays, which are kept in a close compartment or cabinet drier. In solar drier, the air is heated by direct sun rays and carried to the drier from bottom and goes out from the top of chimney. The drying conditions are influenced by the temperature and velocity of the air flow. These solar driers may be with chimney or with amber-coloured glass. For drying vegetables at home, the home drier is ideal.

# Drying schedule for vegetables

| Vegetables | Preparation | Blanching in boiling water (minutes) | Sulphitation in 0.5% potassium metabisulphite solution (minutes) | Drying temp. (°C) | Drying time (hr) | Drying ratio |
|---|---|---|---|---|---|---|
| Beans | Cut into small length | Same as for peas | NIL | 60-65 | 8-10 | 7:1 |
| Bitter gourd | Remove both ends and cut into 6 mm thick slices | Blanch in boiling water for 7-8 minutes | 30 | 66-71 | 7-9 | 16:1 |
| Cabbage | Remove outer leaves and core, cut into 4-8 mm thick shreds | 5-6 | 30 | 60-66 | 12-14 | 18:1 |
| Carrot | Peel, cut into 3-5 mm thick slices | 2-4 (in salt solution of 2-4%) | 30 | 68-74 | 14-16 | 18:1 |
| Cauliflower | Remove stalks, leaves and steam, cut into 10 mm thick pieces | 4-5 | 30 | 60-66 | 10-12 | 35:1 |
| Fenugreek | Sort, remove leaves, stalks and steam and wash thoroughly for 2-3 minutes | Blanch as in spinach | NIL | 60-65 | 8-9 | 20:1 |
| Garlic | Peel the cloves, cut into 6 mm thick shreds | NIL | NIL | 55-60 | 7-8 | 8:1 |
| Knol-khol | Peel, cut into 5 mm thick slices | NIL | 30-40 | 54-60 | 11-13 | 19:1 |
| Onion | Remove tops and tails, peel and cut into 4-8 mm thick shreds | Dip for 10 minutes in 5% salt solution | NIL | 55-60 | 11-13 | 10:1 |

| Vegetables | Preparation | Blanching in boiling water (minutes) | Sulphitation in 0.5% potassium metabisulphite solution (minutes) | Drying temp. (°C) | Drying time (hr) | Drying ratio |
|---|---|---|---|---|---|---|
| Okra | Remove both ends and cut into 6 mm thick slices | 5-6 | 30 | 54-60 | 11-13 | 19:1 |
| Peas | Shelled peas | 2-3 (in boiling water containing 0.5% KMS, 0.1% sodium bicarbonate and 0.1% magnesium oxide) | NIL | 63-65 | 8-10 | 8:1 |
| Potato | Peel, cut into 3-5 mm thick slices | 3-4 | 30 | 60-66 | 7-8 | 7:1 |
| Pumpkin | Peel, remove seeds and soft portions, cut into 6 mm cubes | 2 minutes in hot brine (2% common salt solution) | 30 | 65-71 | 9-11 | 22:1 |
| Spinach | Sort, wash thoroughly in water and cut into 10 mm portions | Same as for peas | NIL | 63-68 | 7-8 | 22:1 |
| Tomato | Peel after scalding in boiling water for 30-60 seconds and make slices of 6-9 mm thickness | NIL | NIL | 60-65 | 9-10 | 27:1 |
| Turnip | Peel and cut into 4-7 mm thick slices | 2-4 | 30 | 52-57 | 11-13 | 28:1 |

The drying time varies with the cultivars, maturity, air velocity tray load and pretreatments. Although sun-drying is more economical but nutritional and organoleptic quality of the dried products are better than dehydrated ones.

## Preservation with sugar

Preserve (murrabba) is made by cooking or heat processing of raw or mature, peeled or punctured and pre-treated whole or cut vegetables in sugar syrup. The vegetable, however, retains normal shape and with wholesome appearance. Vegetables for preserves should be firm, ripe rather than a soft ripe stage. They should be uniform in size or in pieces so as to cook evenly. The manufacture of vegetable preserves involves selection of raw material, peeling, puncturing (to promote sugar penetration), blanching with or without additives to effectively inactivate natural enzymes, sugar addition and cooking or concentration in sugar syrup, control of fermentation by preservatives and then packaging. In preserve making, not less than 45 kg of vegetables are used for every 55 kg sugar and cooking is continued till a concentration of at least 68% of total soluble solids is reached.

## Salient features of industrial processes for preparing preserves

| Preserves | Pre-treatments | Vegetable: sugar (kg) | Days required to process |
|---|---|---|---|
| Ash gourd | Cut into pieces, peeled, priked, soaked in lime water and washed, blanched in alum solution, rinsed with water | 100:125 | 7-15 |
| Carrot | Peeled, blanched and punctured | 100:150 | 6-7 |
| Ginger | Peeled, cut into pieces and blanched | 100:125 | 4-5 |

Candies are made by dehydrating by osmotic pressure of sugar solution (osmotic dehydration). Preserves when drained free from syrup and dried is called *candy* which is prepared largely from petha and ginger. Candies coated with thin transparent coating of sugar crystals are known as *crystalised fruits* which are chiefly prepared from ginger.

## Preservation by salt

The concentration of salt necessary to inhibit the growth of microorganisms in food is related to water content, type of infection, pH, temperature, protein content and presence of inhibitory substances such as acids. The water content is obviously of major importance, since it is the concentration of salt in water phase and not the amount in food as a whole which is significant.

Curing of raw fruits and vegetables in dry salt or brine and subsequent preserving by spices and condiments or in vinegar is known as *pickling*. Spices, condiments and edible oils may also be used to improve their palatability. Turnip, carrot, cauliflower, onion and mixed vegetables are important for pickle making. Various types of pickles of exotic nature such as tender bamboo shoot pickle, chilli pickle, brinjal pickle, karonda and keri pickles are very popular in some parts of the country. Apart from salt, mustard, jaggery, gingelly oil or mustard oil are major constituents of pickles apart from a variety of spices and condiments. A wide variety of traditional pickles are prepared both by organized sector as well as at homes.

For making pickle, wash raw material, cut, peel, mix salt and spices. Fry fine paste of condiments with smoking oil wherever needed and mix with the raw vegetable. Add vinegar, jaggery and oil. Fill in jars and store them. Vegetables are blanched before pickling. Pickles are good appetizers and add to the palatability of a meal.

## Recipes for vegetable pickles

| Ingredient | Quantity of ingredients (g) required for 1 kg of prepared product | |
| --- | --- | --- |
| | Mixed vegetable pickle | Turnip, cauliflower and carrot pickle |
| Salt | 80 | 125 |
| Red chilli | 10 | 20 |
| *Garam masala* | 30 | 30 |
| Turmeric powder | 5 | 20 |
| Rye seeds (mustard) | 30 | 30 |
| Onion (chopped) | 150 | 150 |
| Garlic | 30 | 30 |
| Ginger (green) | 50 | 50 |
| Glacial acetic acid | 10 | 10 |
| Oil | 250 | 250-300 |

Pickles are manufactured on a large scale and exported to other countries also. In India, pickles generally used, are made in either mustard, rapeseed or sesame oil (gingelly oil). Some pickles are made in lime juice or in vinegar. Generally fermented pickles are preserved in vinegar. They are also preserved with sorbic acid or lactic acid.

## Preservation by food additives

A chemical additive or food additive can be defined as a chemical (substance) or mixture of chemicals, other than basic food stuff that is added intentionally either during production, processing, storage or packaging directly or indirectly to improve or maintain nutritional value, enhance quality and consumer acceptability, improve keeping quality and check spoilage caused by microbes and enzymes and facilitate preparation. Common types of intentional food additives are salt, sugars, acids, spices, essential oils, buffers, bleaching agents, emulsifying and thickening agents, food flavour, colour, preservatives, antioxidants, clarifying agents and humectant. Use of food additives is another effective approach/method for preserving vegetables. Some of them in combination can check microbial activity, enzymatic and non-enzymatic browning of the preserved products during storage. Acids lower the pH and act as antimicrobial agents like chemical preservatives. They have bacteriostatic and bactericidal properties. The permitted preservatives used in our country for various products are benzoic acid including salts (sodium benzoate) and sulphurous acid including salts (potassium metabisulphite) or combination of these. They are used either alone (in high concentration) or in combination (in low concentration) to check spoilage in fruits and vegetables and their products as well as increase their shelf-life by keeping proper nutritional value. Preservation by this method is cheap and easy to operate technology, best suited for its application in the developing countries to preserve perishable commodities.

**Limits for permitted preservatives in vegetable products**

| Vegetable product | Preservative | Concentration (ppm) |
| --- | --- | --- |
| Pickles and chutney made from vegetables | Sulphur dioxide or Benzoic acid | 100 250 |
| Tomato and other sauces | Benzoic acid | 750 |
| Dehydrated vegetables | Sulphur dioxide | 2,000 |
| Tomato puree and paste | Benzoic acid | 250 |

In case sulphur dioxide is used as preservative in pickles and chutneys, the product should not be packed in tin containers.

## Chemical Preservation

Seasonal pulps and juices from vegetables such as carrot, tomato etc. can easily be preserved and stored after heating them to $80^0$-$85^0C$ and then by adding chemical preservative.

Potassium metabisulphite is used to preserve most of the pulps/juices, whereas sodium benzoate is used for coloured juices/ pulps. The preservatives are added after dissolving them separately in a small quantity of water. These pulps/ juices can be stored either in jars or glass bottles.

## Processing of sauce, chutney and ketchup

Chutney and sauces are made in Indian homes and also on a commercial scale. Standard recipes have been modified according to consumer's acceptability. Onion, garlic, spices, salt, sugar and acid are added for flavour and to make them more palatable. Vinegar (or glacial acetic acid) also serves as a preservative. Chutney should have at least 50% total soluble solids (sugar) and 1.0% acidity. These can be made from any material either from pieces, shreds or pulps.

Tomato chutney is very popular and also delicious. In the preparation of chutney, spices and vinegar are added. Chutney and sauces possess appetizing properties. Sauces are of two kinds - thin and thick. Thin sauces mainly consist of vinegar extract of various flavoring materials like spices and herbs. Their quality depends mostly on the piquancy of the material used. Mixed vegetable sauce, yellow pumpkin sauce, tomato sauce (tomato ketchup) are very common.

For chutney and sauces both acetic acid and chemical preservatives are essential. Acetic acid having antiseptic property is also used at 0.5% level in pickle preparation. In sauce, ketchup or chutney, boiling with 0.5% glacial acetic acid for 4 minutes checks the spoilage of the products.

## Preservation by fermentation

Fermentation encourages the multiplication of microorganisms and their metabolic activities in food. In this method, food is preserved chemically. Microorganisms are used to ferment sugar either by complete oxidation or partial oxidation in alcoholic fermentation (for wine and fermented beverages), acetic fermentation (for vinegar) and lactic fermentation (fermented pickles) and other minor fermentative actions. Fermentation is a low-cost technology for preservation of vegetables. Lactic acid fermentation is of great importance in food preservation. Lactic acid producing bacteria require small amount of salt for their growth and multiplication. Sodium chloride is useful in lactic fermentation of foods since it limits the growth of putrefactive organisms and inhibits the growth of a large number of undesirable microorganisms.

Fermented foods have many advantages *viz.* prolonged shelf-life, extended seasonal life, less time for cooking and sometimes increased acceptability and digestibility. This also acts as a laxative agent. The vegetables can be preserved by the simple method of lactic acid fermentation which enhances acceptability and nutritional quality of fresh vegetables. By this low-cost technology, some vegetables (individually or mixed) can be preserved safely and popularized.

Fermented spiced beverage juice from black carrot (*kanji*) is very popular in north India. It has a cooling effect especially in summer and it has attractive crimson colour. This beverage can be prepared and preserved for off season use. *Kanji* can be prepared either from black carrot slices in brine or by its juice along with salt, chilly and mustard powder. Here, lactic fermentation is involved and it takes 7-10 days to complete fermentation. The quality of fermented products is better at low temperature (20⁰-25⁰C) than at higher temperatures.

## Cold sterilization/filtration and irradiation

### Filtration

The spoilage in vegetables can also be controlled by mechanical removal of microorganisms by ultrafiltration. It is known as the cold process. It is applied in the treatment of fruit juices. beers and wines. It is of course, applicable only for clear, liquid products.

### Irradiation

In this method, food is preserved by ionizing radiation. The irradiation of food can destroy micro-organisms and enzymes. It is more efficient to employ ionizing radiation to kill microorganisms than enzymes. It may be desirable to inactivate enzymes by other means, in complement to the irradiation action. The sterilization of food with ionizing radiation involves two major considerations, the food product and a suitable radiation source. Since the temperature remains 4⁰-5⁰C, it is also called *cold sterilization technique*. Irradiation technique has been successfully used in controlling the ripening process of vegetables and also for checking sprouting of roots, tubers and bulbs apart from general food preservation.

# SEVEN

## Poly-house Technology for Vegetable Production

The recent advent of plastic has been a unique gift to the modern olericulture and its manifold uses in the form green polythene/plastic sheet commonly known as 'poly-house'. The polyhouse cultivation is possibly the most intensive method of crop production. Cultivation of vegetables in polyhouse involves protection of production stages of vegetables mainly from adverse environmental conditions viz. temperatures, hails, scorching sun, heavy rains and snow. Since vegetables are treated as high value crops and these regular supply in the market in fresh condition is an essentiality, the utility of poly-houses has been found to be maximum with respect to these crops.

The advanced country like Japan is the world leader in poly-house cultivation, having 95% area under poly-houses and out of this about 84% area is occupied only by vegetables.

In India, commercial poly-houses with controlled devices are very few. In early sixties, the Field Research Laboratory (FRL) of DRDO at Leh, attempted solar poly-house vegetable production research and made an outstanding contribution to the extent that almost every rural family in Leh valley posseses a polyhouse these days.

Poly-house technology in India was adopted late but owing to its merits and incentives given by the Central and State Governments, a notable increase in the poly-house area has occurred in the recent years. Indian Petrochemical Corporation Ltd. (IPCL) boosted the polyhouse research and application of raising vegetables by providing UV stabilized cladding film and aluminum poly-house structures. Poly-houses are becoming increasingly popular both in temperate and other regions. In hills, they are being used for raising vegetable nursery, protecting imported germplasm, raising off-season vegetable crops and their seeds. Poly-houses are becoming popular in Ladakh region of Jammu and Kashmir and almost 70% of the total area (approximately 100 hectares) under poly-houses in India is in Ladakh.

## Advantages

- Vegetable crops can be grown under adverse climatic conditions when it is not possible to grow them in open field.

- Certain vegetable crops can be grown round the year in a particular place.

- Poly-house provides an excellent opportunity to produces high quality yields, assured regular supply in huge quantity of vegetables for export.

- Productivity is manifold in poly-houses in comparison to growing vegetables in the open field.

- Poly-houses are ideally suited for raising nurseries of vegetables both sexually and asexually especially in hilly areas.

- Management of weeds, insects, pests and diseases are easier in poly-houses.

- Poly-houses are ideally suited for farmers having small holdings.

- Organic farming of vegetables is easier in poly-houses.

- Poly-houses are ideally suited for production of genetically engineered and micropropagated vegetables, varieties and hybrids.

- Poly-houses helps in saving the valuable time and vegetable crop growth which can be advanced at least 1-2 months in the field.

- Poly-houses are ideally suited for off-season nursery production of high-valued vegetables.

- Poly-houses helps in shortening of the nursery period.

- The produce harvest is neat and clean.

## Disadvantage

The only disadvantage in the development of poly-house technology is the high initial cost. In India, most of the farmers are not economically sound, with the result they are not in a position to establish poly-house structures for taking up vegetable production. Fortunately central/state governments, banks have put in a big way to provide 50% subsidy and other financial assistance for the construction of poly-house units.

## Green-house effect

A poly-house is covered with a transparent polythene. Depending upon the transparency of plastic films major fraction of sunlight is absorbed by vegetable crop plants and the plant material inside the poly-house in turn emit long wave thermal radiations for which covering material has lower transparency. Resulting thereby trapping of solar energy and raising the temperature (10-12ºC) inside the poly-house. This is popularly known as *green house effect*. This rise in temperature in poly-houses is responsible for vegetable forcing in cold climates.

During summer season, air temperature in poly-house is to be brought down by providing cooling device. In commercial poly-houses including temperature, R.H., carbon dioxide, photo-period, soil temperature, plant nutrients etc. facilitate round the year production of desired vegetables. Controlled climate and soil conditions provide an opportunity to the vegetables to express their yield potential.

## Poly-house

The poly-house is a framed structure covered with ultraviolet low density polythene or transparent plastic films in which crops could be grown under controlled or partially controlled environment and which is generally large enough to permit a person to work within it to carry out cultural operations easily.

### Materials used in poly-house

Wood, bamboo, G.I. pipe, conduit pipe, angle iron and aluminium are common materials used for the framework of a poly-house. Durability of the structure depends on the type of material used. Ultraviolet resistant film of 150-200 micron thickness is the commonly used covering material. Such film have 4-5 years life. Air-inflated, double-layered plastic sheet provide more heating inside the poly-house compared to the single-layered sheet.

### Categories of poly-house

For efficient round the year utilization of a polyhouse, some seasonal control is exerted over the environmental factors like air, light, temperature and humidity. Based on environment control mechanism, the poly-houses are categorized as :

**Low cost poly-house :** In low cost poly-house, no equipment is installed to control the environment artificially. In the subtropics during winter poly-house is kept completely closed in the night, while in the day time, depending upon the light intensity, door and /or windows are kept open to allow natural ventilation. Depending upon the ventilation, the temperature inside the poly-house is 5-12°C higher than the outside air temperature. During summer, overheating of the poly-house is a problem which could be avoided by opening the side walls also in addition to door and windows. Besides, use of shading nets and frequent watering are also helpful.

**Medium cost poly-house :** In medium cost poly-house, side walls need not to be flexible. Usually fan-pad cooling system is adopted during summer to keep a check over the temperature rise, besides using shading nets and applying frequent watering through micro-sprinkler. During winter, hot air blowers may be used to maintain higher temperature.

**High cost poly-house :** In high cost poly-house, there is complete control over the environment. For uniform heating, usually propane or electric heater is used. The heat generated is passed through perforated polythene tubing. Heating may also be done by circulating hot water/ stream through pipes. Irrigations in high cost poly-house may be computer-controlled and fully-automated.

**Types of poly-house**

Indian Petrochemical Corporation Ltd. (IPCL) designs, alluminium frames preferably 9m x 3m x 2.7m, mud-brick green house 25m x 4m Leh design, Jorhat design of bamboo framed poly-house and then localized type.

Plastic low tunnels are made of either low density polythene or plastic sheets. They are considered ideal for raising vegetable nurseries and certain vegetable crops. These are called 'miniature poly-house'.

Soil trench is underground solar poly-house. Solar trench of 5' width and 3' depth with different length depending on the availability of land is made for raising vegetables in extreme cold conditions. The ultraviolet stabilized polythene film (Regidex film) is spread over them. These structures harness soil and sun heat for the growth of vegetables. Such structures are the cheapest and quite useful for cold desert regions of the country for raising early vegetable nursery for summer season, production of vegetables and for growing leafy and other vegetables during winter when ambient temperature is subzero for a considerable period. These structures are becoming popular in Ladakh region and are likely to green this frozen barren desert in winters. Soil trench in heavy snowfall areas with detachable dome/tunnel type aluminium/wood/iron bamboo frame cladded tight plastic film are suitable for vegetable cultivation.

## Need of poly-house

### Nursery raising

Under normal weather conditions, it is possible to raise vegetable seedlings in open with ordinary care. But in some crops in certain seasons, the nursery is exposed to adverse weather which results in heavy mortality of the seedlings. Management of vegetable nursery in poly-houses is easier and earlier nursery can be raised. This practice eliminates dangers of destruction of nurseries by hailstorms, rains etc. Protection against biotic and abiotic stresses become easier. So, vegetable nursery raising under poly-houses is becoming popular throughout the regions.

In many subtropical areas the nursery period of early and mid-season cauliflower coincides with the period of heavy rainfalls. The seedling exposed to intense heat of sun under heavily saturated soil conditions suffers due to damping off problem. A properly-ventilated or cooled poly-house may ensure the desired population with less seedling mortality. Similarly, for spring planting the seeds of tomato, brinjal, chilli are sown during November-December, which due to low temperature stress, take unduly long time to germinate and produce transplantable seedlings. Inside a low cost poly-house this time period has been found to be reduced by more than one third, besides increase in the seedling population. Now it is well established that if seeds of cucurbits are sown in small plastic bags during winter and the seedlings are reared in polyhouse, they remain safe from frost injury and become ready for transplanting early in February to give early and remunerative harvest.

Poly-house technique has made it possible to cultivate vegetables like cabbage, cauliflower, knol-khol, tomato and onion in cold desert region of the country (Ladakh). Of late, nursery raising of cucurbits in poly-pouch under poly-houses has become common in most of these areas. Raising nursery in containers (plastic pouch and earthen trays) under poly-house makes it possible to transport them to long distances facilitating transplanting at will.

### Vegetable production in hills

A sizable area in Jammu and Kashmir and Himachal Pradesh comprising Ladakh and Lahaul Spiti is a cold desert. Besides, boosting vegetable cultivation during summer, it has now become possible to grow vegetable under sub-zero ambient conditions of long winters using poly-houses. Spinach, mustard, fenugreek, lettuce, coriander, mint, radish, turnip, knol-khol, green onions and potato can be successfully grown under different poly-house conditions during winters.

In temperate regions of Uttar Pradesh hills, poly-houses have been found suitable for vegetable production during winters when it snows heavily making open field cultivation impossible. Leafy vegetables, peas and root crops are grown successfully in these poly-houses.

During summer cultivation of cucurbits in high altitude in poly-houses has become possible. Cucumber, bottle gourd and bitter gourd can be commercially grown.

In cucurbits, hand pollination in a small poly-house is must. In large poly-houses, colonies of insects pollinators may be used for pollination. Yield of cucurbits mainly depends upon pollination.

Harvesting of poly-house crops is dependent to market demand. Poly-house grown vegetables are handled much more carefully than other field grown vegetables. Exposure to direct sun of poly-house produce particularly that of leafy vegetables should be avoided.

**Vegetables and their varieties suitable for poly-house cultivation**

| Vegetable crops | Varieties |
| --- | --- |
| Amaranth | Green type |
| Bottlegourd | PSPL |
| Brinjal | ARU1, ARU 2C |
| Beans | VL Bonil |
| Celery | All varieties |
| Chinese cabbage | Open-headed |
| Coriander | All varieties |
| Capsicum | HC 202 |
| Carrot | FRL-Sel |
| Chilli | Pant 1C |
| Cucumber | DARL 1 |
| Fenugreek | Kasuri |
| Karamsag | Kashmiri |

| Vegetable crops | Varieties |
| --- | --- |
| Knol-khol | White Vienna |
| Lettuce | Great Lake, Dhum (FRL) |
| Mint | Vegetable types |
| Muskmelon | Hara Madhu |
| Okra | Harbhajan |
| Parsley | All varieties |
| Palak | All Green, Mongol (FRL) |
| Sponge gourd | Pusa Chikni |
| Summer squash | Australian Green |
| Tomato | DARL 303 $(F_1)$, HT6, Marglobe |
| Watermelon | Sugarbady |
| Vegetable mustard | ARU-Black |

The Defence Research and Development Organisation (DRDO) Pithoragarh, using local service units as demonstration centres for cultivation of vegetables under poly-houses. The Indian Council of Agricultural Research (ICAR), State Horticultural Departments (SHDs) and certain Non-Government Organisations (NGO's) are engaged in standardizing poly-houses.

So, this technology is become handy to supply fresh vegetables to troops in land-locked high altitude areas during winter.

**Seed production**

Poly-houses play an important role in breeding of vegetable crops. Elite germplasm is maintained in poly-houses. In limited area, seed of more than one variety of cross pollinated vegetables requiring isolation can be taken up in poly-houses. Breeder seed can be produced by better in poly-houses.

In cold regions, poly-houses are essential for production of cabbage, cauliflower, capsicum, brinjal and okra. Hand or insect pollination in entomophilous, cross-pollinated, vegetable crops is essential under poly-houses.

For hybrid seed production of solanaceous and cucurbitaceous vegetables poly-houses helps to higher seed yield of desired quality. Maintenance of parental lines for hybrid seed production is easier and less cumbersome in poly-houses.

## Micro propagation

Micro propagation of $F_1$ hybrids of vegetables is an emerging field. For hardening of tissue-cultured vegetable seedlings, poly-houses are essential. Management of tissue-cultured vegetable crops in poly-houses, is better than open fields. Poly-house cultivation of vegetable crops using modern biotechnologies, would become common and shall help in achieving the production and productivity targets.

## Off-season vegetable production

Vegetables like capsium, cucumber, bitter gourd, bottle gourd etc. have been grown in off-season during winter under sub-tropical climate producing high yields. In high rainfall areas of the country, the yield of some vegetables like tomato are considerably high under poly-rain shelter. In such areas where no vegetable can be grown during winter months in open, successful crops of lettuce, beet leaf, mustard, radish, turnip and garden beet have been found suitable to grow in the poly-house.

In European countries, tomato yields 63-145 tonnes/ha, the highest yield being in Denmark. In India, where winter temperature does not favour fruit set in tomato, poly-house yields have been found even better. An average yield of 100 tonnes/ha has been obtained from hybrid Naveen grown in medium cost poly-house. Tomato planted in rainy season (July) under plastic rain shelter gave 60-70% higher yield in off-season with maintenance of single stem and reduction of the intra-row spacing by half (50 cm x 25 cm).

Capsicum is a crop of temperate and subtropical climate. In the subtropics in certain areas, the low temperature does not allow proper plant growth and fruit set during winters, hence it is grown in during spring summer. After the first one or two flushes, the fruit size in this crop remains smaller. A yield of 105-149 tonnes/ha could be obtained from mid-September transplanted crop of California Wonder in poly-house, whereas crops planted in open field do not prove successful.

Cucumber seeds fail to germinate if sown in winter in north Indian plains. Cucumber Poinsette gives a fruit yield of about 200 tonnes/ha in poly-house. October is ideal time for sowing of cucumber Poinsette in poly-house.

Cauliflower due to heavy rains during seasons, the soil saturated in certain areas and it becomes difficult to raise successful crop of early varieties of cauliflower. The yield of cauliflower Pusa Deepali, Pusa Katki, Early Chinese Prince, Heavy Silver Plate and Selected Early Dawn transplanted on 24 August in a plastic house-cum-rain shelter is 12.7-16.6 tonnes/ha. Similarly cauliflower Pant Gobhi 4 gives good yield in poly-house.

### Hydroponics

Hydroponic means soilless crop cultivation. A number of vegetables can be grown using hydroponics technology. This technology is better exploited under poly-house. The DARO, Pithoragarh, has standardized techniques of commercial cultivation of tomato, cucumber and capsicum using hydroponics in poly-houses. Crop productivity is increased manifold under hydroponics in poly-houses. Tomato (Marglobe) and cucumber (Long Green) in poly-house under hydroponics gave 1,466 and 980 q/ha respectively compared with 470 and 105 q/ha in open field.

## Constraints

- Lack of standardization of design of poly-house and other structures for different regions of the country.

- Cladding materials are very costly and not affordable by average vegetable grower.

- Plastic films available in the country have very short life.

- Lack of professional poly-house manufacturers.

- Instruments for environment control in the poly-house are not easily available.

- Suitable varieties for different vegetables crops under different types of poly-houses for different zones of the country are not worked out and documented.

- Lack of major research programme of poly-house vegetable culture.

- Lack of awareness among farmers pertaining to potential of poly-house vegetable production.

- Import of poly-houses are not suitable under Indian condition.

## Suggestions

- To reduce poly-house installation by using only the local material.

- For regulating the temperature and humidity in cheap poly-houses, open the poly-ethylene sheet on one side manually when the temperature increases inside and irrigate or sprinkle the beds with water to increase humidity and reduce the temperature.

- To make it more popular among poor and marginal farmers the rate of subsidy already given may be increased so that more farmers can make it use.

- The amount of loan extended to the interested farmers may be increased and the rate of interest may be lowered so that all progressive farmers can get these installed in their land holdings.

- At places, where natural hazards like hails, gales, strong winds, heavy rains etc. are the regular feature, a wire netting may be fixed on the poly-ethylene sheet to protect the same from blowing or tearing and thus its life will increase for a pretty long time.

- The farmers can have their co-operative societies to make best use of the poly-house jointly in arranging materials at cheap rate for their installation and further taking the produce of the market at premium prices for earning lucrative returns.

- The appropriate technology developed at the university or department level for growing vegetables, high-value vegetable and their seedlings and rooting of cuttings should be imparted to all those immature farmers in the initial stages. They should be given the latest technical know-how by training them.

- A suitable situation or place may be selected for installing a poly-house which is near the living house, site should be sunny, free from shade, residual impact of chemicals, disease, insects and weeds, soil should be fertile and well drained with irrigation facility.

- To popularise the poly-house technology, the Directorate of Extension Education of respective SAU's/Department of Horticulture/Agriculture or any other agency should take lead to extend the programme among masses.

# EIGHT

## Off-season Vegetable Production

Importance of vegetables in our diet is tremendous as they have high nutritive value and supply most of the minerals and vitamins which are vital for our body building. Besides, in our present day economy, every individual by growing and consuming more vegetables can save cereals for others and thus help in solving the food problem to some extent. With the increase in population consumption of vegetables has considerably gone up. More consumption needs more production of vegetables and more demand needs to be only met by off-season vegetable production.

### Tips for off-season vegetable culture

#### Site selection

- While growing off-season vegetables, the site selected should have congenial climate for a particular vegetable, variety and hybrid for production.

- The site should be flat with sufficient slope to avoid accumulation of water.

- The site should have a good drainage.

- The site must be free from residual contamination of chemicals etc. from previous crop, harmful diseases and insect-pests.

- The soil should be fertile, add organic manure and fertilizer to the exact necessity.

- A suitable rotation should be followed for growing vegetable year after year on the same field for better results.

In case of cross-pollinated vegetables, a proper isolation distance from other varieties and crops crossable to main vegetable crop should be maintained to have quality produce.

## Soil preparation

⊙ Light loam is the best soil for off-season vegetable production, sandy loam and loam soils are also good.

⊙ Sufficient number of ploughing should be given to ensure good tilth to the soil.

⊙ Soil should be smoothly well levelled to facilitate proper irrigations.

⊙ To know the fertility status, soil should get tested well in advance.

⊙ If the soil is deficient in nutrients, it may be supplemented with heavy doses of manures, fertilizers and micro-nutrients.

## Manures and fertilizers

⊙ Most of the vegetables are heavy feeders and in general 20-30 tonnes of well decomposed F.Y.M./ Compost/leaf mold should be added per hectare.

⊙ Cole crops, potato, tomato etc. require heavy manure dose and leguminous vegetables such as garden peas, beans etc. need less manuring.

⊙ In tuber, bulb and root crops, the application of potassium fertilizer @ 80 kg/hectare is quite essential.

⊙ The nitrogenous fertilizer @ 100 kg N/hectare application is necessary for vegetative growth which leads to higher vegetable yields.

⊙ Application of 80kg $K_2O$/hectare is essential for foliage growth, fruiting and setting of vegetables.

- The whole quantity of manure $P_2O_5$ + $K_2O$ and 50% N should be applied before sowing /planting vegetables and rest half N should be top dressed in two equal splits.

- While applying commercial fertilizer care should be taken to prevent seeds, seedlings and young plants from coming in direct contact.

- Follow a suitable method of applying fertilizers.

- In case of legume vegetables, treat the seed with *Rhizobium* culture @ 25 g/kg seed before sowing.

## Varieties

- Vegetable variety/hybrid widely adopted and which is in great demand should be considered for commercial production.

- The varietal purity should be maintained and should be grown in a particular belt where the vegetables can be raised easily.

- The area should be divided into zones to avoid genetical and physical admixtures.

- Insect-pest/disease resistant and high-yielding varieties should be selected.

- Preference may be given to hybrid varieties which are of uniformly maturing.

## Seed sowing

- The seeds should be obtained from reliable sources *viz.* NSC or Vegetable Research Institutions, SAU's and their sattelite stations or some well-reputed firms.

- Use breeder/foundation/certified seeds only.

- Pure good seeds having high vitality, improved type, free from diseases, suitability for growing conditions and free from foreign matter should be used.

- Healthy seeds have 70-80% germination usually germinate more quickly and uniformly.

- The vegetables such as garden peas, cowpea, beans, radish, carrot, spinach, turnip etc. which are directly sown in the field, the land should be well prepared which is needed for planting the seeds at different depths.

- Seed-sowing depth depends on the soil type, moisture content, season of planting and kind of vegetable variety /hybrid.

- In general, depth of 3-4 times the diameter of thickness of seed is considered appropriate.

- The seeds may be treated before sowing.

- Sowing should be done in rows at it facilitates in subsequent operations.

## Nursery, its preparations and sowing

- Some vegetables such as cabbage, cauliflower, knol-khol, brinjal tomato, chillies, lettuce etc. are first raised in nursery and transplanted later in the main field.

- Normally 5-6 nursery beds will be sufficient for a hectare of vegetable crop.

- In case of small scale/kitchen garden small wooden boxes (50x30x10cm) should be for raising nursery.

- In case of large scale vegetable growing, outdoor nurseries. (300x100x15 cm) should be preferred.

- For raising early and healthy seedlings, green/ply house may be used base on recommended technology in each zone.

- The soil mixture used should be sterilised for killing the disease organisms insect-pests (1 part formalin + 10 part water @ 5 litres/m$^2$ to saturate fully up to 15 cm depth.

- The seeds should be sown in row 5 cm apart in raised beds.

- Thick sowings should be avoided to get healthy seedlings and to eliminate damping-off.

- If seeds are very small, a little sand should be mixed with them for facilitating uniform seed distribution.

- The seeds should be covered with a thin layer of soil mixture and apply water with the sprinkler to maintain optimum moisture in the soil.

- The soil is kept moist till the germination sets in.

- Light shade should be provided to the seed beds by detachable thatch during hot days and rains.

**Transplanting**

- Before transplanting, the seed bed should be water-soaked as this facilitate the removal of seedlings with least root injury.

- When the seedlings attain 4-5 leaf stage in 4-6 weeks should be transplanted.

- The seedlings should be lifted up with little soil attached with roots.

- The seedlings should be transplanted in rows little deeper in the soil. The soil around the roots should be made firm with the help of hands and forefingers.

- Soon after transplanting the field should be given irrigation.

- Transplanting the field should be done preferably either in the evening or during cloudy weather.

- Apply Folidol or Aldrin dust @ 20-25 kg/ha against cut-worms.

- Gap-filling should be done immediately to maintain uniform plant population.

## Irrigation

- A regular and assured irrigation be maintained for successful off-season vegetable culture.

- Light and frequent irrigations are more beneficial than heavy irrigations given after a long span.

- During summer, more frequent watering should be applied during dry spells while in rainy season and winter less watering.

## Interculture

- The vegetable crop should be kept neat and clean through frequent weedings and interculture.

- Shallow cultivation should be given to keep down the weed population and to promote aeration.

- Deep cultivation particularly when the plants have attained a fair amount of growth should be avoided as it may prove injurious as it prunes the roots.

- One earthing-up should be given to the vegetable crop at pre-flowering stage to make it stable.

- In case of poly-type vegetable/hybrids suitable staking may be provided.

- In some vegetables e.g. carrot etc., thinning is needed which may be done at right stage and time.

- Mulching may also be done to increase temperature and conserve moisture.

## Production methods

- For commercial growing, the seed-to-vegetable method should be followed.

- In seed-to-vegetable method, only *breeder/ foundation seed must be used.

- Transplanting, retransplanting and planting methods be used for different vegetable growing.

- Optimum spacing combination may be followed to obtain high quality vegetable growing.

## Isolation distance

- In case of self-pollinated vegetables e.g. garden pea and beans etc., there is no need of isolation.

- The isolation distance should be provided all round the cross-pollinated vegetable field for confirming the varietal/crop standard and have uniform, better and quality produce.

## Roguing and inspection

- Roguing should be done to all undesirable and diseased crop plants whenever observed.

- Any plant which is outside the acceptable limit of varietal variation should be removed.

- The number of roguing depends on the vegetable crop and desired quality produce.

- Field inspection of vegetable crops be done as desired or recommended for checking any disease, insect-pest, weeds, deficiency and physiological disorder, etc.

- The time when inspection should be done varie: with individual vegetable.

 ꘎ Inspection makes the vegetable crop eligible for certification and post-harvest handling, etc.

## Harvesting and grading

 ꘎ The vegetable crop should be harvested as and when it is mature and ready for market.

 ꘎ In some cases, the vegetable crop can be harvested in one lot e.g. hybrids while in others harvesting is completed in 2-3 lots as the crop does not mature at a time.

 ꘎ In some cases, the vegetable crop after harvesting needs curing, selection, trimming, cleaning and preparation etc., before these are sent to the market.

 ꘎ Certain vegetables are ready when they are still at green stage, some a bit over-ripe, others need extra time to mature even after they are picked and some need attention at the height of ripeness.

 ꘎ Those vegetables contain the seeds inside of them such as peppers, melons, tomato, should be harvested depending on the stage and requirement.

 ꘎ Pumpkin and squashes can be harvested and stored for sometimes before these are sent to the market or consumed.

 ꘎ Pod/fruits that are picked directly off the plants such as tomato, garden pea etc. can be cleaned, packed after proper grading and marketed in fresh form.

 ꘎ Lima beans, garden peas etc. are simply picked from their plant by giving a jerk to separate without harming the tender plants.

 ꘎ After proper cleaning and grading the vegetables should be cooled after removing the field, product and container/storage heat before these are packed or stored or marketed.

๏ Chemicals (insecticides, fungicides, weedicides and growth regulators) can influence the quality, self-life of fresh and processing of vegetable in storage and transit. It is very essential to have an appropriate integrated, low cost system of pre and post-harvest technology (including maturity indices) right from the production area to the actual consumer.

## Weeds, insect-pests and diseases

๏ The weeds, insect-pests diseases as and when appear should be carefully examined and got inspected by the nearest scientists (extension staff/pathologist/ entomologist) available and when studied carefully must be sprayed immediately for its control.

๏ For future, the weedicides, insecticides and fungicides should be kept in the ready stock.

๏ Spraying should be done always when it is a sunny day and calm atmosphere for effective control.

๏ Instructions and precautions mentioned on the containers should be strictly followed.

๏ Spraying/dusting should not be done in the morning or at a time when the pollinators are busy in the fields for pollination work.

๏ Always spray safe chemicals so that the pollinators are not killed.

๏ Vegetables should not be harvested immediately after spraying to avoid health hazards.

# NINE

## Soilless Culture and Hydroponics

**S**oilless culture is the art and science of growing plants without soil by feeding them on nutrient solutions. Crops can be raised in the absence of organic matter. Nutrients are provided to the crop plants by artificially prepared nutrient solution.

Hydroponics is a system of soilless culture. The word hydroponics is derived from the Greek words *"hydro"* having to do with water and *"ponos"* means labour. Thus meaning of Hydroponics is working with water. The term hydroponics is often used as a synonym, of soilless culture. Hydroponics, in its true sense strictly means/includes the only water (liquid) or solution culture. The nutrients are provided as per need. In fact, the system is correctly known as 'nutriculture'.

### Historical background

As early as 1666, Boyle who was first to attempt to grow spearmint (*Raphanus aqatica*) in vials containing only water, survived for nine months. Woodward in 1699 in England grew spwarmint in water added with small quantity of soil. However, Liebig (1803-73) and Knop and Sachs (about 1860) initiated the systematic study of plant nutrition in Germany. In fact, soilless culture i.e. crops grow without soil was unknown till 1929. Gericke (1929) was first to promote the commercial potential of liquid culture. He devised a technique for growing plants in tanks of nutrient solution. The plants were planted in layer of sand which was supported on the surface of the solution by netting and canvas through which the roots could pass into the liquid phase. Further, experiments in liquid and aggregate systems were reported in New Jersey and California in the mid-1920s and early 1930s, at a time when greenhouses *per se* (with conventional soil) were becoming common in Europe and the mid-western United States.

Originally, Gericke (1929) defined his method as 'aquaculture' However, in 1940 Gericke coined the word hydroponics (use of liquid culture). Finally, the term hydroponics was proposed by Setchell, based on the Greek hydro (water) and ponos (labour). The static system of Gericke was facing the problem of aeration of the solution as adequate supply of the oxygen to the rootzone is essential for successful cultivation. Firstly, aeration was overcome with the development of nutrients by way of developing aggregate system in which both aeration, and support was available. The technology was used in a few limited applications on Pacific islands during World War II. After the war, Purdue University popularized hydroponics (called *nutriculture*) in a classic series of extension service bulletins describing the precise delivery nutrient solution to plant roots in either liquid or aggregate system.

In the mid-1950s, greenhouses covered with plastic film appeared in Kentucky and evaporative cooling was used in Texas. Air-supported plastic green houses were first used in Washington in 1959. Greenhouse area began to expand significantly in Europe and arid Asia during the 1950s and 1960s and large hydroponic system were developed in the deserts of California, Arizona, Abu Dhabi and Iran in about 1970. In these desert locations, the advantages of the technology were augmented by the duration and intensity of the solar radiation, which maximised photosynthesis activities. In the United States, Japan, Holland, Italy and Russia a majority of greenhouses are used for food production where soilless culture (hydroponics) is common. In India, a few nursery men have initiated work on NFT under protected conditions in which some vegetables are under trial. In near future, more popularity of hydroponics is expected.

## Advantages

- Soilless culture of vegetables may be adopted where ordinary agriculture is not possible.
- Vegetable crops have less chances of soil-borne diseases.
- Quick growth and higher yields are expected.
- Crop remains weed-free.
- Good drainage and no waterlogging.
- Vegetables are obtained without dirt and smell.
- Some crops can be grown out of season, thus fetching higher prices in the markets.
- Addition of organic manures is not essential.
- There is efficient use of water and fertilizers.
- There is no risk of flood, erosion or drought.
- More crops can be raised per year.
- Labour requirement is less.
- More uniform pattern of crop maturation.
- More accurate control over the supply of water, nutrients, pH, root temperature etc.

## Disadvantages

- Initially, more capital investment is required for the construction and maintenance of a hydroponic system.
- Growing crops in restricted volume requires a higher standard of management and skill than for crops growing in the soil.
- Soilless culture is not advisable to adopt when lot of land is available.

## Scope of soilless culture system

Countries having sufficient land under cultivation may not be having much scope for soilless culture. But, where land is the biggest problem for growing crops, the scope of soilless culture in such places is inevitable. Today, hydroponics (one of the three systems of soilless culture) is an established branch of horticultural science. Progress has been rapid, and results obtained during the last few years in various countries have proved it to be thoroughly practical and to have some very definite advantages over normal routine, or conventional methods of horticulture. This technique is important for protected vegetable crops. In Holland, the production of vegetables substrate growing is becoming the normal practice. More than 95 per cent of tomatoes, cucumbers, sweet peppers and brinjals are grown by soilless culture. For other vegetables crops (like lettuce, radish, endive) soilless culture systems are under development.

People living in crowded streets, without garden, can grow fresh vegetables in window-boxes or on house tops. Town residents and workers in industrial settlements often have extensive verandas, backyards or pavements of which they can generally make little use. By means of hydroponics all such places can be made to yield a regular and abundant supply of clean, health-giving greenstuff. Deserts, rocky and stony land in mountainous or barren regions and some wastelands can be made productive at relatively low cost. Certainly the soilless cultivation of plants can give India all the extra food she needs, and its introduction on a commercial-scale would afford lucrative employment to thousands.

## Systems or methods of soilless culture

### Water or solution or liquid culture

This is true hydroponics in which waterproof-troughs or benches of any suitable material are constructed. A wire grid fits closely over the top of tank, which contains the nutrient solutions. This grid serves as a support for the growing plants whose roots descend through the mesh into the liquid or solution below. Aeration of the root system is ensured by adjusting the level of the nutrients so as to leave an air space between its surface and the base of the wire grid. Along with nutrient level, pH is also adjusted.

Solution culture is expensive to instal, because it is essential to make the troughs waterproof, but there is a definite saving in chemicals and fertilizers, since none of the valuable salts are lost. It is, however, difficult to support the plants, and the cost of artificial aeration is high. The prevailing moist atmosphere, together with the high annual monsoon rainfall, the use of hydroponics Gericke method into eastern India appears difficult. Nevertheless, there can be no doubt that this system, combined with a dry sunny climate gives a colossal yield.

### Sand culture

Sand culture is being performed by at least four ways. Description in brief is given below :

**New Jersey method:** In waterproof trough liquid nutrients are forced several times a day into an electrically driven centrifugal pump. Investment is very high initially because culture is possible under glass as rain falling on a hydroponicum of this type destroy the regulated balance of the entire system. However, completely automatic operation is ensured and labour charges are reduced to a minimum.

**Surface watering system:** The plants are grown in improvised beds or pots containing sand, the requisite nutrients and water being supplied in solution from a hosepipe or garden can to the surface of the medium. Free drainage is permitted and water tight troughs are unnecessary. Considering wastage of fertilizers, it is bound to occur with surface watering system. Heavy rainfall might lead to waterlogging of the troughs. In addition, labour cost is high for at least three or four application of the nutrient solution are required weekly.

**Automatic dilution surface watering system:** Nutrient solution is applied by means of a system sprays to the sand bed. A low-pressure, main water supply is needed in addition to an autominor injection pump and 50 litres reservoir. Improved drainage is obtained by a layer of 2-3 cm gravel on the bottom of the troughs underneath the sand.

**Drip culture system:** By means of a feed line, diluted nutrient solution contained in an upper tank is allowed to drip continuously on to a bed of sand in which plants are grown. The solution which percolates through the medium, is collected in a sump and pumped back to the reservoir at intervals. Waterproof troughs are necessary and constant pH testing of the cultural mixture is essential.

**Aggregate culture**

In this type of soilless culture, the nutrient solution is applied to plants via an irrigation system through the media and excess solution is recirculated (e.g rockwool, pumice, perlite, sand culture, gravel culture etc.).

The following are two sub-irrigation system in aggregate culture:

**Sub-irrigation system:** A water-tight trough, filled with 2-3 cm aggregates like gravel, stones, vermiculite, rockwool, peat, perlite, sawdust etc. which are comparatively coarser than sand, to a depth of 15 cm are used for giving support to plants. The substrate is periodically flooded with dilute solution from below, and then allowed to drain. Each bed is built from precast concrete jointed with asphalt; other necessary apparatus consisting of a reservoir sunk below ground level and a centrifugal pump. The entire system is automatic and by virtue of the ebb and flow of the liquid nutrients, the roots of the plants are well aerated. Periodic renewal of the solution is essential.

**The flume system:** This is a device for the use with sub-irrigation consisting of a curved artificial channel down by which the liquid nutrients are directed so that each trough may receive a correct proportion as the solution passes it. Flumes have been employed with great success by growers in Florida.

## Media or substrates for soilless culture systems

A number of media have been used as substrates for soilless culture, of which the most popular are:

- Rockwool (most commonly used)
- Peat (mainly for pot plants)
- Perlite
- Vermiculite
- Sawdust
- Bark chips
- Sand
- Gravel
- Pumice
- Polyurethane mats
- Water

## Future line of work

It is agreed by a majority of people that sufficient knowledge is not available to the common man who can go in for soilless cultivation of vegetable crops. Therefore, it is urgently needed to standardize the technique(s) under local conditions so that in near future this technique becomes helpful to produce extra food without land. Land is limited and it cannot be spared for taking it under plough and hence alternate method of growing crops has to be fully known in advance. Multiple cropping and intensive culture by hydroponics will have profound changes in the social life. Food production and its regular supply will increase employment. Successful soilless culture requires green-house that can control air and root temperature, RH, air movement and composition, pH and EC of nutrient solution, sterilization of root growing media, nutrient solution when recirculated, frequency of cost feeding cycles, etc.

There is need to screen out crops as much as possible and after having confirmed experiments, only then such crops with standardized technique should be made known to public along with constraints.

# TEN

## Container Gardening

In big towns and cities due to population pressure, there is hardly any space available in houses or multistorey buildings to grow any vegetable. In such situation, pots and containers can be used to raise a vegetable garden. This practice is known as container gardening.

### Suitable vegetables and their varieties

All vegetables cannot be grown successfully in containers. Only specific varieties of selected vegetables perform well in containers. Such vegetables, their suitable varieties, sowing or planting time, period of maturity (number of days taken to come to first harvest from the date of sowing or planting) are given in THE table:

### Vegetables and their varieties recommended for growing in containers

| Vegetable | Varieties | Sowing/planting | | Days to first harvest after sowing/planting |
| | | Method | Time | |
| --- | --- | --- | --- | --- |
| Amaranth | Pusa Kirti, Pusa Kiran and Pusa Lal Chaulai | Sowing | Feb.-Mar. Jul.-Aug. | 25-30 |
| Beet root | Crimson Globe and Detroit Dark Red | | Oct.-Dec. Feb. Mar. | 90-100 |
| Bittergourd | Pusa Do Mausami, Pusa Vishesh and Arka Harit | | Feb.-Mar. | 55-60 |
| Brinjal | Pusa Purple Cluster, Pusa Purple Long, Pusa Kranti and Punjab Bahar | Planting | Feb.-Mar. and Jul. | 45-60 |
| Broad bean | Local varieties | Sowing | Oct.-Nov. | 70-75 |
| Broccoli | Pusa Sumeet, Pusa Broccoli, and exotic varieties | Planting | Oct.-Nov. | 90-95 |
| Brussel's sprout | Hilds Ideal, Catskill and Early Dwarf | Planting | Oct.-Nov. | 90-95 |
| Capsicum | California Wonder and Yolo Wonder | Planting | Feb. and Jul. | 55-60 |

## Basic Concepts of Vegetable Science

| Vegetable | Varieties | Sowing/planting Method | Time | Days to first harvest after sowing/planting |
|---|---|---|---|---|
| Chilli | Pusa Jwala, Pusa Sadabahar and locals | Planting | Feb.-Mar. and Jul. | 50-60 |
| Chenopodium | Pusa Bathua 1 | Sowing | Oct-Nov. | 50-60 |
| Cluster bean (guar) | Pusa Sadabahar and Pusa Navbahar | Sowing | Feb.-Mar. and Jul. | 55-60 |
| Cowpea | Pusa Komal and Pusa Do Fasli | Sowing | Feb-Mar. | 60-65 |
| Cucumber | Poinsette, Poona Khira and Pusa Sanyog | Sowing | Feb.-Mar. and Jul. | 45-50 |
| Frenchbean | Pusa Parvati and Contender | Sowing | Jan.-Feb. Sept. | 50-55 |
| Garlic | Agri-Found White and local varieties | Sowing (cloves) | Oct.-Nov. | 100-120 |
| Ginger | Local varieties | Sowing | Apr.-May | 150-180 |
| Green onion | Early Grano, Pusa Red, Pusa Madhvi and local | Planting | Sept.-Nov. | 45-80 |
| Knol-khol | White Vienna and Purple Vienna | Planting | Oct.-Nov. | 70-75 |
| Leek | Exotic varieties | Planting | Sept.-Oct. | 100-110 |
| Lettuce | Great Lakes, Chinese Yellow, Slowbolt | Planting | Oct.-Nov. | 45-50 |
| Methi | Pusa Early Bunching, Pusa Kasuri and locals | Sowing | Sept.-Dec. | 45-50 |
| Mint | Local varieties | Planting (root cuttings) | Mar.-Jul. | 45-50 |
| Mustard | Local varieties | Sowing | Oct.-Nov. | 30-35 |
| Okra (*bhindi*) | Varsha Uphar, Pusa Sawani, A-4 and Parbhani Kranti | Sowing | Feb.-Mar. Jun.-Jul. | 45-50 |
| Onion (green) | Early Grano and Pusa Red | Planting | Sept.-Oct. | 75-80 |
| Palak | Pusa Bharati, Pusa Jyoti, All Green and locals | Sowing | Almost round the year | 50-60 |
| Parsley | Hamburg, Mosscurled and other exotic varieties | Planting | Oct.-Nov. | 80-90 |
| Purslane | Local Varieties | Sowing | Mar.-Jul. | 30-35 |
| Radish | White Icicle, Rapid Red-white Tipped, Scarlet Red and French Breakfast | Sowing | Oct.-Dec. | 25-30 |
| Spinach | Virginia Savoy, and Early Smooth-leaf | Sowing | Oct.-Dec. | 50-55 |

| Vegetable | Varieties | Sowing/planting | | Days to first harvest after sowing/planting |
| | | Method | Time | |
| --- | --- | --- | --- | --- |
| Summer squash | Australian Green and Pusa Alankar | Sowing | Feb. | 50-55 |
| Tomato | Pusa Early Dwarf, Pusa Gaurav, Pusa Hybrid No.1 and Pusa Hybrid No.2 | Planting | Jan.-Feb. Sept.-Oct. | 60-65 |
| Turnip | Pusa Sweti, Pusa Chandrima and Pusa Swarnima | Sowing Sowing | Aug.-Sept. Oct.-Nov. | 40-45 60-65 |

## Types of containers

Containers for raising vegetables can be cement pots, earthen pots and pans, wooden barrels, boxes and crates, plastic jars, cans and buckets, tin boxes, cans and drums of various sizes. These containers should have at least one hole of an adequate size at the bottom as in earthen pots, to drain out excess water. These containers can easily be placed on the terrace, window sills, window boxes, balcony and verandah where sunlight is available for the plants.

## Tools, manures, seeds, fungicides and insecticides

Certain hand tools are the primary need of a gardener. A container garden needs essentially a *khurpi* spade or shovel, watering can, small hand-sprayer, garden hose preferably with a sprinkler, bamboo stakes and string (*sutli*). Good soil, river sand, well-decomposed organic manure (compost or farmyard manure) and nitrogenous fertilizers (urea or ammonium sulphate), insecticides (Malathion or Endosulphan) and fungicide (Captaf) are important inputs.

Quality seed is most important requirement. The seeds can be purchased from the National Seeds Corporation (NSC), agricultural universities, research stations, block development centres and other reliable sources. If one is unable to raise their own seedlings, they may be arranged from reliable nurseries.

The container mixture should be prepared by mixing good soil, river-sand and well-rotten organic manure in equal quantities with the help of a *khurpi* or shovel. The mixture should be free from various soil-borne insects, termites, red ants and cut worms, which generally damage young seedlings. For precaution, add a small quantity of BHC (5%) or Aldrex dust to the mixture before filling it in the containers. After raising a crop for one season the container mixture should be removed and cleaned of roots and exposed to the sun for a few days. This soil could then be reused after mixing one-third the quantity of organic manure and a small quantity of BHC and Captaf.

## Cultivation

### Sowing/planting

Most of the vegetables are raised by sowing their seeds directly in containers. The seedlings of brinjal, chilli, tomato, capsicum, lettuce, Brussel's sprout, broccoli, onion, parsley and leek are transplanted in containers. Their seedlings can be raised in earthen pots or pans. A single, healthy seedling may be transplanted in each container. Several seedlings, each of onion, lettuce, knol-khol, parsley and leek can be transplanted in a container of the same size. Two seeds of summer squash and 4-5 seeds of clusterbean, cowpea, okra (*bhindi*) and Frenchbean are sown in such containers. In radish (table types), turnip and beet root, more number of seeds can be sown in each pot but finally 3-5 seedlings are allowed in a container depending upon the crop. A number of plants can be raised of amaranth, *palak*, spinach, Fenugreek (*methi*), mustard, *bathua, kulfa* and coriander in containers by following thick sowings of their seeds.

## Aftercare

Plants in pots and containers need a lot of care and attention. It is essential to water frequently depending on the season, kind of crop and size of the plant and container. Plants need extra water in dry summer season, so watering should be done twice a day (morning and evening). Too much watering can be as harmful in winter as too little in summer. In the rainy season, proper water drainage is essential. If there is heavy rain, containers should be tilted slightly to drain out the excess water from the top.

Topdressing with nitrogenous fertilizers improves plant growth and yield of vegetables directly. This can be done by applying urea or ammonium sulphate in small quantities. In general 5-10g of urea may be applied in moist soil once a week or 10 days, starting from 3 weeks after sowing or 2 weeks after transplanting. High dose of fertilizer is very harmful since it can kill the plants. If urea or ammonium sulphate is applied in dry soil, the plants must be watered immediately. Plants of cowpea, tomato and bittergourd require staking. Hand-hoeing and weeding with the help of a small *khurpi* should be done periodically to remove weeds. Weeds should be uprooted gently by hand from amaranth, *kulfa, methi, palak*, spinach, *bathua* etc., if thick sowing is done.

Vegetables are attacked by various pests and diseases. Aphids and jassids are small-sucking insects, injuring the plants especially in early stage of their growth. Spraying of Malathion or Endosulphan @ 2ml/litre of water controls these insects. Fruitfly and fruit-borer are serious pests of some vegetable crops. They damage young fruits and make them unfit for consumption. The attacked fruits should be plucked and destroyed. The plants should be sprayed once or twice with Malathion solution @ 1-2ml/litre of water. After spraying, fruits should not be harvested for 7 days for consumption. Fungal diseases (damping off and wilt) and viral diseases affect the plants particularly in the rainy season. Fungal diseases can be controlled by drenching the soil with Captaf solution @ 2g litre of water. Virus affected plants should be removed and destroyed.

## Harvesting and postharvest management

Vegetables harvested at the peak of maturity and used promptly are always superior in nutritional content, flavour and appearance. Leafy vegetables should be picked up frequently when they are most succulent and tender. Root vegetables should be pulled out while still tender as a few days delay makes them pithy, tough and unfit for consumption. Except tomato, all fruit and pod vegetables recommended for container gardening should be picked when they attain proper size and are still tender. Tomatoes are allowed to ripen on plants before harvesting.

Rare vegetables-broccoli, leek, fennel, parsley and purslane (soya)-are not usually available in the market. Most of these are required in a small quantity for consumption. These can be advantageously raised in containers with assured success.

In fact, vegetable container gardening is an interesting hobby and useful method for growing vegetables in urban areas.

# ELEVEN

## Vegetable Marketing

A good farmer has one eye on the plough and the other on the market. But in these days of commercialised agriculture it would be more correct to say that he keeps his hands on the plough and both eyes on the market.

### Essentials of vegetable marketing

Marketing of vegetables includes all the steps from the time of the produce is ready for harvest until it is in the hands of consumer. The prime object of marketing is that the producer realize a suitable net return from his produce. This aim can be achieved through good marketing programme which have the following objectives :

- Vegetable should be present in attractive and saleable form.

- Consumers should get vegetables with quality unimpaired.

- Cost of cultivation of vegetable and other miscellaneous expenses involved, should be kept minimum by adopting efficient scientific management of vegetable farming.

- The vegetable should be sold according to the market prevailing rates to get a good price.

### Quality parameters

The quality of vegetables is a prerequisite for success in marketing. The following factors may be considered:

### Appearance

The vegetable should be of normal shape, good colour and free from dirt and blemishes.

## Texture

According to the types of vegetables, their texture may be hard or soft, granular or smooth, crisp or flabby, fibby, fibrous or non-fibrous. However, these textural property may vary according to the stage of maturity, culture and varietal conditions e.g., crispness in cucumbers, mealiness in potatoes and juiceness in melons.

## Flavour

Vegetable should be of normal flavour and odour. These may be due to the presence of essential of volatile oils, acids and other chemical compounds. The quality of the vegetable can be kept unimpaired, till they reach the consumers.

## Nutritive value

This is known after undertaking tests. High chemical constituents leads to good nutritive value of vegetables.

# Vegetable transportation

Timely and speedy delivery of vegetables with minimum damage and deterioration at minimum cost are essential in vegetable marketing. Good transportation facilities have made it possible to extend the area of production of perishable vegetables. In India, marketing of vegetables is mostly governed by the nearness of the market.

## Transportation methods

### Rope ways

Pulley base trolley system mainly based on mechanical and gravitational force mechanism. It is the most suitable method to transport vegetable produce in the hilly areas in undulating topographical situations.

### By road

The transportation of vegetables may be done by trucks, public vehicle, bullock-carts, tractor-trolley and any other possible means, depending on situations, but the best way for transportation is to involve the truck/tractor unions and transport the produce through them, which will be assured timely, cheap and economic to producers. If the material with high perishable nature like tomato, *palak, methi*, etc. are transported by road to long distances, they are subjected to spoilage quickly in comparison to onion, potato etc. need special care while handling during transit.

## By rail

Wherever possible, rails (goods train) are important means of transportation for most of the commodities e.g. root vegetables, onion, potato etc. However the time taken by rail is sometimes more than by the roads but the cost of transportation in generally cheaper in case of goods' trains.

## Air transportation

This method is followed only in case of high value/rare vegetables, because of the very high cost of transportation. Sometimes, there are unnecessary delays due to disturbances in flight. It is noted that the produce is handled carelessly on the airport and also space available in cargo are not enough. Now-a-days, the things have improved and perishable and high value vegetables which need quick disposal, air transportation will be the only solution.

## Water ways

This method is used among growers whose fields are situated near on bank of river or lakes. In India, this transportation system is used only in Kashmir, Kerala and some parts of Andhra Pradesh and West Bengal. It is yet to be developed further for the quick and easy disposal of perishable vegetables from the fields which are situated far away from big markets but near or on the banks of river or lakes.

## Assembling agency

### Producers

They collect vegetables from their own fields and from the fields of several other small producers. They bring the so-collected vegetables to the market or transport it to the distant markets or sell it to commission agents.

### Village merchants

They collect non-perishable vegetables like potatoes, sweet potatoes from the producers of the locality directly or through exchange with other articles of daily use such as salt and kerosine oil etc., sell them in the mandies directly or bigger merchants.

### Itinerant merchants

These merchants travel from one place to another and purchase the vegetables when it is harvested or whole vegetable fields on the conditions that it will be irrigated by the owner. Keep watch over it and sell vegetables in the market as and when they are ready for harvesting. These are professional merchants locally known as *Chaudhari* or *Kunzras* etc.

### Whole merchants and commission agents

They purchase vegetables from the producers through their agents or *dalals*, collect and transport them to those markets, where they are in great demand. Onions at Nasik or in Bellary are collected in this way and transported to other cities and town in Northern India.

### Producer's co-operative societies

These societies collect the vegetables from the producers who are members of the society, so that they may be sold at better prices and producers may not undergo any loss. These societies also store the vegetables, when they are cheap, for selling them in future at high rates, provided storage facilities exist with them.

## Vegetable marketing methods

### Private negotiation

In this method, the purchaser makes his offer personally to the commission agent and the bargain is settled. This method is common all over the country.

## Open auction

In this method, almost all buyers assemble at a particular time in the market or at shop of the commission agent. Each lot of vegetables is kept for auction separately by the auctioneer. The highest bid is accepted for approval of the seller.

## Under cover

Producer makes standard packing in local-made baskets with a cover of gunny cloth in such a way so that vegetable(s) is not seen. The prepared baskets are handed over to the commission agent. The commission agent calls for bids from buyers, the produce is given to the bidder who bids the highest price.

## Sales channels

1. To consumers    a) House to house
   b) Roadside markets
   c) Weekly markets
   d) Small markets
2. To retailers    a) Mandies
   b) Selling at the farm
   c) Auction market
3. Selling to wholesale merchants
4. Co-operative selling

## Constraints

o Non-availability of proper transport facilities.
o The vegetables production areas are located far from road head.
o The roads are still *kutcha* and rough and not metalled.
o Labour is not available and more so very costly.

o  Lack of properly organised marketing, transport and distribution systems.

o  Vegetable growing done by the small and marginal farmers who are at the mercy of middle-men for the disposal of their produce.

o  No proper packaging and grading of the produce.

o  High transport charges.

o  Inadequate organisational setup for marketing of vegetables at district and state levels.

o  Lack of adequate extension services to impart knowledge of various channels involved in vegetable marketing.

o  Lack of credit facilities.

o  Lack of adequate infrastructure for prevention of huge post-harvest losses of vegetables due to improper handlings.

o  Lack of marketing intelligence service.

o  Old system of under-cover auction is still prevailing.

## Recommendations

o  The market intelligence should be properly organised as it is quite essential both for the interest of producers and consumers or wholeseller.

o  The price of vegetables and their supply should be properly regulated for the mutual benefits of producers and consumers.

o  Secret method of sale of vegetables should be prohibited and auction of vegetables should be regular and uniform in all cases.

o  The grading, packing and processing of vegetables should be properly organised by Agricultural Department in main vegetable producing areas.

O There should be proper and timely inspection of vegetables on arrival and spoiled vegetables should not be sold in the market unless properly certified for disposal.

O There should be priority and preference for vegetable transportation rails or roads.

O There should be separate market boards for the vegetables.

O There should be proper refrigeration arrangements for quick and timely delivery of perishable vegetables without any deterioration in their quality.

# TWELVE

## Breeding Techniques in Vegetable Crops

Vegetable breeding is a science based on principles of genetics and cytogenetics. It aims to improve the characteristics of vegetable crops so that they become more desirable horticulturally and economically. Thus the chief objective of vegetable breeding is to develop such improved varieties of vegetable crop that will be commercially successful. Generally, "a successful variety is one with a total balance of traits that makes it more profitable for growers than any other one they might choose. This is why breeders worry about emphasising one trait to the exclusion of others".

### Objectives of vegetable breeding

- High yields
- Improved quality
- Wider adaptability
- Acceptable to grower and consumer
- Disease and insect resistance
- Frost tolerance
- Change in maturity duration
- Photoinsensitivity
- Synchronous maturity
- Non-shattering characteristics
- Determinate growth
- Varieties for new seasons
- Suitability for processing
- Moisture stress and salt tolerance
- Elimination of toxic substances
- Better ability for seed production

## History of vegetable breeding research in India

1940     Successful attempt of seed production of temperate vegetables at Quetta (now in Pakistan)

1947     Sanctioning of a nucleus 'Plant Introduction Scheme' at the IARI, New Delhi.

Simultaneous start of ad hoc schemes by the ICAR in different states like Punjab, Uttar Pradesh, West Bengal, Maharashtra, Himachal Pradesh, J & K and Tamil Nadu.

1949     Establishment of a vegetable breeding station at Katrain in Kulu Valley, Himachal Pradesh for production of seeds of temperate vegetables.

1955     Transfer of vegetable breeding station, Katrain to IARI, New Delhi to undertake research on temperate vegetable crops, standardisation of seed production technology and to produce seeds of improved varieties of temperate vegetable crops.

1956     Creation of Division of Horticulture at the IARI, New Delhi.

1960     Start of establishment of state agricultural universities (SAUs) under which the G.B. Pant University of Agriculture and Technology, formerly known as the Uttar Pradesh Agriculture University (UPAU), Pantnagar was the first agriculture university to be established on land grant pattern in 1960.

The establishment of state agricultural universities on the pattern of land grant colleges/universities of U.S.A. having fullfledged and separate departments of horticulture and/ or vegetable science started from 1960 onwards.

1968 Establishment of Indian Institute of Horticultural Research (IIHR), Bangalore with strong focus on vegetable improvement among other things.

1970 Initiation of All-India Coordinated Vegetable Improvement Project (AICVIP) with headquarters at IARI, New Delhi, headed by a Project Coordinator, Dr. Vishnu Swarup being the first one who joined the project in 1971.

1984 Recommendation by quinquennial review team (QRT) of ICAR to upgrade the AICVIP at the level of Project Directorate of Vegetable Research (PDVR).

1987 Start of Project Directorate of Vegetable Research during the Seventh Five Year plan by upgrading the erstwhile AICVIP, with headquarters at IARI, New Delhi.

1991 Joining of PDVR by the first Project Director (Dr. G. Kalloo).

1992 Shifting of the headquarters of PDVR from New Delhi to Varanasi.

1994 Initiation of All-India Coordinated Research Project under National Seed Project (NSP) for production of breeder seed of vegetable crops.

1995 Initiation of ICAR research network on promotion of hybrid research in vegetable crops (ad hoc project) for three years, spread over different vegetable research centres/state agricultural universities.

## List of State Agricultural Universities having independent department of vegetable science

| Sl. No. | State Agricultural Universities | Year of establishment |
|---|---|---|
| 1. | G.B. Pant University of Agriculture and Technology, Pantnagar, Uttar Pradesh | 1960 |
| 2. | Punjab Agriculture University, Ludhiana, Punjab | 1963 |
| 3. | Jawaharlal Nehru Krishi Vishwa vidyalaya, Jabalpur, Madhya Pradesh | 1964 |
| 4. | Chaudhary Charan Singh Haryana Agricultural University, Hissar, Haryana. | 1970 |
| 5. | Tamil Nadu Agricultural University, Coimbatore, Tamil Nadu. | 1971 |
| 6. | Rajendra Agriculture University, Pusa, Samastipur, Bihar | 1972 |
| 7. | Kerala Agricultural University, Vellanikkara, Kerala | 1972 |
| 8. | Gujarat Agricultural University, Sardar Krushinagar, Dantiwada, Gujarat (With College of Agriculture at Anand, Navsari, Junagadh, Sardar Kushinagar) | 1972 |
| 9. | Narendra Deo University of Agriculture and Technology, Narendranagar, Kumarganj, Faizabad, Uttar Pradesh | 1975 |
| 10. | Himachal Pradesh Krishi Vishwa vidyalaya, Palampur, Himachal Pradesh | 1978 |
| 11. | Y.S. Parmar University of Horticulture and Forestry, Solan, Himachal Pradesh. | 1984 |

From the above, it is clear that the number of independent departments of vegetable sciences presently in India are 11. Pantnagar University is the latest in the chain where an independent department of vegetable science was established in January, 1995 after the bifurcation of the existing Department of Horticulture.

# Mode of reproduction

The breeding procedures that may be used with a particular vegetable crop species are determined by its mode of reproduction. The different modes of reproduction applied in vegetable crop are:

**Asexual:** In asexual reproduction, the vegetative parts of the plant *viz.* rhizomes, corms, tubers, bulbs, cuttings etc. whole or part thereof are utilized to produce new individuals. The typical examples are potato, sweet potato, pointed gourd, artichoke, colocasia, ginger, turmeric, elephant's foot, etc. There are some vegetable crops which can be reproduced through both vegetative part as well as seed e.g. asparagus.

**Sexual:** Sexual reproduction involves fusion of male and female gametes to form zygote which develops into an embryo. In vegetable crop plants male and female gametes are produced in specialised strucutre known as *flower.*

## Floral biology in vegetables

The floral biological studies relating to such vast diverse vegetable plant wealth is an immense task and need to be undertaken fully before proceeding to breeding work and improvement of the vegetable crop in a particular area or zone of the country.

Floral biology : Includes morphology and anthesis of flowers. Anthesis includes the opening of a floral bud and the time of fertilization. Floral biology of a vegetable is a variable and depends upon place, variety, nutrition and environmental conditions like temperature and humidity. Overall floral biology includes-flowering duration, emergence of floral branch reproductive organs-sequence of blooming in a floral branch, duration of blooming period, daily blooming, opening and closing of individual spikelets, anther dehisence, stigma receptivity and viability of the pollen.

Normally flower opening signals the maturity of reproductive organs. When the stamens are mature, the anthers burst and shed their pollen grains. When the pistil is mature the stigma becomes receptive, fresh and is coated with sugary syrup.

Dehisence of the anthers usually occurs within few hours after the unfolding of the petals, depending upon rainfall, temperature and humidity. Transfer of pollen can be affected either by hand, wind or insect-pollinator to the stigma.

## Bud formation

Bud is an embryonic or unelongated shoot. The plants more or less continuously increase in size and develop new organs at least intermittently throughout their life history is one of the most evident of natural phenomena popularly known as *growth*. The origin of terminal bud present on most stem axis, lateral buds which are essentially rudimentary side branches develop in the axil of leaves. These first appear as mound-like meristems in the axils of the embryonic leaves. In temperate zone in most species both terminal and lateral buds are encased in bud scales which are shed only when and if growth of the stem tip is resumed. The buds of most herbaceous plants are devoid of bud scales. A great many more axillary buds form on most plants than ever develop into lateral branches. Some vegetative apical stem meristems continue to grow as such independently but sooner or later some of them become transformed into reproductive meristems. The length of time which a given apical meristem remains in the vegetative state before undergoing transformation into a reproductive meristem differ greatly from one kind of plant to another and from one meristem to another on a given plant, being conditioned in part by genetic and in part by environmental factors. Under certain environmental conditions, not exactly the same for any two kinds of plants, a plant may remain indefinitely in vegetative condition. Delayed transformation of a vegetative into reproductive meristem can be seen in any herbaceous plants which bears a terminal inflorescene at the end of a leafy shoot which has first elongated for some weeks or months before the initiation of flower primordia begins. The first steps in the transformation of a vegetative to reproductive meristem are invisible physiological changes resulting in metabolic conditions within the meristematic cells which completely after the differentiation pattern of the meristem.

In some species of plants the transformed meristem becomes in effect on inflorescence primordia from which an inflorescence bearing a number of flowers develops whereas in the other species only a single flower becomes differentiated from the transformed meristem. The first microscopically visible change in the transformation of a vegetative into a reproductive meristem is a change in configuration. Growth of the central portion seemingly is inhibited, and the meristem becomes flattened on top instead of more or less conical. Small protuberances develop from this modified meristem in a regular spiral or whorled arrangement. These, mound-like protuberances are the primordia from which flower parts develop in a manner analogous to that by which leaves develop from similar protuberances on a vegetative meristem. A marked difference in the development of the two meristems, however, is that there is no elongation of the axis between successive floral primordia such as usually occurs between successive leaf primordia. Although details of flower development differ considerably one species to another, the fundamental patterns followed is similar in all species.

Anthesis includes the opening of flower bud. At first the floral parts of most flowers are tighly enclosed within the

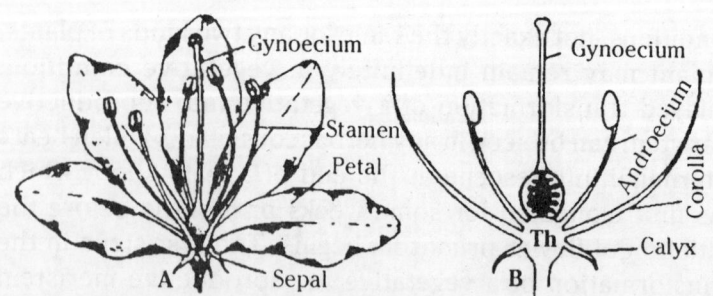

*A - Part of a flower; B - A flower in longitudinal section showing the position of the whorls on the thalamus (Th)*

overlapping sepals, constituting a flower bud. Subsequently expansion of the flower bud into the opened flower occur, a stage in flower development called *anthesis*.

While sepals or both sepals and petals are present in the majority of flowers the only essential parts are stamen and

## Flower parts and their function

| Flower | | | |
|---|---|---|---|
| **Accessory organs** | | **Essential organs** | |
| **Sepals (Calyx)** | **Petals (Corolla)** | **Stamens (Androecium)** | **Pistils (Gynoecium)** |
| • Lower most organ | • Present inside calyx | • Male part | • Female organ |
| • Leaf like | • Brightly coloured | • Produce pollen grains form male cells or sperms | • Ovary containing ovules-immature seeds |
| • Meant to protect the bud till it takes the form of flower | • Possess glands and nectaries sweet liquid called *nectar* is produced and secreted | • Long stalk at the top bears pollen sac or anther | • Slender style or tube |

- Produced odours of essential oils which attract insects, humming birds, people and other creatures for pollination

- Consists of filament, anther and connective

- Stigma which receives the pollens

- The surface of the stigma secretes substances which may provide optimum conditions for pollen germination
- The pollen tubes traversing the style secretes pectinase which dissolves intercellular substances of the style tissue
- After traversing the style the pollen tube enters embryo sac of the ovule.

the pistils. The majority of vegetables (plants) are bisporangiate (bisexual) species and have both of these vital organs present in each flower. In some vegetables,

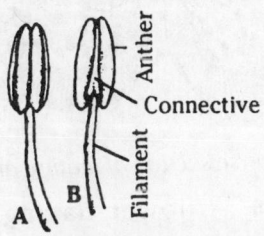

*Two stamens. A - Face of the anther showing four pollen sacs; B - Back of the anthers showing connective*

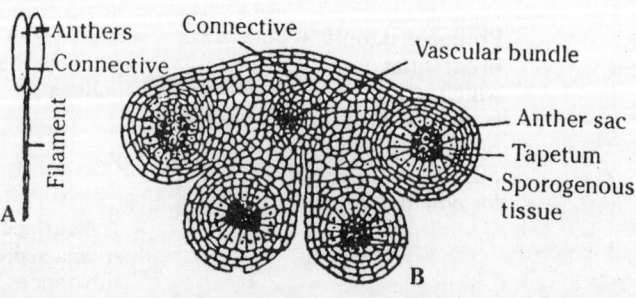

*A - Stamen; B - A transverse section of anther internal structure*

monoecious species - some flowers are staminate and other pistillate, but both organs are borne on the same plant e.g. cucurbits, sweet corn etc. In dioecious species, some plants bear only staminate, other plants only pistillate flower e.g. spinach.

It is the existence of flower borne by the plants including vegetables which is ultimately responsible for the formation of seed. Within the flower two organs are essential to complete the whole process of seed initially supported by the accessory organs (sepal and petals) which are not directly involved in sexual reproduction. Some flowers do not have them.

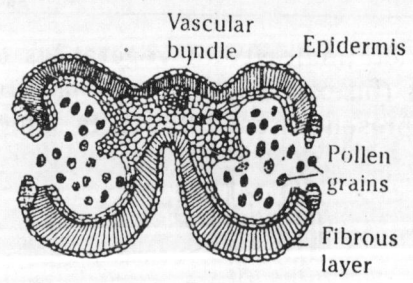

*Transverse section of mature anthers*

**Complete flowers**- A flower having all four organs, **Incomplete** - One of them missing, **Perfect** - When both male and female organs are present, **Imperfect** - When either of the essential organs are missing, **Monoecious** - The reproductive organs are formed in separate flowers on them same plant e.g. cucurbits and sweet corn and **Dioecious** - When the essential organs are situated on different plants e.g. spinach, pointed gourd, asparagus.

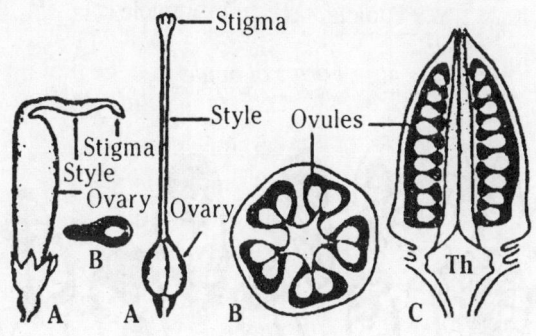

*Pistil, A - Simple, one chambcred ovary; B - Five chambered ovary trans-section; C - Longitudinal section.*

## Anthesis

A marked difference in the development of two meristems, however, is that there is no elongation of the axis between successive floral primordia such as usually occurs between successive leaf primordia. Although details of flower development differ considerably from one species to another, the fundamental pattern followed is similar in all species. At first the floral parts of most flowes are tightly enclosed within the overlapping sepals, constituting a flower bud. Subsequently flower development called *anthesis.*

Young developing flowers are centres of a myriad of metabolic changes. The hormones play a role in the initiation of flowers. Respiratory activity, an indirect index

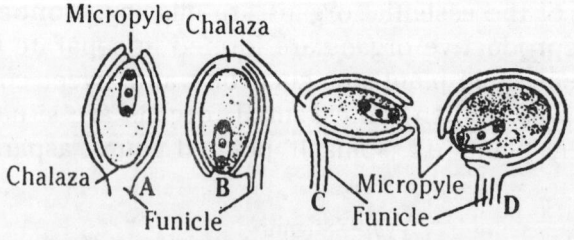

*Forms of ovule*

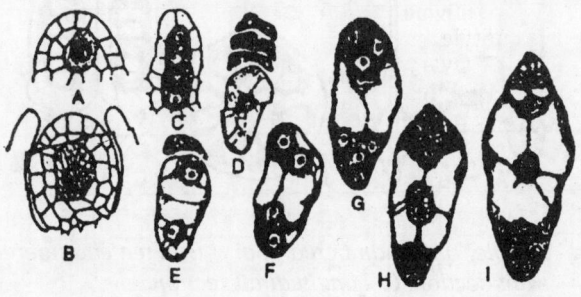

*Development of embryosac A-H, I-fully developed embryosac*

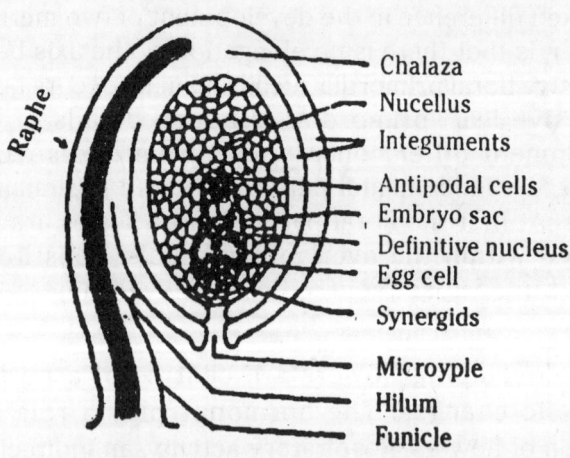

Chalaza
Nucellus
Integuments
Antipodal cells
Embryo sac
Definitive nucleus
Egg cell
Synergids
Micropyle
Hilum
Funicle

Raphe

*Ovule in longitudinal section*

of metabolic activity is always high in young floral meristems, but this is also true of other meristems. Assimilatory rates are also high and there is a continuous translocation towards a developing flower of food, water compounds containing mineral elements and hormones.

## Dehiscence

The dehiscence of anther usually occurs within few hours after the unfolding of the petals, depending upon rainfall, temperature and humditiy. However, the anthesis (opening of the flower) which signals the sexual maturity of the sexual organs.

### Parts of the ovule

**Funicle:** Each ovule is attached to the placenta by a slender stalk known as *funicle*.

**Hilum:** The point of attachment of the body of the ovule to its stalk or funicle is known as *hilum*.

**Raphe:** In the inverted ovule, the funicle continues beyond the hilum alongside the body of the ovule forming a sort of ridge, this ridge is called the *raphe*.

**Nucellus:** The upper end of the raphe which is the junction of the integuments and the nucleus is called the *chalaza*. The main body of the ovule is called the *nucellus*.

**Integuments:** Necellus is surrounded by two coats is termed as *integuments*.

**Micropyle:** A small opening is left at the apex of the integuments is called the *micropyle*.

**Embryo sac:** A large, oval cell lying embedded in the nucellus towards the micropylar end is called as *embryo sac*, and is the most important part of the ovule.

## Reproductive characteristics of some common vegetables

| Family | Vegetable | Genus | Species | Sub-species | Flowering type | Selfed/ or crossed | By | Life cycle | Variety type | Remarks |
|--------|-----------|-------|---------|-------------|----------------|--------------------|-----|-----------|--------------|---------|
| Chenop-odaceae | Garden beet | *Beta* | *vulgaris* | | Perfect | C | W | B | OP | All forms crosses sugarbeet. |
| | Swish chard | *Beta* | *vulgaris* | | Perfect | C | W | B | OP | Cross with other beets |
| | Beet leaf | *Beta* | *vulgaris* | | Perfect | C | W | B | OP | Crosses with other beets |
| | Spinach | *Spinacia* | *oleracea* | | Dioecious | C | I | A | OP or H | Contains perfect flowered plants |
| Compo-sitae | Lettuce | *Lactuca* | *sativa* | | Perfect | S | - | A | PL | Seed may carry lettuce Mosaic |
| | Globe artichoke | *Cynara* | *scolymus* | | Perfect | C | I | P | C | Poor type from seed |
| | Jerusalem-artichoke | *Helianthus* | *tuberosus* | | Perfect | C | I | P | C | Seedlings variable |
| | Chicory | *Cichorium* | *intybus* | | Perfect | C | I | B | OP | Crosses with wild type |
| Convolv-ulaceae | Sweet potato | *Ipomea* | *batalas* | | Perfect | C | I | P | C | Propagated by rooted tubers |

| Family | Vegetable | Genus | Species | Sub-species | Flowering type | Selfed/ or crossed | By | Life cycle | Variety type | Remarks |
|---|---|---|---|---|---|---|---|---|---|---|
| Cruci-ferae | Cabbage | *Brassicca* | *oleracea* | capitata | Perfect | C | I | B | OP or H | All vegetables in this group |
| | Cauliflower | *Brassicca* | *oleracea* | botrytis | Perfect | C | I | B | OP or H | Cross readily; are often |
| | Broccoli | *Brassicca* | *oleracea* | italica | Perfect | C | I | B | OP or H | Self-incompatible |
| | Brussel's sprouts | *Brassicca* | *oleracea* | gemm-iiera | Perfect | C | I | B | OP or H | |
| | Knol-khol | *Brassicca* | *oleracea* | gongylod es | Perfect | C | I | B | OP | |
| | Kale | *Brassicca* | *oleracea* | acephala | Perfect | C | I | B | OP | |
| | Collards | *Brassicca* | *oleracea* | | Perfect | C | I | B | OP | |
| | Turnip | *Brassicca* | *rapa* | | Perfect | C | I | B | OP or H | |
| | Rutabaga | *Brassicca* | *napobrassica* | | Perfect | C | I | B | OP | |
| | Chinese cabbage (Heading) | *Brassicca* | *pekinensis* | | Perfect | C | I | A | OP or H | |
| | Non-heading | *Brassicca* | *chinensis* | | Perfect | C | I | A | OP | |
| | Radish | *Raphanus* | *sativus* | | Perfect | C | I | A | OP | Crosses with wild forms |
| | Rat tail radish | *Raphanus* | *sativus* | | Perfect | C | I | A | OP | Crosses with wild forms |

| Family | Vegetable | Genus | Species | Sub-species | Flowering type | Selfed/ or crossed | By | Life cycle | Variety type | Remarks |
|---|---|---|---|---|---|---|---|---|---|---|
| Cucurbitaceae (Gourd) | Cucumber | *Cucumis* | *sativus* | | Monoecious | C | I | A | OP or H | Does not cross with muskmelon |
| | Muskmelon | *Cucumis* | *melo* | | Andromonoecious or monoecious | C | I | A | OP or H | Includes netted (cantaloupe) honey dew, casaba and mangomelon |
| | Bitter gourd | *Momordica* | *charantia* | | -Do- | C | I | A | OP or H | |
| | Bottle gourd | *Leginaria* | *siceraria* | | -Do- | C | I | A | OP or H | |
| | Ridge gourd | *Luffa* | *acutangula* | | -Do- | C | I | A | OP or H | |
| | Sponge gourd | *Luffa* | *cylindrica* | | -Do- | C | I | A | OP or H | |
| | Snake gourd | *Trichosanthes* | *anguina* | | -Do- | C | I | A | OP or H | |
| | Pointed gourd | *Trichosanthes* | *dioica* | | -Do- | C | I | A | OP or H | |
| | Ash gourd | *Benincasa* | *hispida* | | -Do- | C | I | A | OP or H | |
| | Coccinia | *Coccinia* | *cordifolia* | | -Do- | C | I | A | OP or H | |
| | Long melon | *Cucumis* | *melo* | utilissimus | -Do- | C | I | A | OP or H | |
| | Tinda | *Citrullus* | *vulgaris* | fistulosus | Monoecious | C | I | A | OP or H | |

| Family | Vegetable | Genus | Species | Sub-species | Flowering type | Selfed/or crossed | By | Life cycle | Variety type | Remarks |
|---|---|---|---|---|---|---|---|---|---|---|
| | Water melon | *Citrullus* | *lanatus* | | Monoecious | C | I | A | OP & H | Seedless vegetables are sterile triploids |
| | Summer squash | *Cucurbita* | *pepo* | | Monoecious | C | I | A | OP & H | |
| | Winter squash | *Cucurbita* | *maxima* | | Monoecious | C | I | A | OP & H | |
| | Pumpkin | *Cucurbita* | *moschata* | | Monoecious | C | I | A | OP & H | |
| Graminae | Sweet corn | *Zea* | *mays* | | Monoecious | C | W | A | OP & H | All corn types (pop, flint, dentflour, sweet) cross |
| Leguminosae | Bean | *Phaseolus* | *vulgaris* | | Perfect | S | - | A | PL | Seed can carry serious diseases |
| | Lima bean | *Phaselous* | *vulgaris* | | Perfect | S | - | A | PL | |
| | Garden pea | *Pisum* | *sativus* | | Perfect | S | | A | PL | |
| | Field pea | *Pisum* | *sativum* | arvense | Perfect | S | - | A | PL | |
| | Cowpea | *Vigna* | *sinensis* | | Perfect | S | - | A | PL | |

| Family | Vegetable | Genus | Species | Sub-species | Flowering type | Selfed/ or crossed | By | Life cycle | Variety type | Remarks |
|---|---|---|---|---|---|---|---|---|---|---|
| | Soyabean | Glycene | max | | Perfect | S | | A | PL | Includes black eye and cowder peas blackeye bean |
| Liliaceae | Asparagus | Asparagus | officinalis | | Dioecious | C | I | P | C | Propagated by seed |
| | Onion | Allium | cepa | | Perfect | C | I | B | OP & H | Includes shallot (usually propagated by division) |
| | Garlic | Allium | satiuum | | Perfect | C | | P | C | Never produce seed |
| | Welsh onion | Allium | fistulosum | | Perfect | C | | B | OP | Does not cross with *Allium cepa* |
| | Leek | Allium | ampelo-prassum | | Perfect | C | I | B | OP | |
| Solanac-eae | Tomato | Lycoper-sicon | esculentum | | Perfect | CS | - | A | PL or H | |
| | Capsicum | Capsicum | annuum | | Perfect | CS | I | A | PL or H | |
| | Chillies | Capsicum | frutescens | | Perfect | CS | I | A | PL or H | |
| | Brinjal | Solanum | melonegena | | Perfect | S | I | P | C | |

| Family | Vegetable | Genus | Species | Sub-species | Flowering type | Selfed/ or crossed | By | Life cycle | Variety type | Remarks |
|---|---|---|---|---|---|---|---|---|---|---|
| Malvaceae | Bhindi | *Abelmoschus* | *esculentus* | - | Perfect | C (often) | I | A | OP or H | |
| Solanaceae | Potato | *Solanum* | *tuberosum* | | Perfect | S | | P | C | TPS and tubers |
| Umbelle ferae | Carrot | *Daucus* | *carota* | sativa | Perfect | C | I | B | OP or H | Crosses readily with wild carrot |
| | Celery | *Apium* | *graveolens* | | Perfect | C | I | B | OP or H | |
| | Celeriac | *Apium* | *graveolens* | | Perfect | C | I | B | OP | |
| | Parsley | *Petroselinum* | *hortense* | | Perfect | C | I | B | OP | |
| | Parsnip | *Pastinaca* | *sativa* | | Perfect | C | I | B | OP | |

⋀ Perfect-male and female in the same flower; dioecious-male and female flowers on separate plants; monoecious-separate male and female flowers on the same plant; andromonoecious-male and prefect flowers on the same plant.

⋀ S = self-pollinated; C = Cross-pollinated; CS = both crossed and selfed

⋀ Pollinated by I = Insect and W = wind

⋀ A = Annual; B = Biennial and P = perennial

⋀ PL = Pureline; C = Close; OP = Open-pollinated and H = hybrid.

## Pollen

Pollen grains differ greatly in size, ranging from about 5 to 200 μ in diameter and also in configuration from one plant to another, but their physiological role is similar in all species. The quantity of pollen produced also differs greatly from one kind of plant to another. Some plants produce pollen only sparingly; other with great prodigality. A single sweet corn plant, for example, is estimated to release 50,000,000 pollen grains. Only about 1000 pollen grains are required if fertilizations of all of the egg cells in all the ovules of one sweet corn plant is to be accomplished.

## Pollination

Pollination refers to the transfer of pollen grains from anthers to stigma pollen from an anther may fall on to the stigma of the same flower leading to *self-pollination* or *autogamy*. When pollen grains from flowers to one plant are transmitted to the stigmas of flowers of another plant, it is known as *cross-pollination* or *allogamy*. A third situation, geitonogamy, may resultswhen pollen from the flowers of one plant falls on the stigmas of other flowers of the same plant. The genetic consequence of geitonogamy are the same as those of autogamy.

## Self pollination

Self pollinated species are believed to have originated from cross pollinated ancestors. These species, as a rule, must have hermaphrodite flowers, but in most of these species, self pollination is not complete and cross pollination is not complete and cross pollination may occur up to 5%.

## Mechanism promoting self-pollination

There are various mechanisms that promote self pollination, which are generally more efficient than those promoting cross-pollination. They are briefly summarised as:-

O    Flowers do not open at all and is termed as *cleistogamy*. This ensures complete self-pollination e.g. lettuce.

O    The flowers open, but only after pollination has taken place and is termed as *chasmogamy*. This ensures self-pollination e.g. capsicum, brinjal, tomato.

O    In vegetable crops like tomato and brinjal, the stigmas are closely surrounded by anthers. Pollination generally occurs after the flower opens. But the position of anthers in relation to stigma ensures self-pollination.

O    In vegetable crop like pea, flowers open but the stamens and the stigma are hidden by the two petals forming a keel.

O    In vegetable crop like lima bean, stigma is ruptured by bees which leads to its receptivity.

## Genetic consequence of self-pollination

Self pollination leads to a very rapid increase in homozygosity. Therefore, populations of self-pollinated species are highly homozygous. Self-pollinated species do not show inbreeding depression but may exhibit considerable heterosis. Therefore, the aim of breeding methods generally is to develop homozygous varieties.

## Cross pollination

In cross-pollinating species, the transfer of pollen from a flower to the stigmas of the others may be brought about by wind (anemophily e.g. spinach, sweet corn, beet leaf, garden beet etc.), water (anemophily e.g. water cress), gravity (earth's attraction e.g. sweet corn), insect (entomophily e.g. cole crops, carrot, onion etc.). Many of the crop plants are naturally cross-pollinated. In many species, a small amount of (up to 5-10%) of selfing may also occur.

## Mechanism promoting cross-pollination

| Mechanism | Brief description |
|---|---|
| 1. Dicliny | Dicliny or unisexuality is a condition, in which the flowers are either male (staminate) or female (pistillate). |
| 1a. Monoecy | When male and female flowers are separate but present in the same plant, it is known as *monoecy*, e.g. cucurbits, sweet corn, cassava. |
| 1b. Dioecy | When male and female flowers are present on different plants, it is called *dioecy*, e.g. pointed gourd, asparagus, spinach |
| 2. Dichogamy | It refers to maturation of anther and stigma of the same flower (hermaphrodite) at different times. |
| 2a. Protandry | When anthers mature before pistil, it is known as *protandry*, e.g. carrot, sugar beet, onion. |
| 2b. Protogyny | When pistil matures before anthers, it is known as *protogyny*, e.g. *Brassica* spp. |
| 3. Heterostyly | When styles and filaments in a flower are of different length, it is called *heterostyly*. |
| 3a. Heteromorphic | Existence of styles of three lengths (long, medium and short) e.g. brinjal. |
| 3b. Dimorphic | Flowers with long and short filament of anther e.g. tomato. |
| 3c. Trimorphic | a) Flowers with long style, medium and short anthers.<br>b) Medium style, long and short anther.<br>c) Short style, long and medium anthers e.g. *Brassica* spp. |
| 4. Herkogamy | Hinderance to self-pollinations due to some physical barriers such as presence of hyline membrane around the anthers is known as *herkogamy*. |

## Genetic consequences of cross pollination

Cross-pollination preserves and promotes heterozygosity in a plant population. Cross-pollination species are highly heterozygous and show mild response to severe inbreeding depression and a considerable amount of heterosis. The breeding methods in such species aim at improving the crop species without reducing heterozygosity to an appreciable degree. Usually hybrids or synthetic varieties are the aim of vegetable breeder wherever the seed production of such varieties is economically feasible.

### Often cross-pollination

In many crop plants, cross pollination is 5% and may reach 12% such species are generally known as *often cross-pollinated crop*. The architecture of such crops is intermediate between those of self-pollinated and cross-pollinated species. Consequently, in such species breeding methods suitable for both of them may be profitably applied. But often hybrid varieties are superior to others e.g. okra, brinjal, chilli, lima bean.

### Breeding systems in vegetables

The vegetable crops are divided into two groups from the standpoint of the vegetable breeder, depending upon whether they are predominantly self-pollinated or largely cross-pollinated. This distinction of one of primary importance because of the methods of breeding are applicable to self-pollinated vegetables as for the most part distinct from those that apply to cross-pollinated species.

All the plants in the population of out crossing species are highly heterozygous and enforced inbreeding results in determinate true in general vigour and other adverse effects. Heterozygosity appears to be an essential feature of commercial varieties of these species and as a result it must be maintained during the breeding programme or

## Vegetable crops needing pollination

| Vegetables | Pollinators | Nature of pollination | Seed set increase (%) | Conditions favouring cross-pollination |
|---|---|---|---|---|
| **1. Cruciferous vegetables** | | | | |
| Cole group, radish, turnip, Chinese cabbage | Bees, Diptera, Blow Flies, Solitary bees, other insects (Leaf cutting bees, mining bees, bumble bees) | Highly cross-pollinated (some wind-pollination in turnip) | 14-22 | i) Protogynic condition of the flowers.<br>ii) Dehiscence of anther of the long stamens which are in the level of stigma outwards and short stamen inwards<br>iii) Self-sterility |
| **2. Cucurbitaceous vegetables** | | | | |
| Pumpkin, gourds, cucumber, melon etc. | Honey bees | Cross-pollinated | | i) Monoecious occurs singly<br>ii) Large and sticky pollen carried by insect and not by wind |
| **3. Umbelliferous vegetables** | | | | |
| Carrot, parsley, celery, parsnip, fennel, coriander, celeriac, dill, chirvil, anise and caraway | Honeybees, solitary beet etc. | -do- | 50 | i) Protandry condition |
| **4. Solanaceous vegetables** | | | | |
| Tomato, sweet pepper, chillies, brinjal and potato | Honey bees, and other insects | Self-pollinated (intensification of the pollination process) | Brinjal 16-20, pepper 60 | i) Pendent position of the flowers facilitate pollination |

| Vegetables | Pollinators | Nature of pollination | Seed set increase (%) | Conditions favouring cross-pollination |
|---|---|---|---|---|
| | | | | ii) Projection of stigmas beyond the cone of anthers (in some varieties) |
| | | | | iii) Style does not elongate and grows the anthers tube until the time of anther dehiscence or afterwards |
| **5. Lilliaceous vegetables** Onion, garlic, leek, asparagus, shallot, welsh onion and chive | Honeybees, solitary bees, and diptera | Cross-pollinated (Honey bees are also used to produce hybrid seed) | | i) Male sterility ii) Dehiscence of anthers of the inner whorl of stamen at irregular intervals before those of outer whorl and the process is completed between 24-36 hours and before stigma become receptive. |
| **6. Compositacious vegetables** Lettuce, endive, salsify, chicory, dandelion and globe artichoke | Wild bees honeybees and other insects | Self-pollinated | | i) Varied flowering peaks occur. |
| **7. Chenopodiaceous** Garden beet, sugarbeet beetleaf, spinach, Swiss chard and orach | Wind, diptera | Wind-pollinated | | i) Shedding of pollen over a period of weeks. |

| Vegetables | Pollinators | Nature of pollination | Seed set increase (%) | Conditions favouring cross-pollination |
|---|---|---|---|---|
| | | | | ii) Tetramorphic flower present in spinach |
| | | | | iii) Presence of dioecy |
| | | | | iv) Monoecious plants |
| **8. Malvaceus vegetables** | | | | |
| Bhindi | Bees | Often-cross-pollinated | 8.75 | i) Sticky pollen grains |
| **9. Leguminaceous vegetables** | | | | |
| Lima bean, Broad bean | Beetle and thrips | Self-pollinated but some cross-pollination | 30 | Pollination through beetle and thrips by making holes at the base of nectary |
| **10. Graminaecious vegetables** | | | | |
| Sweet corn and popcorn | Wind and gravity | Wind pollinated (2500 pollen in each anther is produced) | | i) Monoecious plants |
| | | | | ii) Location of male and female organs at different parts on the same plant |

restored as a final step of the programme. The population of self-pollinated vegetables usually consists of mixture of many closely related homozygous lines and remain more or less independent of one another in reproduction. With these vegetable species the goal of most breeding programmes is pure line. When vegetative parts of some vegetables usually can be used to produce new individuals a sexual pattern becomes possible in these vegetable/plant breeding programmes. There are more than 32 different kinds of vegetables grown in India which are cross-pollinated in nature.

The breeding method in any vegetable crop depends upon its breeding system and genetic architecture resulting from natural selection as well as artificial by human selection during the course of cultivation. Broadly speaking, the breeding methods for all the pollinated vegetable crops are the same since they have similar breeding system. However, the breeding methods differ in some respects in these vegetables, particularly those which have self-incompatibility, dioecious or monoecious sex form and biennial or perennial growth habit. The extent of loss of vigour due to inbreeding, a general characteristic feature of cross-pollinated vegetables, will also necessitate modification of breeding procedures to some extent.

The genetic architecture and the pattern of inheritance of characters are important considerations while determining the most appropriate breeding procedures applicable to any particular vegetable crop. The inheritance of the characters may be simple if these are governed by a few genes only such as, colour and shape of fruit in cucurbits -bottle gourd, colour of pod in bhindi. But not so in polygenically controlled characters. Resistance to diseases in some cases may by polygenic (controlled by several genes) vegetables. Several economic characters in vegetable crops are polygenically inherited. These include

## Breeding systems, sex form and growth habit of different cross-pollinated vegetable crops

| Breeding system | Sex form | Growth habit | Vegetable crops |
|---|---|---|---|
| Highly cross-pollinating | Hermap-hrodite | Annual | Amaranth*, Spinach beet (Palak)*, Chenopodium*, |
| | | Biennial | |
| | | i) Self-incompatible | Cabbage, cauliflower, (late varieties), knol-khol, Brussel's sprout, sprouting broccoli, kale, radish, turnip, garden beet |
| | | ii) Self-fertile | Carrot, onion, cauliflower (some varieties) |
| | Monoecious | Annual | Bottle gourd, bitter gourd, sponge gourd, ridge gourd, snake gourd, Ash, gourd, cucumber, pumpkin, squash, pointed gourd, Indian squash, long-melon, watermelon, muskmelon**, sweet corn* |
| | Dioecious | Annual | Spinach* |
| | | Perennial | Asparagus, pointed gourd |
| Often cross-pollinating | Hermaph rodite | Annual | Bhindi (Okra), Lima bean, Chillies |

* Anemophilous (wind-pollinated), others are entomophilous (insect-pollinated).
** In muskmelon most of the varieties are andromonoecious, while some are monoecious.

such as metric traits as yield, size, weight, shape and compactness of head and maturity in cabbage and cauliflower, shape size and number of fruits in bhindi, tomato and brinjal, size, weight and shape of roots in radish, turnip, carrot and garden beet or bulbs in onion, size and weight of fruits and thickness of flesh in melons, cucumber and pumpkin and size of leaf in spinach and beet leaf. A few other quality characters like carotene content in carrot, ascorbic acid content in tomato or cabbage, sugar content in melons, capsicum content in chillies are also governed by polygenes. Similarly resistance to diseases in certain cases is also polygenically inherited, such as yellow disease in cabbage unlike the qualitative characters which have a simple inheritance, the polygenic characters are greatly influenced by environment and hence in these cases a phenotype is often not a true indicative of its genotype. The variation existing in a breeding population is of great importance to a vegetable breeder. The phenotypic variability is the result of the genotypic value, the effect of environment and the genotype-environmental interaction. In a quantitative characters, the cumulative effects of the individual genes and their interactions, both allelic or dominance and the non-allelic or epistasis, determine the genotypic value.

## Gene and genotype frequency

The gene frequency is the number of that particular gene (allele) in a population relative to the total number of genes at the same locus. The genetic constitution of a population is commonly described in terms of gene and genotype frequencies. The genes carried by a population have continuity from generation to generation but the genotype in which they appear do not in the transmission, the genotype of parents are broken down and new ratio of genotypes is constituted in the progeny from genes transmitted gametes. The law governing the frequencies of genes and genotypes in the population was formulated by Hardy in England (1908) and Weinberg in Germany. The law states that in large random maturing population both gene and genotype frequencies are constant from generation to generation to the absence of migration, mutation and selection. The population is said to be equilibrium in cross-pollinated vegetable crops and sum of A and a is always unity. If the frequency of A = 9 and a = 1 and (the frequency of genotype) AA + 2Aa + aa then $(.9)^2$AA 2 (0.9) + $(.1)^2$Aa + $(.1)^2$aa = .81AA + 1.8Aa and 0.01 aa-1.00. Among the temperate vegetables/varieties falling under cole group are important and all widely grown. Next to these are European types of carrots, radish, turnip and garden pea. All these vegetables are cross-pollinated. Pollination is being carried out mostly by insects notably honey bees. From the breeding point of view cabbage and allied vegetable crops have various important characteristics in common.

- **O** Most botanical varieties of *Brassica oleracea* are biennial, flower after the winter (overwintering) because a period of low temperature (0-7°C) is required for flower formation. However, most varieties of sprouting broccoli and cauliflower are normal, only the late winter varieties are biennial in nature.

- **O** Cole vegetables are usually cross-pollinated only in some cauliflower varieties, seed setting is partly brought about by selfing.

- **O** It is possible to judge the quality etc. before flowering and this increases the effectiveness of selection. For most vegetables, production-cycle followed is as grown as usual, selection-vegetative phase, maintenance-overwintering (throughout the winter)-bolting-flowering- pollination by insects (after the end of winter).

The breeding procedures for cole vegetables are not much different from other cross-pollinated vegetables, however, the varying degrees of self-incompatibility plays a great role in breeding programmes.

## Self-incompatibility

Self-incompatibility means where both male and female parts are capable of producing normal gametes but the mating as a result of selfing fails to set seed. The phenomenon of self- incompatibility is very widely spread in plant kingdom and it enforces out breeding in many plant species. The advantages of self-incompatibility has been taken in the production of hybrid seed in vegetable crops. Also it is important to know the mechanism of self-incompatibility if the various vegetables are to be minimized through self-crossing.

### Advantages

O    There is no need for the maintainer lines.

O    The seed set takes place on both the parents.

### Disadvantages

O    The incompatibility mechanisms are complex.

O    The incompatibility mechanisms are influenced by environment.

### Types of self-incompatibility mechanism

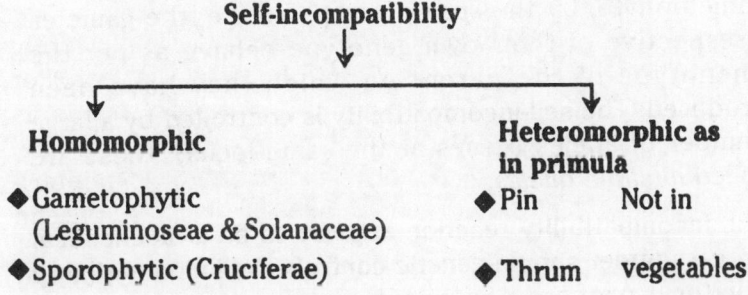

**Self-incompatibility**

**Homomorphic**

◆ Gametophytic
   (Leguminoseae & Solanaceae)

◆ Sporophytic (Cruciferae)

**Heteromorphic as in primula**

◆ Pin          Not in

◆ Thrum        vegetables

### Homomorphic

There is no morphological difference in the male and female organs which hinders their compatibility, within the same flower/plant. This system is prevalent in vegetable crops.

### Heteromorphic

As an example it is found in evening primrose where pin and thrum two types of anthers are found such that pin crosses with thrum and thrum with pin. Not important in vegetables.

### Gametophytic

Under the homomorphic system, gametophytic system is fairly widespread. It means that the incompatibility reaction is right in the gametophyte itself i.e. the pollen (or for that reason any generation which is gametophyte i.e. that has "n" number of chromosomes). Under the gametophytic system the genotype of the gametophyte determines the incompatibility reaction irrespective of the parent plant (sporophyte) from which it is produced. This system is prevalent in some plants of the family solanaceae and leguminosae. Also called *operational factor hypothesis*.

**Sporophytic:** This is one of the most important types of self-incompatibility found in vegetable crops especially crops belonging to the family cruciferae. In this the incompatibility reaction is imparted upon on the gametophyte, by the *sporophyte* parent i.e. the gametes irrespective of their own genotype behave as per the phenotype of the parent on which they have been produced. The self-incompatibility is controlled by a large number of alleles (genes at the same locus), these are called *multiple alleles*.

The incompatibility reaction appears to be a biochemical process under simple genetic control. The incompatibility (system) process can operate at any stage between pollination and fertilization. In cabbage and radish incompatible pollen usually fails to germinate and if it does germinate, the pollen tube fails to penetrate the stigma. In some species of this type, the inhibition is lost if the stigmatic surface is removed. In other species incompatible pollen germinate and the pollen tube grows down the style, but slowly that rarely reaches the ovary in time to effect fertilization. The rate of growth of compatibility pollen tubes can be included by the background genotype in which the major gene(s) governing the incompatibility (system) reaction operates.

In the gametophytic system, the incompatibility is controlled by single gene "S", which is usually characterised by the very large number of allelic forms in which it exists. The pollen tube growth is usually very slow in a style that contains the same allele of S. The gametophytic system gives rise to three main types of pollination (i) fully incompatibility ($S_1S_2$ x $S_1S_2$) in which both alleles are common, (ii) half the pollen compatible ($S_1S_2$ x $S_1S_3$) in which one allele is different and (iii) all the pollen is compatible ($S_1S_2$ x $S_3S_4$) in which both alleles are different. The sporophytic system is similar to the gametophytic system in that the genetic control is by a single gene with multiple alleles. It differs from the gametophytic system in that incompatibility reaction is imparted to the pollen by the plant upon which the pollen is borne.

## Inheritance of incompatibility

In nature cross-incompatibility is infrequent owing to the great number of factors that exist. In selfed $F_1$ generation of a plant with a genetical constitution $S_1S_2$ consists of 25% $S_1S_1$ plants, 50% $S_1S_2$ and 25% $S_2S_2$ plants. If $S_1S_2$ is crossed with $S_3S_4$ the $F_1$ will consist of 25% $S_1S_3$, 25% $S_1S_4$, 25% $S_2S_3$ and 25% $S_2S_4$. If $S_1$ is dominant over $S_2$, the cross $S_1S_2$ x $S_2S_3$ is compatible, whereas it would be incompatible if the 'S' factors were to act independently. Odland and Noll (1959) suggested a double cross in cabbage. Four homozygous self-incompatible but cross compatible plants were used as shown below. The $F_1$ hybrid is a double cross.

Var. A

$$S_1S_1 \times S_2S_2 \qquad\qquad \times \qquad\qquad S_3S_3 \times S_4S_4$$
$$\downarrow \qquad\qquad\qquad\qquad \downarrow \qquad\qquad\qquad\qquad \downarrow$$
$$S_1S_2 \qquad\qquad S_1S_2S_3S_4 \qquad\qquad S_3S_4$$

Var. B

(Double cross $F_1$ hybrid)

## Factors influencing incompatibility

The incompatibility is strongest in fully opened flowers. When young buds are selfed, seed setting is much better than at later stages and may be as good as in cross-pollination. In breeding, use is made of bud pollination for the production of inbred lines. The best results are usually obtained when the buds are from 5-8 cm length. At 15°C they reach this size about 2-3 days before the flowers open, whereas at lower temperatures, they do so some days earlier. If the buds are pollinated too early, seed setting decreases as the flowers are not yet sufficiently fertile; moreover, very young buds are, more easily damaged in such operations than the older ones. In older buds, the incompatibility gradually increases. Some days after opening of the flowers the incompatibility decreases again. Thus, at temperature over 15°C, selfing of flowers that are from 2-4 days old gives the same seed setting as cross-pollination, whereas at lower temperatures, at which the flowers bloom longer, the decrease in incompatibility starts later. This senile compaction is used in selfing parents with the aid of insects and in this manner it is possible to increase inbred lines on a large scale.

## Male sterility

One of the outstanding achievements in plant (vegetable) breeding over the last four decades has been the utilization of male sterility in the production of hybrid seed. Discovery of male sterile clones in onion by Jone and Emsweller (1936) led to a search of male sterility in other crop species for their possible use in hybrid seed production. So as to capitalize male sterility for hybrid seed production in vegetable crops, two factors are of prime importance, first how is male sterility inherited, second its manifestation.

## Impact of mode of inheritance on utilizing male sterility

The mode of inheritance is vital as it will not only determine the breeding methodology in maintaining the male sterile stocks every year but will also indicate whether or not male sterility can be used at all in a vegetable crop for production of hybrid seed within economic feasibilities. This is more evident if we look at the patterns of inheritance of male sterility in plants (vegetables). Basically only three systems of inheritance of male sterility have been recognised (i) cytoplasmic (ii) genic and (iii) cytoplasmic genic. In cytoplasmic male sterility, the sterile reaction is imparted upon the individual through its cytoplasm, which is mainly contributed by the mother which results in that all the individuals produced by a male sterile, mother will be male sterile, consequently all the $F_1$ hybrid seed produced through this type of male sterility is male sterile and fail to produce seeds or even fruits. Cytosterile plants shall always need a pollinator for their maintenance and also for seed production in the hybrids. This type of male sterility is thus most useful in vegetable crops where any part of the plant or fruit is of commercial value. The genic type of male sterility is practical in fruit and seed vegetables like tomato, lettuce, cowpea, lima bean, watermelon and muskmelon. The male sterility in these crops is controlled by a single recessive gene which is maintained by a heterozygous pollinator, like as female x SS male, (self sterile male). In every generation fifty per cent of the plants have to be rogued out from the female nursery stocks as these are male fertile. The roguing in such situations can, however, be facilitated with the help of marker genes as in case of tomato where two recessive genes that of potato leaf and green stem linked to male sterility have been used by Currence and Nickelson (1956). Yet another way to maintain such male sterile stocks is resorting to vegetative propagation as in tomato and cucurbits. Use of partial male sterile material as in carrot has also been advocated by Hensche and Gabelman (1963) which facilitates the maintenance of such stocks, but captalizing of such an unstable type of male sterile in hybrid breeding programme is rather dangerous as it can break down with a change in environment and may thus cause contamination in the hybrid seed. The most compelling practical consideration, therefore, to use genic male sterility is to obtain hybrid seed which is fully fertile.

Turning to the cytoplasmic genic male sterility, it behaves just as cytoplasmic type of sterility and has the same practical bearings like rendering the $F_1$ seed male sterile. But an important difference lies in the interaction of cytoplasmic and nuclear genes, for the cytoplasm to cause male sterile in this case it has to be accompanied by recessive male sterile gene(s) in the cytoplasm. Thus, this type of sterility has been made use in production of hybrid seed on very large scale in onion, where it is controlled by sterile cytoplasm and a recessive nuclear gene (Smsms, Jonnes and Emsweller, 1936) and in garden beets where it is controlled by sterile cytoplasm and two recessive genes (SXXZZ, Owen, 1942). Thus it has to be maintained in homozygous condition by a maintainer line (pollinator). Here the pollinator has a special qualifications, that is, it should be able to give 100 per cent sterile progeny when crossed to female male sterile line. In garden beet too these pollinators are referred to as "O" type pollinators. This type of male sterility has also been found in broad bean (Bond *et al.*, 1964) and carrot (Banga *et al.* 1964). However, its use in the production of hybrid seed in these vegetable crops has been very little. This is mainly because bean and capsicum are fruit vegetables and male sterile $F_1$ is uneconomical and its stability in carrot is unreliable.

## Importance of mode of manifestation of male sterility

The way by which male sterility is manifested in the plant species (vegetables) can have a far reaching consequences on its capitalization in the production of hybrid seed. One of such outstanding mode of manifestation of male sterility are found in what is called, "*functional*" or "*positional*" type of male sterility, as is found in tomato (Rover, 1948) and brinjal (Jasmin, 1954). In this type of male sterility, the pollen are normal and functional but the anthers fail to dehisce. Advantage of the situation can be exploited in maintaining male sterile stocks in homozygous state through selfing without the aid of pollinators. Moreover, as with other genic types of male sterilities where either pollen are sterile or stamens are malformed. This kind of male sterility rules out roguing out 50 per cent population from the male sterile nurseries. In tomato, this desirable type of male sterility was linked with green shoulders, but, Mittal and Thomas (1961) found a mutant which though functional male sterile, was free of green shoulders.

Male sterile flowers produce no pollens and this feature which is often heritable is used in breeding of hybrids. In cauliflower, brussel's sprout and broccoli, a form of male sterility has been found which is determined by one recessive factor. The male sterile flowers are distinguished from normal flowers by their shrivelled anthers and filaments.

## Future outlook

So far functional types of male sterility holds a great promise and efforts should be made to locate the same in other vegetable crops, where cytoplasmic or cytoplasmic-genetic type of sterility has to be used, stress should be laid on finding out more diversified sources of sterile cytoplasm rather than capitalizing on one or few sources. This is needed to save the hybrid varieties from the ravages of diseases and insect-pests.

## Gametocides

More recently efforts have been made to cause sterility with the help of chemicals. Such attempts have been made in many vegetable crops like cucurbits and onions but the practical utility on commercial scale is still doubtful.

## Poly cross method

The selected plants are allowed to flower together, care being taken that every plant is pollinated with the same pollen mixture. The seeding plants are harvested separately and the progenies are compared following which work is continued with the best parent plant which must be maintained vegetatively for the purpose. If necessary the procedure can be repeated for some generations to allow selection of the genetically superior plants.

## Diallel or paired crosses

In diallel, all possible crosses between the selected plants are made. When diallel crosses are made, it is easy to find the heterozygous plants and the selection is only continued with progenies of AA x Aa plants, when there is a need to eliminate an undesirable character which is determined by a single recessive gene. In paired crosses, a selected plant is crossed with another selected plant a x b, c x d, e x f etc. with sporophytic control of the pollen, two alleles are normally present at the time of gene action. Thus, there has been strong selection for the type of allele interaction operating in the pollen. It has been shown the gametophytic system where only one allele is present at the time of pollen determination that a breakdown of allele activities occur when two different alleles are present in pollen of tetraploid. Clearly the negative effects of polyploidy in the sporophytic system is to be expected and confirms the confusions obtained with the gametophytic systems. There will be possibilities of new dosage relationship in autopolyploids with sporophytic system and because weak and complete dominance are features of the system. It is possible that such new dosage relationship will result in an occasional change in gene action.

## Polyploidy

Most of the vegetables are diploids i.e. they have "2n" number of chromosomes but some like potato, bhindi, sweet potato are natural polyploids.

### Triploids (3n)

Polyploidy has been used as a tool in genetic deseeding where the final produce is required to be devoid of seed. Since the triploids are generally sterile and they do not produce seed but may set fruits (watermelons, cucumbers). Triploid seeds are produced by crossing 2n x 4n = 3n. Triploid seeds are given to the farmers along with pollinator (2n). So, production of triploid seeds calls for maintaining diploids and triploid lines for producing triploid seed every year.

### Tetraploids (4n)

Tetraploids produced from their diploid progenators are generally sterile/less fertile but may be more vigorous vegetatively. Tertraploids have been made in spinach (*Beta vulgaris*) and are useful in leafy vegetables.

### Aneuploids

It means not true polyploids. Aneu mean 'false', ploid = 'levels' (folds) such as n + 1 ; 2n ı 2 etc. These have been used to cause sterility and produce seed less watermelons

## Parthenocarpy

It means setting of fruit without fertilization. Some gynoecious lines in cucumber are fully gynoecious and male flowers are produced by the spray of $GA_3$ 1500 ppm at 3-4 leaf stage.

## Maintenance of variety

(i) By mass selection and (ii) production of nucleus seed in cages (not less than 25 plants to avoid loss of vigour).

## In breeding

Varies from vegetable to vegetable. It is drastic in carrot, while in onion, it is not to that extent, and in cabbage, it is still less than that in Onion.

| Carrot | < | Onion | < | Cabbage |
|---|---|---|---|---|
| (Not more than 2 years) | | 2-3 years | | 3-4 years |

In cauliflower, the loss in vigour due to inbreeding varies with the varieties depending upon the proportion of self-compatible and self-incompatible plants occurring in the variety.

## Activities in vegetable breeding

The desired changes in genotypes of vegetable crop species and the consequent benefits to farmers are brought about by a series of interrelated and largely interdependent activities. These activities are as follows :

(1) Creation of variation.

(2) Selection.

(3) Evaluation.

(4) Multiplication, and

(5) Distribution.

For an efficient vegetable crop improvement programme the breeding activities have to be properly coordinated and efficiently geared to maximise the outputs from a programme.

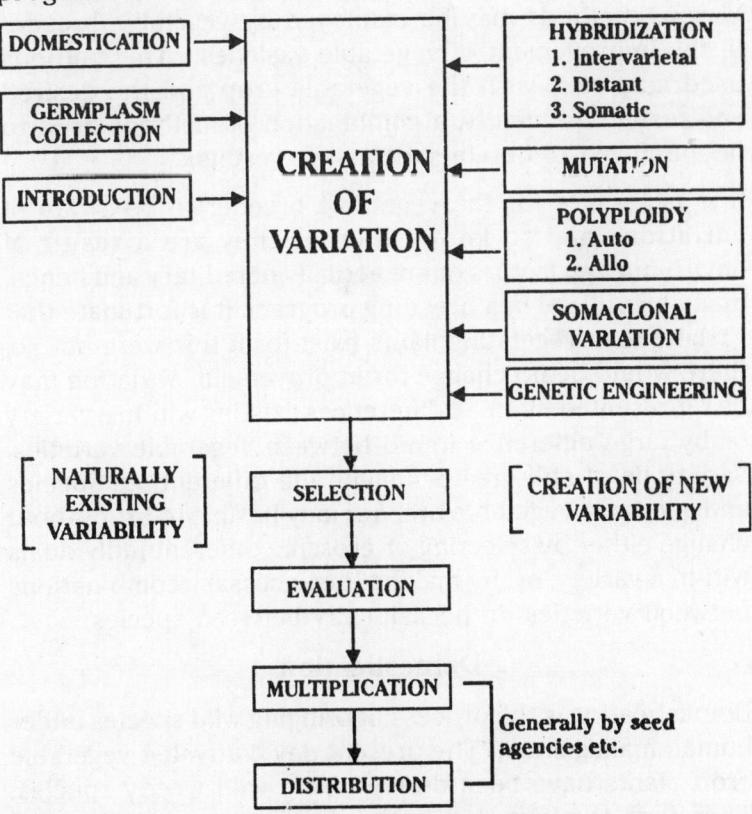

*Activities in breeding*

A deficiency or slackness in any one of these steps will definitely reduce the efficiency of the programme. This will lead to a wastage of valuable resources, which no nation, particularly a developing one, can afford. Therefore, a regular and critical review of the progress and the problems has to be made, the deficiencies in the various activities have to be identified and appropriate remedial measures have to be taken to ensure continued progress in vegetable breeding programmes.

## Breeding methods

Various methods may be employed by vegetable breeder in the improvement of vegetable varieties. The methods used may vary with the vegetable crop and the desired end result. Frequently, a combination of methods need to be employed to obtain satisfactory results.

It is necessary for the vegetable breeder to be aware of variations and to know whether they are a result of environmental factors or are actually hereditary and hence, many be utilized in a breeding program. It is fortunate that variations in vegetable plants exist for if this were not so, there would be no change for improvement. Variation may be represented by small differences existing within a variety or by large difference found between vegetable varieties. Variations of still greater magnitude differentiate species and genera. A vegetable breeder may be able to bring about change either by selecting or crossing different individuals within a variety, or by making the necessary combinations between varieties on occasionally between species.

## Domestication

Domestication is the process of bringing wild species under human management. The present day cultivated vegetable crop plants have been derived from wild weedy species. Therefore, the first step in the development of cultivated vegetable crop plants were their doemstication.

## Germplasm collection

A germplasm collection of a crop species consists of a large number of lines, varieties and related wild species of vegetable crops. Such collection are called *gene banks* when a germplasm collection is sufficiently large to include entries or accesions from all over the world, it is called *world collection* some of the world collections are listed:

## Some important vegetable germpalsm world collections

| Germplasm world collection | Vegetable crop |
| --- | --- |
| Cambridge, U.K. | Potato |
| International Centre for Potato, Lima, Peru | Potato |
| New Zealand | Sweet potato |
| Wisconsin, U.S.A. | Potato |
| Royal Botanical Garden, Kew, England | All vegetable crops |
| Institute of Plant Industries, Leningard, USSR | All vegetable crops |
| International Institute for Tropical Agriculture (IITA) , Ibadan, Nigeria | Cowpea, Lima bean, Cassava, Sweet potato |
| International Centre for Tropical Agriculture (CIAT), Palmira, Colombia | Beans |
| International Centre for Agriculture Research in Dry Areas (ICARDA), Lebnon, Syria | Beans |

In India, NBPGR (National Bureau of Plant Genetic Resources, New Delhi) maintain large collections of horticultural crops.

Germplasm collections maintained in India at institutions other than NBPGR, New Delhi.

| Institutions | Germplasm collection |
| --- | --- |
| Central Plantation Crops Research Institute (CPCRI), Kasaragod, Kerala | Plantation crops |
| Central Tuber Crops Research Institute (CTCRI), Thiruvananthapuram | Tuber crops (except potato) |
| Indian Institute of Horticultural Research (IIHR), Bangalore | Horticultural crops |
| Central Potato Research Institute (CPCRI), Shimla, Himachal Pradesh | Potato |

## Plant Introduction

Plant introduction consists of taking a genotypes or a group of genotypes of plants into a new environment where they were not being grown before. Thus, the plant introduction may involve new varieties of a crop already grown in the area, wild relatives of crop species or a totally new crop species for the area.

For several centuries A.D., in India the agencies of vegetable crop introduction were invaders, settlers, traders, travellers, explorers, pilgrims and naturalist. In the 16th century A.D., Portuguese introduced chillies, potato, sweet potato. The East India Company brought cabbage, cauliflower and other vegetables from the Mediterranian.

## Introduction of vegetable/varieties

| Vegetable | Introduction |
| --- | --- |
| Asparagus | Perfection, Selection-841 |
| Bean | Contender, Kentucky Wonder |
| Brussel's sprout | Hills Ideal, Brussel's Dwarf |
| Cabbage | Golden Acre, Pride of India, LLDH |
| Cauliflower | Snowball-16, Improved Japanese |
| Carrot | Nantes, Chatenay |
| Celery | Wrightly Grove Giant |
| Chillies | Hot Portugal, Hungarin Wax. |
| Garden pea | Arkel, Lincoln. |
| Garden beet | Crimson Globe, Detroit Dark Red |
| Knol-khol | White Vienna, Purple Vienna |
| Lettuce | Great Lake, Almo, Chinese Yellow Simpson |
| Onion | Early Grano |
| Parsley | Mosscurled |
| Radish | White Icicle, Japanese White, Chinese Pink, Rapid Red White Tipped |
| Turnip | Purple Top White Globe, Snowball, Golden Ball |
| Watermelon | New Hempshire Midget, Sugarbaby, Charleston Grey, Skipper |

## Selection

Selection in each generation of a number of individuals most nearly approaching the desired type for progenitor of succeeding generations. However, in terms of genetics, discrimination among individuals in the number of offsprings contributed to the next generation is known as *selection*.

## I. Selection made in introduced varieties not desirable for some traits but suitable for few characters

(a) The varieties introduced are used in the hybridization program i.e. for crossing with the local existing material (used as one parent).

| | | |
|---|---|---|
| Carrot | Pusa Kesar] | Nantes was used as one parent |
| | Sel-233] | for their development |
| Turnip | $L_1$ ] | PTWG as one parent |
| | Pusa Kanchan ] | |
| Radish | Sel 271 ] | Japanese white as one parent |
| | Pusa Himani | |
| | Pusa Reshmi | |
| | Punjab Sufaid | |

(b) Some introduced material have been used in inter specific hybrids

Potato-*Solanum demissum* have been used as resistant source for late blight disease.

Tomato-*Lycopersicon piruvianum* against TMV and Root knot nematode resistance

*L. hirsutum*- Leaf mold and TMV.

*L. pimpinellifolium*- Leaf mold, wilt, Buck eye rot.

(c) Some standard varieties have been used for improving sugars and ascorbic acid content

*Abelmoschus manihot*- yellow vein mosaic.

## II. Selection made in the directly introduced material (germplasm) which is heterozygous

### Selection made from the introduced material

| Vegetable | Variety | Selection from | Country |
|-----------|---------|----------------|---------|
| Cabbage | Pusa Drum Head | EC-6774 | Japan |
| Tomato | Punjab Tropic | Complex parentage | |
| | Keck Ruth | Selection | Bulgaria |
| | Sel-120 | -do- | Hawaii |
| | Lal Mani | EC-4532 | - |
| | Solan Surkha | EC-4552 | - |

## Inbreeding

The mating of individuals more closely related than individuals mating at random is known as *inbreeding*. The lines produced by continued inbreeding are known as *inbred lines*. Self-fertilization is the most intense form of inbreeding. In plant breeding nearly homozygous lines are produced by continued self-fertilization accompanied by selection for 5-6 generations. This can be used as the method of breeding only in those crops which do not show any loss of vigour due to inbreeding like cucurbit. The important uses of inbreeding in cross-pollinated crops are: (i) to obtain uniformity in plant characters, (ii) to improve yield, etc. by individual plant selection as in cucurbit in which there is no inbreeding depression and (iii) to combine suitable inbred lines in production of hybrids and synthetics.

## Selection and inbreeding

Selection and inbreeding of individual plants as a method of breeding is limited, it is at the same time an essential part of any breeding program. Whenever a desired type may be obtained by selection; it is not only less expensive but may also require less time than when hybridization is necessary. The material should have broad genetic base from where selections can be made from time to time in future programs.

## Mass selection

Mass selection is a form of selection in which individual plants are selected and the next generation continued from the aggregate of seeds of the selected plants. This is achieved by simply roguing out the undesirable plants from the field and harvesting the remaining plants in mass. Another approach is to tag or mark desirable plants and harvest them in mass. A refinement of mass selection is to harvest the best plants separately and grow individual plant progeny rows. At maturity seeds from similar and superior lines are bulked. As a matter of fact, nucleus seed production in self-pollinated vegetables constitutes this form of selection. Progress in the improvement of varieties by mass selection method is slow and is partly dependent upon the ability of the vegetable breeder to select the best individuals for seed parents. The method, however, has considerable value and has accounted for many vegetable varieties. Seed producers utilize mass selection for the purpose of maintaining the stocks true to type. Change hybrids are frequently the most likely plants to be selected, as they will usually have the best appearance.

## Stratified mass selection

This method was proposed by C.O. Gardener in 1961. In this method, the source population is grown in isolation at low plant density to allow better phenotypic expression by each individual plant. The whole area is divided into sub-plots or grids. Selection for superior plants is carried out in each grid. Seeds from selected plant across the grids are composited. This method is also called as *modified mass selection*.

## Single plant and pure line selection

Genetically pure lines developed by making repeated selections. Some variations may be due to mutations in nature but very rare or natural out-crossing.

A strain homozygous at all loci, normally obtained by successive self-fertilization in vegetable breeding is known as *pure line*. W. Johannsen, a Danish biologist was the first to detail its genetic basis. Pure line breeding generally include three main steps, *viz.* (i) sufficient number of selections need to be made from the original available source population, which is normally a land variety (ii) growing of progeny rows from the individual plant selections obtained in the first step. This evaluation for initial few years/generations is usually based on visual observations and during this process the lines/plants with apparent defects are eliminated. In case of selecting resistant lines, epiphytotic conditions could be created during this stage and (iii) the final stage consists of critical evaluation of remaining lines in the replicated trials to compare the selections among themselves and with established commercial variety. The outstanding line is finally selected and seeds are bulked for further increase. The time taken for pure line selection shall depend on prevailing circumstances, usually it may take at least three years. The breeder has to make his individual selections with little concern about natural crossing. On the other hand, in cross-pollinated species, there is dire need to prevent crossing between different plants. This is accomplished by protecting the flowers during the pollination season with suitable cages. In cabbage and related species it is possible to use the bud-pollination technique. In asparagus, by using progeny test is possible to determine whether or not the high yielding ability of plants is inherited. It is necessary to test out the combining ability of the male parent plants with all of the desirable female parent plant if one is to expect satisfactory results. In spinach there are male, female and hermaphroditic plants and these latter, if caged individually, can be self-pollinated, thereby making it possible to eliminate undesirable types more readily.

## Pedigree method

This method consists of selection among individual plants and their progeny during inbreeding following crosses between selected donor parents. The parents are normally crossed in a single cross system. The $F_2$ populations are subjected to selection for apparent desirable traits *viz.* days to maturity, plant type, plant height, resistance to diseases and other easily identifiable traits. The $F_3$ progenies of the $F_2$ plants are grown in rows. The most desirable $F_3$ progenies' rows are selected/marked followed by selection and harvesting of superior individual plants from each of the selected rows. This practice of selection among rows followed by selections within rows continued till $F_5/F_6$ generation in which desirable rows are bulked within themselves separately and these bulks/new breeding lines are evaluated in the replicated trial for further selection.

## Mass pedigree method

In this method, best individuals with desired characters are selected on the basis of phenotypic performance in a source population. Open-pollinated seeds of the selected individual plants are divided into two halves. Second year replicated progeny row trial is conducted using one set of half seeds from each plant. On the basis of progeny performance, the best parental individuals are identified. The remnant half seeds from the superior parental plants are mixed and grown in isolation for random mating during the third year. This method is called *line breeding* when selection is based on progeny tests and a group of progeny lines is composited. The name, mass pedigree method was given by S.S. Rajan.

## Single seed descent method

It is also called *modified pedigree method* according to C.A. Brim (1966). This method involves single seed procedure, multiple seed procedure and single hill procedure. In this method all the plants of $F_2$ generation are advanced to the next generation by harvesting and growing one seed from each plant. This process continues till $F_5/F_6$ in which superior individual plants are finally selected. The advantages of this method are (i) all the $F_2$ plants get a chance to be represented in the subsequent generations, (ii) adequate level of variability is maintained by in all the generations, (iii) no potentially superior plants are eliminated in the early generation, (iv) two generations can be grown per year as the generations could be advanced in the off-season as plant growth and development is of no consequence in this selection scheme because the productivity of the plant does not affect the genetic makeup of the population advanced by single seed descent as long as plant produces at least one seed and (v) it is cost effective as less resources are required to follow this scheme of selection.

## Bulk method

Growing of genetically diverse populations of self-pollinated vegetables in a bulk plot with or without mass selection followed by single plant selection in $F_5/F_6$ generation is known as *bulk breeding*. According to Nilsson- Ehle (1908), in this method plants of segregating population are harvested together and sample of the seed is used to plant the next generation in $F_5/F_6$ selected individual plants are harvested and planted in individual plant progeny rows, from which superior rows are selected and handled as per procedures outlined in pedigree method of breeding. The advantages of this method are (i) large populations can be grown in each generation, thereby increasing the probability of more gene combinations, (ii) little effort is required to handle anyone cross permitting several crosses to be carried forward, (iii) selections from later generations would breed true as any comparable method of breeding and (iv) more generations can be grown per year involving off-season nurseries since there is a great role of natural selection in bulk breeding.

## Back-cross method

A system of breeding whereby recurrent back crosses are made to one of the parents of a hybrid, accompanied by selection for a specific character (s) is known as *back cross breeding*. In case separate recurrent parents are utilized, then it is called *modified back cross*. In this, the character to be transferred should be simply inherited and retain intense expression in the successive transfers. The only defect observed is that the production potential of the improved variety is fixed at the level of the recurrent variety. Moreover, it excludes the combination of genes of the parental cultivars which is a common feature and advantage inherent in pedigree, bulk methods, etc. The advantages are (i) the improvement through this method is in a stepwise manner and thus, is to a greater extent regulated and not random, (ii) there is no role of environment on this program, (iii) yield evaluation of back cross derived lines is not essential as the derived lines are *at par* with the recurrent parent for most of the characters except the introduced character from non-recurrent parent, (iv) it requires a small number of plants to be handled, (v) the performance of the derived lines is predictable (almost equal to that of recurrent parent) and (vi) it is repeatable.

## Double back-cross

This method of breeding is normally adopted to break the undesirable association between two characters. The $F_1$ is back-crossed with both the parents for 3-4 generations. The back-crossing is accompanied by selection for the character received from the non-recurrent parent. This method was adopted in tomato for combining large fruit size and earliness.

## Multilines breeding

This method is an extension of backcross method and can be called *multilateral back crossing*. It consists of simultaneous back cross program to derive isogenic lines for resistance to diseases in the background of the same recurrent parent. Therefore, each isogenic line will be similar to the recurrent parent but will be differing for resistance to various physiological races of a disease. Mixture of these isogenic lines is called *multiline variety*. In order to avoid drastic changes within a multiline variety over years, the individual back cross derived isogenic lines may have to be separately increased annually and multiline variety synthesized afresh. Multiline component lines should be evaluated individually and in various combinations. The inferior lines should be rejected.

## Mutation breeding

A sudden heritable variation in a gene or in chromosome structure is known as *mutation*. The discovery of effective mutagenic agents like X-rays, gamma-rays, beta-rays, chemical mutagens, etc. The mutations may be natural or artificially induced and have been made use of in vegetables. Now, mutation induction has become an established tool in plant breeding to supplement existing germplasm and to improve cultivars in certain specific traits. Many improved varieties have been released to farmers for many different crop species, demonstrating the economic value of the technology. Limitations arise mainly from the large size of the mutanized population to be screened and from the unsatisfactory selection methods. Both limitations may be eased to some extent by advances in techniques of plant *in vitro* culture. Different varieties react differently to mutagens. Therefore, optimum dose is determined by trial runs if such information is not available. Normally, seed, seedling or any other plant part particularly growing point is irradiated or treated with specific mutagenic agent to induce mutation. Various steps in mutation breeding are as follows: Plants grown from the treated seeds are referred as $M_1$ generation (M=mutation), seed from each $M_1$ plant is grown as single $M_2$ row, consisting of about 20 plants/row, $M_2$ can be grown as bulk also taking 2-3 seeds from each $M_2$ plant. $M_2$ generation is grown under spaced planting and under the conditions that will permit the best expression of the desired character.

## Polyploidy

The development of polyploid varieties, or those with more sets of chromosomes than the two found in diploids, is another method of improvement of some vegetables which offers a certain amount of possibility to the vegetable breeders. This is especially true now that the induction of tetraploid types is relatively easy. Tetraploids are larger than diploids. The improvement by the development of polyploids seem to be gained most easily in cross-pollinated vegetables having a low basic chromosome number. Many vegetables would fit into the category. Generally, the autotetraploids (double diploids) forms are quite sterile and produce much less seed than the normal diploids. The role of polyploidy is better understood and has been used in many vegetable crops e.g. production of seedless watermelon, in this crop tetraploid formed from a homozygous diploid is usually quite fertile and when crossed to a normal diploid produce a triploid type which rarely produce seed. In order to get viable triploid seed, it is necessary to use tetraploid as the female parent. The triploid is more vigorous either the diploid or tetraploid. In onion, the Beltsville Bunching variety resulted from a cross of onion (*Allium cepa*) and Welsh onion (*A. fistulosum*). The number of chromosomes in $F_1$ hybrid were doubled, thereby producing the new variety. This same phenomena has occurred many times in nature as, for example, with leeks and rutabagas which are amphidiploids. It should be pointed out that the mere production of polyploids does not necessarily mean that a better type of plant variety will result. In most cases it is necessary to improve the polyploid type by recombination and selection, thereby increasing its vigour and fertility. The vegetable breeder, then should consider this method as simple tool for use in vegetable crop improvement.

## Composite cross

A mixture of simple crosses is called a *composite cross*. This is a very economical method of handling a large number of crosses. Once such composite is made, it represents a valuable segregating gene pool for placement at stations with limited breeding facilities. However, the original scheme did not allow crossing between crosses, but now the composite crosses can be made by crossing pairs of $F_1$ creating double crosses and still higher order compound crosses.

## Recurrent selection

A method of breeding designed to concentrate favourable genes scattered among a number of individuals by selecting in each generation among progeny produced by matings *inter se* of the selected individuals or their self progeny of the previous generation. Based on the ways in which plants with desirable characters are identified, recurrent selection has been classified into four types:

i)  **Simple recurrent selection or recurrent selection. for phenotype:** In this a number of plants are self-pollinated in a source population in first year. At maturity, superior plants based on phenotypic performance are selected. In the second year, seeds produced by self-fertilization of the selected plants are planted and crossed in all possible combinations and the produce is bulked. This completes original selection cycle. Since selection is based on the phenotype of the plant, it is useful only for characters with high heritability. In case of cabbage and cauliflower, etc. where it is not possible to identify the desired selections before flowering, inter-crosses of selections may be made in the first year of each cycle and the second year may be eliminated from each cycle. Thus, strictly speaking, selfing is not an integral component of simple recurrent selection, rather it is done to prevent crossing from the inferior pollen grains before the plants reach to selection stage.

ii)  **Recurrent selection for general combining ability (gca):** Here a three-year cycle is involved. In first year, a number of plants are self-pollinated and crossed to a broad-based heterozygous tester stock to identify the so-called plants with good general combining ability. In second year, the crosses are evaluated to identify those that are superior. Selfs of first year are kept in reserve. In third year, the reserve selfed seed are grown out, inter-crossed in all possible combinations, and a composite of inter-crossed seed is used to establish an improved population for further selection.

iii)  **Recurrent selection for specific combining ability (sca):** This method is the same as that of recurrent selection for general combining ability except that the tester selected is a narrow base inbred line. The recurrent selection for general and specific combining ability is equivalent to half sib progeny.

iv) **Reciprocal recurrent selection:** Aims at simultaneous improvement of two heterozygous and heterogeneous population, one acts as a tester for another and *vice versa*. This method is as effective as recurrent selection for sca when non-additive effects are of major importance.

## Synthetic variety

A variety which is maintained from open-pollinated seed following its synthesis by hybridization in all combinations among a number of selected genotypes which have been tested for combining ability. The components of a synthetic variety may be inbreds (normally), clones, mass selected populations or various other materials. The component units are maintained so that the synthetic may be reconstituted at a regular intervals. The inbreds to be used as a component lines are chosen on the basis of combining ability tests. The component inbreds are crossed in all possible combinations. This inter-crossed seed is called as $S_0$. Equal quantity of seed from all crosses is composited and the mixture is allowed open-pollination in isolation and seed is harvested. This becomes $S_1$ generation. In absence of reconstitution of a synthetic at regular intervals, the population becomes an open-pollinated variety.

## Composite variety

A number of open pollinated varieties (which are properly tested lines/varieties) or lines are allowed to cross themselves. Composite varieties are generally derived from the varietal crosses in advanced generation. These are usually developed from open-pollinated varieties or other heterozygous populations or germplasm which have originally not been subjected to inbreeding or have not been elaborately tested for their combining ability. Usually they involve open-pollinated varieties, synthetics, double crosses, etc. selected for yield performance, maturity, resistance to diseases and pests. These composites often a high order heterosis in $F_1$'s when widely diverse populations are crossed. Advanced generations of such heterotic crosses often show stabilized yields. GCA and additive gene effects play predominant role in exploitation of these populations.

## Hybridization

Often sufficient progress cannot be made by selection alone, so it becomes necessary to combine characteristics of varieties, sub-species or even species to bring about the improvement that is desired. The vegetable breeder needs to ascertain his objective and then to acquaint himself with the material that is available. Variation will come from different varieties, but possible use of different species should not be overlooked. It should be realized that the less diverse the parent material, the easier will be the selection following the hybridization program. Many plant characters are rather simple in their inheritance, since they are determined by a single pair of factors. On the other hand some characters are dependent upon the interaction of several factors. Yield is controlled by a combination of factors. Regardless of the character, it is beneficial for the vegetable breeder to know its inheritance to conduct a more intelligent breeding program. However, much vegetable breeding is done without adequate knowledge of the inheritance of many of the characters.

Each species, and often each variety, has its own morphological and physiological peculiarities which require a different method of crossing. The pollination technique for many vegetables has already been worked out, whereas in others the breeder may need to determine this for himself. It is essential that the flowering periods of the parents overlap, for normally the length of time the pollen is viable and stigmas are receptive is relatively short. Whenever there is no overlapping, the treatment given the parent material may need to be varied or manipulated by adjusting the sowing time or enhancing or delaying by the use of growth regulators to bring the pollen maturity and stigma receptivity to coincide at the time of pollination.

As a result of gene combination in the progeny following a cross, it is possible to select desirable segregates, possessing a combination of useful genes from two or more parents involved in the cross. In view of the procedural difficulties encountered sometimes in making a cross, the large area required for growing the hybrid material and considerable time in evolving a variety, it is not advisable to initiate a hybridization program if the desired objective can be achieved through other breeding methods like introduction and selection. Moreover, starting hybridization, it is important to get fully acquainted with the performance and chief attributes of different vegetable varieties and related species available so that the most appropriate parents could be chosen for crossing, if required. A well-defined objective, a complete knowledge of the performance and chief characteristics of the parents and the pattern of inheritance of the characters under study, the correct choice of parents, an appropriate size of segregating population to be grown, a careful selection in the hybrid material and adequate testing of promising hybrid selections will generally assure the success of a hybridization program. The main steps followed are :

o    Selection in the hybrid material and bulk method.

o    Pedigree method.

o    Size of population and number of families selected.

o    Observations to be recorded.

- Final evaluation and selection.
- International crosses in the improvement of yield.
- Location testing for performance.
- Release of the variety if appropriate.

## General combining ability

The average performance of the line (inbred x variety). In case of cross-pollinated vegetable crops it is known as *top cross*. This is due to the additive effects (intra-type-fixable) and additive x additive (interallelic-pallially-fixable). The elite base varieties can be then tested for their gca in diallel crosses or partial diallel crosses or by top cross with the best indigenous variety or a varietal complex as a tester.

## Specific combining ability

Inter-allelic and non-fixable crossing will be allowed in open-pollinated lines only. For this the vegetable breeder will have to see which combination is better. So the breeder should select a best combination that can make specific. combination.

**Diallel crosses:** Cole and root vegetables *viz*. radish and turnip.

**Partial diallel crosses:** Cole vegetables.

**Poly-crosses :** Cole vegetables.

**Top-cross :** Garden beet, *palak*, beet leaf and spinach.

**Both diallel and top crosses:** Cucurbit.

## Open-pollinated varieties

These are perpetuated by open-pollination and the genetic constitution remains the same i.e. constant from generation to generation.

## Complexes

If the varieties of inbreds are not tested and allowed to mix together are called complexes.

## Hybrids

The term hybrid variety is used to designate $F_1$ population that are used for commercial planting. The $F_1$'s are obtained by crossing genetically unlike parents.

## Hybrid varieties

2, 3 or 4 inbred lines or varieties are involved. The lines are tested for their general combining ability and then the crosses are made.

### Single cross

Two inbreds are crossed A x B=AB (hybrid). A hybrid progeny from a cross between two unrelated inbreds.

### Three way cross

Three way cross is the hybrid progeny from a cross between a single cross and an inbred e.g. (A x B) x C.

### Double cross or double hybrid

Two inbreds are crossed, similarly another two are crossed and then their $F_1$ hybrids are crossed e.g. (A x B) x (C x D) = (AB x CD) = ABCD.

### Modified single cross

A modified single cross is the hybrid progeny from a three-way cross which utilizes the progeny from two related inbreds as the seed parent and an unrelated inbred as the pollen parent e.g. (A x A) x B.

### Double modified single cross

It is the hybrid progeny from two single crosses, each developed by crossing two related inbreds e.g. (A x A') (B x B').

**Triple cross**

Two inbred A x B are crossed to produce AB $F_1$ hybrid (0) will be further crossed with another inbred (C as 0) gives ABC.

**Modified three way hybrid**

It is the progeny of a single cross as female parent and another single cross between two related inbreds e.g. (A x B) x (C x C).

**Top cross**

Inbred line A x tester (a variety) or inbred line x variety tested for its gca.

**Double top crosses**

To inbred lines are crossed first and then the $F_1$ hybrid crossed with a tester variety e.g. A x B = $F_1$ x tester variety.

**Advanced generation hybrid/cross**

(A x B) = $F_1$ (C x D) = $F_2$, (F, x $F_2$) = $F_1$ $F_2$

**Varietal cross**

Variety x variety or Inbred x Inbred.

**Multiple cross**

Crossing of a large number of inbred lines or varieties.

**Synthetic**

Combination of so many inbred lines followed by a successive pollination.

**Poly-cross**

In case of inbred lines which cannot be crossed easily. Planting is done in such a way to get equal chance to cross with each other. Maintain the lines by vegetative propagation.

**Back-cross**

A x B = $F_1$ x A.

**Single back-cross**

[(A x B) (C x D) x C]

## Breeding objectives and methods in vegetables

| Vegetable | Breeding objectives | Methods |
|---|---|---|
| Tomato | Increased fruit yield, earliness, fruit quality suitable for fresh market, suitable for home garden with a regular continuous growing, suitable for processing, indeterminate cultivars for green-house production, resistance to diseases, insects and abiotic stresses-drought, salt and herbicide tolerance, low temperature, germination and growth, chilling injury tolerance and cold or hot set varieties, ability to withstand marketing pressure. | Introduction, pure line selection, pedigree method, bulk method, single seed descent method and back-cross method. |
| Garden pea | High green pod yield, long attractive green pods with more seeds/pod, sweetness, high shelling percentage, specific maturity-early, mid and late, suitable for freezing and canning, resistance/tolerance to abiotic factor-frost, resistance to diseases-downy mildew, powdery mildew rust and wilt, resistance to insects-leaf miner, aphids, pod borer and pea stem fly. | Pedigree method, bulk method, single seed descent (SSD), back cross method, mutation breeding. |
| French bean | High pod yield, non-stringy long pods flat or round in shape, early pod harvesting, bush/pole plant type, high number of pods/plant, high number of pod cluster/plant, free from interlobular space, photo- insensitivity, wider adaptability, resistance to diseases-bean common mosaic, virus, *Fusarium* root rot, white mold, anthracnose, angular leaf spot, resistance to insects-pod borer, pea stem fly, tolerance to environmental stress. | Pedigree selection, single plant selection and bulk breeding. |

| Vegetable | Breeding objectives | Methods |
|---|---|---|
| Vegetable Cowpea | High green pod/seed yield, dual purpose (vegetable and pulse), earliness, appropriate plant type-erect, determinate, wider adaptability, photo-insensitive, short, tender pods for whole pod for processing, long tender and stringless pods for fresh consumption, varieties suitable for inter- cropping, resistance to diseases-anthracnose, cercospora leaf spot, powdery mildew, *Fusarium* wilt, ascochyta blight, bacterial blight bacterial postules, cow pea yellow mosaic virus, resistance to insects-hairy caterpillar, leaf hopper, aphids, thrips, bruichids pod borer and pod sucking insects. | Pedigree method, back-cross method, combination of both and mutation breeding. |
| Methi | Earliness, high yield, better quality leaf characters-dark green colour, tender, better flavour, quick rejuvenation, resistance to diseases and insects. | Introduction, selection, bulk selection and hybridization. |
| Lettuce | High leaf yield, leaf shape, size, colour, texture and crispness, quality in both non-heading and heading types, butter head type, bolting resistance, richness in nutritive value, uniformity, adaptation to specific environment, resistance to diseases-mildew, big vein, sclerotinia drop, lettuce mosaic, tip bum, resistance to insects-cabbage looper, resistance/tolerance to abiotic stress-salinity. | Introduction, selection, inbreeding, back-cross method, pedigree selection and mutation breeding. |
| Cauliflower | High yield, non-ricey, compact curds with retentive cream/white colour, suitable varieties for off-season in hills, better seeding ability, heat tolerance for producing curds in August- September, self-incompatible but cross-compatible inbred to produce hybrids of tropical types, resistance to diseases-black rot, *Sclerotinia* rot, *Alternaria* blight, *Erwinia* rot. | Mass selection, family selection, simple pedigree method, bulk method, back-cross method and hybrid breeding. |

| Vegetable | Breeding objectives | Methods |
|---|---|---|
| Cabbage | High yield, longer staying capacity in field after head foundation, narrow, short and soft core, short stem, cultivars suitable to grow under mild summer, self-incompatible but cross-compatible inbred to be used in development of hybrids, resistance to disease-black rot, cabbage yellows, resistance/tolerance to insects-cabbage butter fly, caterpillar, aphids and diamond black moth. | Mass selection, inbreeding, hybrid breeding, (single, double and top cross). |
| Knol-khol | High yield, spherical, swollen bulb like stem, quality-soft, mature bulbs at edible stage, free from early bolters, tolerance to frost, resistance to *Fusarium* wilt, *Xanthomonas campestris* and club root. | Mass selection, line breeding, family selection & hybrid breeding. |
| Sprouting Broccoli | High yield, suitable varieties both in green and purple colour heads, better seeding ability, resistance/tolerance to diseases, insects and abiotic stress. | Mass selection, family breeding and bulk method. |
| Chinese Cabbage | Earliness, high green and seed yield, quality-leaf shape, size, deep green colour and crispness, more number of leaves producing capacity per plant and cuttings, better cooking quality, resistance to diseases, insects and abiotic stress. | Mass selection, inbreeding and hybrid breeding. |
| Radish | Early rooting, high yield, quality-white, long/stump roots with thin root and non-branching habit, non-pithy roots, pungency of roots as per consumer's preference, slow bolting habit, resistance/tolerance to abiotic stress-heat, drought, wetness, resistance to *Alternaria* blight, white rust, radish mosaic virus and tolerance to aphids. | Mass selection, hybrid breeding-single, double and three way crosses. |

| Vegetable | Breeding objectives | Methods |
|---|---|---|
| Carrot | High root yield, scarlet/orange colour of roots, high carotene content in roots, uniformity in root shape and size, thick fleshy roots, thin and self coloured core in roots, broad shouldered, cylindrical, uniformly tapering or stump rooted carrot with non-branching habit, early rooting, cracking free roots, higher sugar and dry matter in roots, slow bolting habit, smooth root surface, resistance to *Alternaria* blight and *Cercospora* leaf blight. | Mass selection, and pedigree selection, inbreeding depression and hybrid breeding. |
| Turnip | Earliness, root colour as per consumer's preference-white, purple types more liked in India, stump rooted varieties with their thin tap root and non-branching habit, appropriate dry matter (8-9%) in roots, resistance to club root, powdery mildew, turnip mosaic virus, white rust, phyllody, cabbage root fly and turnip root fly. | Mass selection and hybrid breeding. |
| Celery | Earliness, commercial potential for salad and spice, quality-colour, flavour, taste, length of petiole, tenderness, crispness, non-pithiness, lack of ribbing cross section, shape and flavour, high yield, volatile oil in seed, resistance/tolerance to diseases-early blight, late blight, *Fusarium* yellow, pink rot and leaf curl. | Introduction, selection and bulk method. |
| Parsley | High yield, earliness, better quality for salad and spice, flavour, taste, tenderness, green colour, crispness, resistance to diseases and insects. | Introduction and selection. |

| Vegetable | Breeding objectives | Methods |
|---|---|---|
| Onion | High yield, longer bulb storage life, bulb quality-size, shape, colour, pungency, firmness, dormancy, dormancy is important for prolonged storage life, high TSS is important for processing industry-chips, powder, etc. pungency-S-alkyl cysteine Isulfoxide precursors and the enzyme allinase contribute to the yield of sulphur compounds in onion bulbs, resistance to diseases-purple blotch, basal rot, stemphylium blight, bacterial storage rot, resistance to thrips, resistance to abiotic stresses-moisture stress, high temperature, salinity and alkalinity. | Mass selection, modified mass selection, family selection and hybrid breeding. |
| Leek | High yield, better quality. resistance to bolting, dark green leaf blade, freedom from bulging, long shaft (the blanched pseudo-stem formed by the young folder leaf blade), resistance to leek rust and yellow stripe virus. | Mass selection and heterosis breeding. |
| Cucumber | Early fruiting, high female to male sex ratio, attractive green or dark green fruits with smooth surface ‹ and without prominent spines or prickles, uniform, long, cylindrical shape without crook neck, fruits free from carpel, separation without hollow spots, fruits free from bitterness, less seeds at edible maturity, resistance to powdery mildew, downy mildew, anthracnose, cucumber mosaic virus, resistance to fruit fly and abiotic stresses. | Inbreeding, back-cross method, pedigree method, use of sex inheritance and chemicals in breeding, hybrid breeding. |
| Muskmelon | Attractive round/spherical fruit shape, thick flesh with attractive colour, small seed cavity, quality fruit-sweet, juicy, musky flavoured fruits, TSS not less than 10%, tough netted skin of fruit, high early and total marketable yield, resistance to diseases-powdery mildew, downy mildew, virus, resistance to insects-red pumpkin beetle, fruit fly and aphids. | Controlled inbreeding, pedigree method, back-cross method and hetcrosis breeding. |

| Vegetable | Breeding objectives | Methods |
|---|---|---|
| Watermelon | Earliness, pistillate flowers at lower node, number, tough skinned, fruits for long distance transportation, TSS content not less than 10%, fruits with smaller and fewer seeds or seedless with attractive deep red flesh, intermediate fruit shape between typical long and round, high yield, resistance to diseases-virus, *Fusarium* wilt, anthracnose, powdery mildew, resistance to cucumber aphid, fruit fly, cucumber beetle, red pumpkin beetle. | High breeding. (3x and 2x interplanted-stimulative parthenocarpy-seedless) |
| Squash and Pumpkin | High fruit yield, early fruiting, first pistillate flower at early node, number, high female to male ratio, yellow or mottled skin of fruit, non-ridged fruit surface, resistance to diseases, insects and abiotic stresses. | Inbreeding, hybrid breeding (i) manual pollination/ use of insects, (ii) use of chemicals and interspecific hybridization. |
| Bottle Gourd | High yield, greater fruit number, more fruit weight, earliness (appearance of pistillate flowers at early node number), high female : male flower ratio, round, long club-shaped fruit, sparse hairs persisting on skin, non-fibrous flesh at edible stage, non-bitter fruit, attractive, green fruit, resistance to powdery mildew, resistance to red pumpkin beetle and fruit fly. | Inbreeding, pedigree method, bulk method, back-cross method, heterosis breeding and resistance breeding. |
| Bitter Gourd | Earliness, high female to male sex ratio, whitish green to glossy green fruit colour depending upon consumer's preference, less ridged fruit surface, thick fruits particularly suitable for stuffing, fruit size variation as per consumer's preference, Immature seeds for longer period during green edible stage, high yield (in terms of both fruit weight and number), resistance to red pumpkin beetle and fruit fly. | Inbreeding, pure line selection (in segregation generations) and heterosis breeding. |

| Vegetable | Breeding objectives | Methods |
|---|---|---|
| Ridge and Sponge Gourd | Earliness, high female to male sex ratio, uniform, thick cylindrical fruits, free from bitterness, tender non-fibrous fruits for longer time, high fruit yield in terms of both weight and number, resistance to powdery mildew, downy mildew and insects. | Inbreeding, selection, crossing of complementary parental lines and handling of segregating generations through pedigree, bulk, back-cross and single seed descent method & heterosis breeding |
| Beet leaf | High leaf and seed yield, delayed bolters, high quality-leaf and stem colour, deep green, enlarged shape and tenderness, quick rejuvenation, resistance to diseases-damping-off, *Cercospora* leaf spot, mildew and rust, resistance to insects-aphids, caterpillars and beetles. | Introduction, selection, mass selection and heterosis breeding. |
| Spinach | High green and seed yield, quality-green, broad and thick leaves, better rosette of leaves both number and size, delayed bolting, resistance to diseases, insects and abiotic stress. | Introduction, mass selection and hybridization. |
| Vegetable Amaranth | Earliness; high leaf and seed yield, non/slow' bolting habit, low content of oxalates and nitrates, leaf, stem and petiole colour, green/purplish green, tenderness, plant habit, short internodal length, wider adaptability and resistance to amaranth leaf webber. | Mass selection |
| Garden Beet | High root yield, quality-dark red, uniformly coloured roots, uniform root shape, slow bolting habit, absence of internal white rings in roots, monogerm seed, resistance to downy mildew and powdery mildew. | Inbreeding, mass selection, progeny breeding and mutation breeding. |

| Vegetable | Breeding objectives | Methods |
|-----------|---------------------|---------|
| Okra | High pod yield, quality-dark green, tender, thin, medium, long, smooth, 4-5 ridged pods at marketable stage, pods free from conspicuous hairs, early and prolonged harvest, short plants with more number of nodes, short internodes, pods suitable for processing industry and export market, resistance to diseases-YVMV, *Fusarium* wilt, *Cercospora* leaf spot, fruit rot, resistance/tolerance to insects-fruit and shoot borer, jassids and fruit fly. Tolerance to abiotic stress-low temperature, excessive rains, saline and alkaline soils. | Introduction, pure line selection, pedigree method, mutation breeding and heterosis breeding. |
| Brinjal | High yield, earliness, fruit shape, size and colour as per consumer's preference, low proportion of seed, soft flesh, lower alanine content, upright, sturdy plant, free from lodging, resistance to diseases and insects. | Pure line selection, pedigree method, bulk method, modified pedigree method (single seed descent), combination of bulk and pedigree methods, back-crossing and heterosis breeding. |
| *Shimla Mirch* | Earliness, increased yield, desirable (oblate or round) fruit shape and size, superior fruit quality-pleasing flavour, high sugar/acid ratio, high pigment content and vitamin C, resistance to diseases fruit rot, *Cercospora* leaf spot, powdery mildew, bacterial leaf spot, *P. hytophthora* root rot, root knot nematode, common TMV, resistance/tolerance to insects-thrips, mites, aphids, fruit borer, resisance tolerance to abiotic stresses-heat, water stress and salinity. | Pure line selection, pedigree method, back-cross method, heterosis method and mutation breeding. |

| Vegetable | Breeding objectives | Methods |
|---|---|---|
| Chillies | Earliness, desirable fruit shape and size-long fruits, superior fruit quality-high capsaicin (a fat soluble flavourless, odourless and colourless compound), high oleoresin, resistance to diseases (as in *Shimla mirch*), resistance/tolerance to insects (as in *Shimla mirch*), resistance to abiotic stresses-heat, water stress & salinity. | Pure line selection, pedigree method, back-cross method, heterosis breeding, mutation breeding. |
| Paprika | Earliness, high yield, increasing productivity, shortening the vegetation period, high colour content, high pigment content, low or no pungency, resistance to diseases-anthracnose, fruit rot, powdery mildew, bacterial leaf spot, virus complex, resistance to insects-aphids, thrips, mites, fruit borer or hairy caterpillars, resistance to blossom end rot, sun scald and abiotic stresses. | Introductions, pure line selections, heterosis breeding and pedigree method. |
| Lima bean | Earliness, high pod yield, quality pods-tender, stringless, wider adaptability, tolerance to abiotic stresses-high temperature and frost, resistance to diseases and insects. | Pedigree method, single plant selection and bulk method. |
| Potato (Tuber) | Higher tuber yield, earliness, photoperiod insensitivity, responsive to fertilizer, better keeping quality (resistance/tolerance) against shrinkage, rottage, accumulation oil sugars especially reducing sugars and reasonable dormancy, better quality tubers-(i) round, medium sized with shallow eyes and free from greening for general consumption, (ii) high vitamin C and protein content, (iii) high specific gravity (dry matter content), suitable for French fries, chips and dehydrated products, (iv) low sugar content for chips and French fries to avoid browning, resistance to diseases-Late blight, early blight, charcoal rot, wart common scab, bacterial wilt, soft rot, viral diseases and nematodes (cyst and root-knot nematode), resistance/tolerance to aphids and potato-tuber worm resistance, tolerance to heat, drought, frost and soil salinity . | Selection, induced mutations, di-haploid breeding and biotechnological techniques. |

| Vegetable | Breeding objectives | Methods |
|---|---|---|
| TPS | Avoid bulky tuber transportation, avoid diseases common in seed tubers, elimination of storage losses in seed tubers. | Hybrid breeding. |
| Cassava | High yield (>35 t/ha fresh root), high starch (>25%), high harvest index, responsive to additional inputs, unbranching or late branching plant type, low IICN content, good cooking and eating quality, early harvestability, better root storage quality, wider adaptation, compact branches, compact root system, shade tolerance for use as an inter-crop under coconut etc., resistance to major diseases-cassava bacterial blight, anthracnose, brown leaf spot, tolerance to adverse soil and climatic conditions. | Introduction, selection, intervarietal hybridization, heterosis breeding, polyploidy, induced mutations and biotechnological techniques. |
| Sweet Potato | High yield, high quality (high dry matter, orange flesh), good storage characteristics, good field performance, resistance to field (*Fusarium* wilt and root knot nematode) and storage diseases, high sprout production, resistance to pests-sweet potato weevil and white grubs. | Introductions, selection, mass selection, pedigree breeding, polyploidy, somatic hybridization & poly-cross method. |
| Colocasia | High yield, increase in nutritive value of corm, petiole and leaves, quality green leaves, shape, size and colour of corms, acidity free varieties, resistance to diseases *Phytophthora* leaf blight, mosaic, *Pythium* rot, resistance to insects-leaf aphids, thrips, spider, mites, leaf eating caterpillars. | Selection and mutation. |
| Yam | High yield, quality flesh white or yellow colour, tuber shape-oval or irregular, less acidity, more tuber cluster/plant, resistance to diseases-leaf spot/blight, resistance to insects-scale and tolerance to frost. | Selection and induced mutation. |

| Vegetable | Breeding objectives | Methods |
|---|---|---|
| Elephant's foot (Zimikand) | High yield, resistance to abiotic stress-waterlogging, resistance to diseases-*Sclerotium* rot, bacterial spot. | Selection and mutation breeding. |
| Asparagus | High spears yield, quality of spears-tenderness, shape and size, less branched spear, wider adaptation, seed production ability, production of male hybrids, resistance to diseases-*Fusarium* wilt, rust resistance to insects and abiotic stressees. | Introduction, selection, induced mutation, breeding male asparagus hybrids and biotechnology. |

# Centre of origin

Each of the world's basic vegetable crop originated in a relatively confined geographic region. There is considerable evidence that the cultivated vegetable crops were not distributed uniformly throughout the world. Even today, certain areas show far greater diversity than others in the forms of certain cultivated crops and their wild relatives. N.I. Vavilov (1951), a Russian scientist surveyed the world and divided the whole world into eight main centres of origin. Origin in simple words mean how a particular vegetable has been originated and domesticated for the welfare of living beings. A centre or origin is the place of maximum dominant factors exists in the population of the crop/vegetable and maximum diversity of that particular vegetable. The recessive genes are on the periphery of that centre of origin. These were the main convections followed. The recessive genes arise by mutation in nature. This is also called '*primary centre of origin*'. When two or more species crossed with each other and artificial natural selections occur in the population is called '*secondary centre of origin*' for example, several species of *Brassica* and *Allium*. Harlan (1954) called these pockets or valleys small local areas where maximum diversity plus dominant genes existed as '*microcentres*'.

## Primary and secondary centre of origin of vegetable crops

| Gen centre | Primary centre | Secondary centre |
|---|---|---|
| Chinese-Japanese | Brinjal, wax gourd, Chinese cabbage | Water melon, amaranth |
| Indo-Chinese | Sponge gourd, ridgegourd, bitter gourd, sword bean, winged bean, taro, chayote, cucumber | Chinese cabbage, bottle gourd, yam bean, amaranth |
| Hindustan Centre | Brinjal, wax gourd, cucumber, ridge gourd, sponge gourd, bitter gourd, hyacinth bean, drumstick, okra | Water melon, rosella, bottle gourd, amaranth |
| Central Asia | Onion, garlic, carrot, spinach | Brinjal, watermelon, cauliflower, okra |
| Near East | Onion, garlic, leek, garden beet | Okra |
| Mediterranean | Cabbage, cauliflower, broccoli, radish | Sweet pepper, garlic, orka |
| Africa | Brinjal, watermelon, bottle gourd, cowpea, okra, rosella | Onion, lima bean, amaranth |
| Central America and Mexican region | Tomato, hot pepper, pumpkin, squashes, yam bean, sweet potato, common bean | - |
| South American region | Tomato, hot pepper, cassava, pumpkin, lima bean, chayote | Common bean |
| North American region | | Tomato, brinjal, watermelon squashes, onion, lettuce, lima bean |

Moreover, the evolutionary processes proceed at more rapid rate than the larger geographical areas/regions. About 400 species constitute the global diversity in vegetable crops.

The regions overlap for a number of crops. But one major and three minor centres in the old world (Africa, Asia and Europe) and new world (America) respectively, have been identified as the areas of the origin and diversification of the vast majority of vegetable crops.

## Centres of origin of vegetable crops

| Centre of origin | Vegetable crops |
|---|---|
| Ethiopia | Bottle gourd, Celery, Cucumber, Cowpea, Fenugreek, Garden pea, Okra, Onion, Watermelon, West Indian Gherkin. |
| Mediterranean | Asparagus, Cabbage, Beets, Cauliflower, Celery, Garlic, Kale, Knol-khol, Leak, Lettuce, Onion Turnip. |
| Asia Minor | Beans, Beets, Sprouting broccoli, Cabbage, Carrot, Kale, Lettuce, Pea, Brussel's sprouts. |
| Central Asiatic | Bottle gourd, Broad bean, Carrot, Afghanistan, Garlic, Onion, Pea, Spinach, Turkistan Turnip, Welsh onion. |
| Indo-Burma | Amaranthus, Ash gourd, Brinjal, Bitter gourd, Cluster bean, Colocasia, Cowpea, Cucumber, Dolichos bean, Elephant's foot, Indian spinach, Long melon, Muskmelon, Pointed gourd, Ridge |

| | |
|---|---|
| | gourd, Round melon, Smooth gourd, Snake gourd, Snap melon, Spinach beet, Winged bean, Yam. |
| Siam, Malaya, Java | Ash gourd, Colocasia, Elephant's foot, Yam. |
| China | Cowpea, French bean, Turnip, Muskmelon, Radish. |
| Mexico-Guatemala | Amaranthus, Chilli, Chow-chow, French bean, Jack bean, Lima bean, Squash, Sweet potato, Tomato. |
| Peru, Ecuador, | Chilli, French bean, Lima bean, Bolvia Potato, Pumpkin, Squash, Tomato. |
| South Chile | Potato. |
| Brazil, Paraguay | Tapioca. |
| United States | Globe artichoke, Jerusalem artichoke. |

# THIRTEEN

## Vegetable Biotechnology

Plant biotechnology has recently created unprecedented opportunities, not only for the manipulation of plant biological systems for the benefit of human beings, but also under studies for better understanding of its fundamental life process. Consequently it has become the fastest and rapidly growing technology in the world. It has the potential to increase food productivity, reduce dependency of agriculture on chemicals, lower the cost of raw material and reduce the negative environmental impacts associated with traditional production methods.

### Definition

Plant biotechnology may be defined as, "generation of useful products or services from plant cells, tissues and, often organs (very small organ explants)".

### Uses of plant biotechnology

- Production of plants rapidly on large scale by tissue culture technique.

- Production of disease free material by meristem culture.

- Somaclonal variation to broaden the gene base for the isolation of resistance at cell level to disease, soil salinity and acidity etc.

- To add desirable genes to recipient species through vector like plastids virus, plasmids etc.

- Gene cloning of desirable materials for incorporation in the common cultivars.

- The somatic hybrid in case of distant species.
- Embryo rescue culture for making distant crosses.
- Industrial use.

## Techniques of plant biotechnology

### (A) Tissue culture

The growth of tissues of living plants in a suitable culture medium (*in vitro*) is known as tissue culture or *in vitro* technique.

**Classification of tissue culture techniques in vegetables**

| Types of tissue culture | Vegetable crops |
| --- | --- |
| Callus culture | Lettuce, sweet potato, cucumber |
| Cell culture | Garlic |
| Organ culture | Lettuce, tomato, brinjal, sweet potato, cucumber |
| Meristem culture | Onion, cauliflower |
| Embryo culture | Tomato, okra, French bean, *Brassica*, pumpkin |
| Anther or pollen culture | Potato, tomato, peas, asparagus, French bean, *Brassica*, capsicum |

### Callus culture

This is the process where by small pieces of living tissues (explants) are isolated from an organism and grown aseptically for indefinite period on a medium. The usual explants are buds, root tips, nodal segments and germinating seeds and these are placed on suitable culture media where they grow into an undifferentiated mass known as *callus*. Once established it can be propagated indefinitely by subdivisions.

### Cell culture

Regeneration of a plant from a single cell in nutrient medium. It may include somatic cell or germinal cell pollen.

Callus is transferred to a liquid medium and agitated, the cell mass breaks up to give a suspension of isolated cells, small clusters of cells and much larger aggregates. Such suspensions can be maintained indefinitely by·subculture but, by virtue of the presence of aggregates they are extremely heterogeneous. Then placed in a suitable medium isolated single cells from suspension culture are capable of division.

## Organ culture

Regeneration of a plant from an organ which has separate identity such as anther, ovule, embryo and bud. Apical shoot tips are surface sterilized and placed on growth medium lacking plant hormones, they will develop into single seedling. If the medium supplemented with cytokinins, axillary shoots will emerge from their normal positions in the leaf axils and produce a shoot cluster. These clusters can be subdivided into smaller clumps of shoots which will, in turn, form similar culture also include

## Regeneration of plantlets in vegetables

| Vegetable | Organs |
| --- | --- |
| Potato | Shoot, buds, roots |
| Brinjal | Hypocotyl, embryo |
| Sweet potato | Shoot, buds, leaf, stem, shoot tips, root |
| Ginger | Buds from rhizomes |
| Tomato | Cotyledons, stems internode, leaf, shoots, apical meristem, hypocotyle |
| Capsicum | Cotyledon, internode, shoots, stems, apical meristem, hypocotyle |
| Asparagus | Stem segments, shoot tips |
| Garden pea | Shoot apices, epicotyle segment |
| Bean | Root, hypocotyl, cotyledon |
| Onion | Axillary and adventitious shoots |
| Garlic | Shoot, buds |
| Knol-khol & Brussel's sprouts | Stem section, axillary buds |
| Cauliflower | Curd, leaf veins, lamina and cotyledon |
| Cabbage | Shoot, buds, meristem tip |
| Chicory | Leaf vein segment, storage root (chicon) |
| Cucumber | Main lateral apices |

embryo, anther, ovary culture on nutrient medium.

## Meristem culture

Regeneration of a plant from tissues of an actively dividing organ like stem tip, root tip or vegetative bud. A septic culture of shoot apical meristem on nutrient media for the production of complete plant. Apical meristem are excised from the shoot tip and placed on media. Meristem is not capable or growing independently unless some leaf primordia are retained below it. By this method plants can be rapidly multiplied, regenerated plants are genetically similar to the donar plants and pathogen free plants especially viruses can be produced.

## Embryo culture

Removal of developing young embryos from seeds and their cultivation *in vitro*. Exchange of genes between cultivated and wild species and transfer of resistant gene(s) from wild species to cultivated varieties is hampered due to certain barriers in sexual hybridization. Therefore, the *in vitro* method has been successfully used to overcome such barriers in distant hybridization. In this case, hybrid embryo are rescued prior to their abortion and cultured directly into artificial media. The embryo culture technique has been practically utilized for proper development of fertilized embryos in following cases :

- *Lycopersicon esculentum* x *L. peruvianum*
- *Capsicum pendulum* x *C. annum*
- *Phaseolus vulgaris* x *P. acultifolius*
- *Cucurbita pepo* x *C. ecuadorensis*
- *Cucurbita metulifenus* x *C. zeyheri*
- *Abelmoschus esculentus* x *A. moschatus*

● *Abelmoschus esculentus* x *A. ficulneus*

## Anther/pollen culture

Culture of anthers (or pollen grains) on a suitable medium (*in vitro*) for the production of callus and/or haploid plants. This technique has been used to obtain haploid in many vegetable crops e.g. potato, brinjal, tomato, capsicum, pea, French bean, muskmelon and sprouting broccoli.

## Protoplast culture

Regeneration of a plant from naked single cell in a culture medium. Protoplasts are ideal material for genetic transformation because their naked plasma membrane allow entry of foreign DNA, cell organelles, bacteria or virus particles.

## Somatic hybridization

Somatic hybridization allows fusion of complete cytoplasm fusion of two parents. The cytoplasm mix, obtained offers the opportunity of producing hybrids. Somatic cell fusions have been successfully employed for resynthesis of *Brassica napus* by fusing protoplasts of *B. oleraca* and *B. compestris*. Cytoplasmic male sterile gene(s) can also be transferred in vegetables like onion and carrots to produce hybrid seed cheaply. Hexaploid hybrids resulting from protoplast fusion of *Solanum brevidem* x *S. tuberosum*, which show resistance to leaf roll virus and race 'O' of late blight. Also protoplast fusion of *Solanum melongena* x *S. sisubrifolium* show resistance to phomopsis blight. Fusion between sexually incompatible member of the same family have produced hybrid plants that retain some chromosomes numbers of both parents. The best known example is the 'pomato' which is tomato-potato somatic hybrid.

## Applications of tissue culture

### Micropropagation

Axillary shoot cultures of potato when cultivated in presence of appropriate level of cytokinins and gibberellic acid they form large number of very small tubers. These mini tubers can be sown directly in the field and will generate normal plants.

### Recovery of virus free stocks

The basic method of obtaining virus free plants is culture of apical meristems. If a small enough piece is taken and cultured preferably less than the first half millimeter of growing tip, *in vitro* and then used to regenerate virus free plants.

### Germplasm conservation

Tissue culture can be used to maintain the germplasm in the plants which are unable to produce self-incompatibility mechanisms. This has advantages, such as it occupies a small space and has the possibility of elimination of insect-pests and diseases. The conventional storage of germplasm is very expensive, time consuming and occupies large space (potato, root and other asexually propagated vegetable crops), the problem can be solved by tissue culture techniques. Two approaches are being adopted for the *in vitro* germplasm conservation i.e., freeze preservation of meristems and cells in liquid nitrogen at-196°C (*cryopreservation*) and *slow-grow cultures* where meristem culture can be maintained without subculture.

### Germplasm exchange

The commonly used propagules in most of the clonal vegetable crops, particularly in case of those that do not produce seeds, are rather bulky. Further, there is the necessity of careful quarantine. Seedlings in test tubes obtained from apical meristems are extremely useful in germplasm exchange of such vegetable crops. Such seedlings are usually free from pathogens and insect-pests and are easier to handle.

## *In vitro* selection for stress resistance

### *Salt tolerance*

In order to identify a salt tolerant genotype, generally the inhibitory salt solutions are incorporated in the medium. The cells surviving and regenerating may be tested for suitable genetic salt tolerance. Salt tolerant lines have been isolated in many vegetable crops by the tissue culture techniques e.g., carrot, tomato and capsicum.

### *Metal ion toxicity*

Selection for tolerance to toxic level of cadmium and aluminium ions has been reported for tomato cell line. Tolerant to $Al^{+3}$ and $Mn^{2+}$ ions have also been isolated in carrot.

### *Cold tolerance*

The pollen selection technique is an economical method to identify a genotype for cold tolerance. Genotypes which tolerate low temperature at gemetic level are also tolerant at somatic level. The pollen germinating at a low temperature contributed a gene for cold tolerance. Several lines have been isolated in tomato, potato, capsicum, etc. to cold tolerance.

### *Drought and flood tolerance*

Tolerance to drought is a polygenic trait, water-stress tolerant lines have been successfully isolated in tomato using polyethylene glycol (PEG), non-penetrating osmotic solute. Crop injury due to flooding results from anaerobic respiration in roots. As a consequent, alcohol accumulates due to catalytic activity of alcohol dehydrogenase enzyme

(ADH). Selection for ADH deficient mutation could help to develop cell lines and plants tolerant to flooding.

## Disease resistance

Host selective toxins produced by certain pathogens are used in the medium for selecting resistant plants. Host cells resistant to this toxin may be resistant to the diseases. This has been extensively studied in potato against late blight disease and resistant material was selected with this method. The resistant plants selected at cellular levels have also been resistant under field conditions and resistance is a heritable characteristic.

**Tissue culture systems used to study disease resistance for various host pathogen introduction.**

| Host | Pathogen/Elicitor | Phytuberin |
|------|-------------------|------------|
| Potato | *Phytophthora infestans* | Rishitin & Phytuberin |
| Tomato | *P. infestance* | Rishitin & Phytuberin |
| Sweet potato | Elicitor not added | Furanoterpines |
| Garden pea | Elicitor not added | Pisatin |
| French bean | *Botrytis cinerea* | Phaseolin |

### Generation of variability

### Somaclonal variation

The genetic modification which occurs in plants regenerated from callus cultures of somatic explants is called as *somaclonal variation*. Somaclonal variation has been extensively exploited for the improvement of asexually propagated vegetables. These have been observed for a large number of attributes in potato, tomato, onion, letuce, etc. In potato, early blight resistant clone could only be identified by inoculating leaves of regenerated plant with toxin derived from *Alternaria solani*. In sexually propagated crops, chromosomal rearrangements sometimes cause infertility. Somaclonal variation is neither organ nor explant specific in occurrence, e.g., in potato somaclonal variation has been observed in plants regenerated from leaf discs, rachis or petiole explants. Since plant regeneration from somatic explants is relatively early compared to either gemetic cells or protoplasts, somaclonal variation can play an important role in breeding of superior vegetable varieties or hybrids.

## Some useful plant variations obtained through somaclonal variation in vegetables

| Vegetable crops | Improvement |
|---|---|
| Carrot | Improved snacking characteristics (sweetness, crunchiness, crispness). |
| Lettuce | Leaf shape and leaf colour improved snacking characteristics. |
| Celery | Improved snacking characteristics. |
| Potato | Resistance to early and late blight, tuber shape, growth habit and tuber colour. |
| Sweet potato | Variety 'Scarlet' developed, which is resistant to *Fusarium oxysporum* f. *batatis* and give high tuber yields. |
| Tomato | Increased solids content, jointless pedicel and male sterility. |
| Onion | Bulb shape, size and plant height. |

## *Protoclonal variation*

The variation which is observed among the plants which are regenerated from callus cultures of protoplast is referred to as *protoclonal variation*. The range of variability among protoclones is significantly higher compared to callus culture. This increased variation is attributed to stress imposed by the process of cell wall denudation and its subsequent synthesis. Over 1000 protoclones have been screened for potato variety Russet Burbank for the traits growth habit, maturation period, tuber uniformity, skin colour, etc. Some of the protoclones were resistant to symptoms caused by *Alternaria solani* toxin. Lines to late blight were also recovered.

## Haploid production

Anther or pollen cultures are useful in obtaining haploid plants, which are useful in genetic studies in vegetable breeding. Haploids are of considerable value in obtaining homozygous diploids in a short period of time. Haploids obtained from pollen grains are being used for the improvement of tomato, potato, capsicum, cole crops and cucurbits.

## (B)   Genetic Engineering

Plant genetic engineering is identified as the isolation, introduction and expression of foreign DNA in the plant. In other words, it refers to direct introduction of foreign (DNA) into a plant's system by micromanipulation at cellular level.

### Ways of gene transfer

### 1.    Via plasmid

Plasmids is an extra chromosomal genetic material which is found within bacteria cell and replicates independently of chromosomal DNA. Plasmids are used as cloning vectors in genetic engineering. The recombinant DNA is inserted into plant cell via plasmid. The plasmid can enter the cell and then combine with genetical material of host cell, resulting in transformation.

### 2.    Micro-injection

In this method, the recombinant DNA is directly injected into the cell nuclei of host plan.cell or recipient plant species.

### 3.    Direct uptake

In this method, the DNA is absorbed directly through the cell membrane. Such treatment is given in the cell culture. After introduction of gene into new cell, it gets incorporated into the recipient cell and expresses its effect in the new environment which can be easily observed by change in morphological or physiological change.

## Recombinant DNA technology

The DNA which contains genes from different sources and can combine with DNA of any organism is called *recombinant DNA*. Since genetic engineering utilizes recombinant DNA, it is also known as *recombinant DNA technology*. Recombinant DNA is obtained by special techniques rather than simple breeding procedures.

## Gene cloning

It is a technique of genetic engineering by which a gene sequence with many identical copies is replicated. Identical gene sequence is isolated by using restriction endonuclease enzyme. This can also be obtained by making a complementary DNA from mRNA templete using reverse transcriptase. It is then inserted to cloning vector i.e., a plasmid or bacteriophage. The hybrid is used to infect a cell (plant or bacteria) and replicated within the cell. Gene cloning is used for identification of molecular structure of genes.

## Gene sequencing

Gene sequencing refers to the determination of the order of bases of a DNA molecule making up a gene. The DNA is purified and broken at a specific point using restriction endonuclease enzyme. Thus all strands have one identical end. These strands are then broken at a random distance from this end so that there are strands ending on every base present. These strands are then separated, their end base identified and put in order of fragment size to determine the entire sequence.

## Gene splicing

In genetic engineering, the enzyme catalysed joining of DNA fragments is referred to as *gene splicing*.

## DNA probes

The small segments of DNA with known base sequences, origin and function are called *DNA probes*. DNA probes can be obtained either through DNA templet or can be produced by gene cloning technique. DNA probes are extremely useful in determining the nucleotide sequence in vegetable crops.

## Applications of genetic engineering

### Insect resistance

The crystal protein (*cry-proteins*) is encoded for the *cry* gene of *Bacillus* thuringiensis. *Cry proteins* are toxic to a specific range of target insects. The *cryIA* gene has been successfully transferred (using Ticplasmid vector) in tomato and potato to create resistance against insects.

### Resistance to viruses

Transfer of cDNA (complementary DNA) of satRNA (satellite RNA) that reduce the severity of symptoms produced by RNA viruses have been transferred into the genomes of their host in the hope of conferring upon them resistance to infection by that virus. For example Cucumber Mosaic virus (CM), Cauliflower's Mosaic Virus (CaMV) and Gemini virus have been successfully used as vectors for plant transformation.

### Herbicide resistance

The strategy that pertains to the transfer into the crop plants of genes that code for enzyme capable of detoxyfying the concerned herbicides. Such gene (*bxn*) is present in bacteria *Klebsiella pneumoniae* has been transferred into tomato which detoxifies herbicide bromoxynil. Another bacterial gene *bar* (it codes for phosphinothricin acetyl transferase; PAT, which detoxifies the herbicide L-phosphinothricin, PPT) from *streptomyces* spp. has been transferred into tomato and potato.

## Development of transgenic plants

A plant in which a gene has been transferred through genetic engineering is called a *transgenic plant* and the gene so transferred is called *transgene*. A transgenic tomato 'Flavr Savr' suitable for processing have been developed by using antisense RNA. It is bruise resistant, exhibiting delayed ripening and increased shelf life.

### Molecular markers

Plant DNA, digested with restriction enzymes, generated fragments of varying lengths. Restriction fragment length polymorphisms (RFLPs) from different sources reflects variation in the distribution of restriction sites. RFLP markers can be used for detecting, mapping and monitoring genes controlling quantitative and for speeding up breeding of multigenic traits. The usefulness of the approach has been demonstrated in identification and interogation of chromosomal segments of *Lycopersicon chmielewsk* associated with increased contents of soluble solids into cultivated tomato.

## Biotechnological achievements

List of varieties released by using biotechnological approaches for commercial cultivation

| Crop | Variety | Remarks |
|------|---------|---------|
| **Somaclonal** | | |
| Sweet potato | Scarlet | Darker and more stable skin colour |
| Tomato | | Bacterial wilt resistant |
| **Transgenic** | | |
| Potato | New leaf | Resistant to Colorado potato beetle |
| Squash | Freedom | Resistance to viruses |
| Tomato | Flavr Savr | Delayed fruit softening, delayed ripening, delayed ripening, thicker skin and altered pectin content |

# FOURTEEN

## Vegetable Crop Improvement

The improvement denotes either an addition or an alteration that augments the value of a crop plant or a production resource etc. In vegetable crops, improvement specifically would mean the amelioration in their quantitative as well as qualitative productivity and associated traits. Improvement of a vegetable crop is done through appropriate use of the principles of heredity and variation. Its tools and techniques constitute basics of plant breeding which denotes an art as well as a science of identifying and isolating more productive (than the existing one) hybrid or variety (cultivar). A perfect vegetable crop variety or hybrid is yet to be developed. All the existing ones have one or the other deficiency which precludes the expression of the maximum of in-built potential productivity in terms of quantity and quality except under rare conditions. A vegetable breeder endeavours to correct such a deficiency or deficiencies by evolving plant types embodying higher yield capability.

### Alliaceae

#### Keys to the genera

Leaves well developed, herbs without distinct above ground stems, flowers in heads or umbels, sheathed by one or more spathes : *Allium*

### Onion

Onion *Allium cepa* L. is an important vegetable crop grown throughout the world. It is a cool season vegetable. It is grown for its bulbs. In India, the most important onion growing states are Maharashtra, Tamil Nadu, Andhra Pradesh, Bihar and Punjab. Onion is one of the crops which

India exports. A global review of area and production of major vegetables shows that onion ranks second in area and third in production in the world. About 25.40 million tonnes of onion is produced in the world from 1.79 million hectare of land. India ranks first in area and second in production (0.29 million hectare and 2.45 million tonnes).

## Origin

The Russian scientist Vavilov reported Central Asia as the primary centre of origin and near east as the secondary centre of origin. Onion originated from the region comprising north-west India, Afghanistan, the Soviet Republic of Tazik and Uzbek and Western Tiensens. Western Asia and the area around Mediterranean sea are its secondary centres of origin. It is now cultivated throughout the world.

## Floral biology

The flowering structure is called an *umbel* which is aggregate of many small inflorescences (cymes) of 5 to 10 flowers, each of which opens in a definite order causing flowering to be irregular and to last for two or more weeks. Each individual flower contains 6 stems, 3 carpels united into one pistil and 6 perianth segments. The pistil contains 3 locules each of which has 2 ovules. The flower also contains nectaries which secrete nectar to attract insects for cross-pollination. The flowers are protandrous and anthers shed pollen over a period of 3-4 days prior to the time when full length of style is attained. Anthesis occurs in earily morning (6-7 hr). Anther dehiscence is between 7.00 and 17.00 hr and on next day also with peak between 9.30 and 17.00 hr. Pollen fertility is maximum on the day of anthesis. Thus, stigma becomes receptive 3-4 days after shedding of pollen grains and.protandary leading to cross-pollination is favoured.

## Breeding goals

- High yield.

- Long bulb storage life.

- Resistance to diseases (purple blotch, basal rot, stemphyllium blight, bacterial storage rot).

- Resistance to insect pests (thrips).

- Resistance to abiotic stresses (moisture stress, high temperature, salinity, alkalinity).

- Bulb quality (size, shape, colour, pungency, firmness dormancy, amount of soluble solids).

## Cytogenetics

The *Allium cepa* species are diploid with basic chromosome number of $x = 8$ ($2n = 16$). Occasional tetraploids have also been reported. There are morphological and cytological similarities between the species of section *cepa*, but still strong crossing barriers exist between them. This prevents gene flow between the two even where sympatric distribution of two species occurs. Introgression of genetic material from wild to cultivated species is also difficult. Low success has been recorded in several interspecific hybridizations.

## Breeding methods

Introduction, mass selection, selfing and massing, inbreeding, hybridization and heterosis breeding have been used for improvement of onion.

### Plant introduction

Early Grano is an introduction developed by this method. In long-day types, Brown Spanish has also been developed.

## Mass selection

Common in cross-pollinated crops. Most of the onion varieties in India have been developed by mass selection.

## Selfing and massing

Suggested by Jones and Mann (1963). This method is very good for improvement in a.crop, where inbreeding depression is common. Improvement in cultivar can be affected by selfing followed by massing. The procedure is as follows:

**First year (bulb crop)** : Select 100 best bulbs of desired type.

**Second year (seed crop)** : Grow selected bulbs, self one or more umbels per plant to initiate a separate line each.

**Third year (bulb crop):** Grow the progenies of each inbred line separately. Discard the poor performing lines during the growing season, at harvest or in storage. Select at least 25 best lines and keep 15-20 bulbs of each for selfing and open pollination for next year.

**Fourth year (seed crop)** : Self pollinate 1-2 umbels in each plant and allow others to open pollinate. Mass open pollinated seed and increase for large-scale production for more than two generations to avoid much inbreeding depression.

**Fifth year (bulb crop)** : Grow the selfed progenies separately. Select again the best 25 lines and 15-20 bulbs of each lines as above in the third year.

**Sixth year (seed crop)** : Composite and plant bulbs of all selections in a field or in a cage for free open pollination in between the unrelated lines. The open pollinated seed can be massed and increased as foundation seed.

## Inbreeding

In any onion improvement programme a considerable amount of inbreeding of selfing is necessary. Brown paper bags and three ring muslin cloth bags are used for selfing.

## Hybridization

Used when we want to introduce characters from other varieties.

(i)   **Intervarietal** :   Very common

(ii)  **Interspecific** :   Very rare

Utilized in *A. cepa* x *A. fistulosum*.

## Heterosis breeding

The $F_1$ hybrids are high yielding with uniformity in bulb size, the two most desired characters. One of the main component for exploitation of heterosis in onion is isolation of male-sterile lines. The male-sterile lines have been isolated in Pusa Red at IARI. Male-sterile lines have been also isolated at IIHR, Bangalore, but there is no hybrid in commercial production yet; work is in progress to isolate the best combiners.

Heterosis breeding involves three important stages, namely:

(i)    Production of inbred lines.

(ii)   Testing combining ability of inbred lines.

(iii)  Production of the seeds of $F_1$ hybrids.

*A line =*   male sterile line (S msms) caused by sterile cytoplasm (S) and nuclear recessive gene (msms).

*B line =*   maintainer line (N msms)

*C line =*   male parent (N MsMs or N Msms or N msms or S MsMs or S Msms)

Maintenance AxB 1:1, $F_1$ production AxC 4:1 or 8:2.

## Salient breeding achievements

- The varieties Italian Red and Local Brazilian are observed resistant to purple blotch.

- Granex and Excel 35 are reported to be resistant to pink root.

- Some promising $F_1$ hybrids identified are Abundance, Aristocrat, Banaza, Champion, and Contender.

- $F_1$ hybrids developed in India namely, Purple x Red, Purple x White and Red x White have more weight of plants and increased bulb diameter.

- Yellow-red colour of bulb was linked with resistance to *Colletotrichum arienans*.

- Punjab Selection and N 2-4-1 were good for dehydration. Pusa Red and Punjab Red Round were best suited for storage at room temperature (up to 5 months).

## Future prospects

- Development of varieties with improved quality and high yield suitable for different agro-climatic conditions.

- Breeding varieties resistant to major disease and insect pests, such as purple blotch, basal rot, stemphyllium blight and thrips.

- Heterosis breeding for development of short-day, high yielding $F_1$ hybrids in red, white and yellow types with longer storage life.

- Development of varieties of multiplier onion with bright red colour and higher bulblets.

- Standardization of improved package of practices particularly for new varieties and hybrids.

## Cruciferae

### Keys to the genera

Pods linear and dehiscent, beak of pods cylindrical or conical, seeds I-seriate, flowers yellow not aquatic: *Brassica*.

## Cauliflower

Cauliflower (*Brassica oleracea* var. *botrytis* sub var. *cauliflora*). In India, cauliflower is grown both in hills and plains and from 11°N to 35°N. Important cauliflower growing states in India are U.P., Karnataka, West Bengal, Punjab, and Bihar. It is also commonly grown in northern Himalayas and in Nilgiri hills in South. Cauliflower is harvested from August or early September to late February or early March in north Indian plains and from March to November in the hills.

## Origin

Cauliflower has descended through mutation and selection from a common ancestor, the wild cabbage (*Brassica oleracea* L. var. *sylvestris* L.) which is still found growing in western and southern Europe and North Africa, Cyprus and areas around Mediterranean coast are considered the primary centres of origin of cauliflower.

## Floral biology

The buds open under the pressure of the rapidly growing petals. This process starts in the afternoon and usually the flowers become fully expanded during the following morning. The anthers open a few hours later, the flowers being slightly protogynous. The flowers are pollinated by insects, particularly bees, which collect pollen and nectar. The nectar is secreted by two nectaries situated between the bases of the short stamens and the ovary. Situated outside the bases of the pairs of long stamens are also two nectaries, but these are not active. Flowers are borne in recemes on the main stem and its branches. The inflorescence may attain a length of 1-2 m, but the slender pedicels are only 1.5-2 cm long.

## Breeding goals

- High yield.
- Non-ricey, compact bract-free protected curds with retentive cream/white colour.
- Heat tolerance for producing curds in August/ September.
- Suitable varieties for curd formation in summer and rainy season in the hills.
- Better seeding quality.
- Self-incompatibility but cross-compatible inbreds to produce hybrids of tropical types.
- Resistance to diseases (black rot, sclerotinia rot, alternaria blight, erwinia rot).

## Cytogenetics

The somatic chromosome number of cauliflower is 2n=18. Studies on secondary association have been interpreted as showing a basic chromosome number x = 5, making cauliflower a modified amphidiploid from a cross between two primitive and five chromosome species with subsequent loss of one pair of chromosomes. In another study, three groups of two bivalents each and three groups of one bivalent each have repeatedly been found. From this secondary pairing of chromosomes, it has been concluded that they all derive from a basic species with x = 6. The six chromosomes are designed by the letters A, B, C, D, E and F. Later it was indicated that the minimum number of secondary association groups which included all chromosomes was three. This implied that the basic chromosome number is three.

# Breeding methods

## *Mass selection*

Mass selection has been widely used for improvement of cauliflower in India. However, this method is not so effective for the improvement of polygenic attributes, also it is slow and far from an ideal option. Consequently, modification like mass pedigree method and family selection method that are based on progeny testing have been found better than mass selection.

## *Recurrent and reciprocal recurrent selection*

It was found that recurrent selection is useful to improve compactness, diameter, depth and weight of cauliflower and improvement in yield from 18-47% over the original material.

## *Inbreeding*

Inbreeding resulted in uniformity of characters under selection and also increased the yield. Many inbred lines which were derived by inbreeding, had much better yield than the original varieties.

## *Hybridization*

Intervarietal hybridization can also be attempted to combine useful characters of two varieties like compactness of curd from one and high harvest index from other variety and in later generations selection is done for both the characters in the segregating populations.

## *Heterosis breeding*

In recent years, a number of $F_1$ hybrids have been developed in cauliflowers. For economical hybrid seed production, mechanism like self-incompatibility or male sterility can be used. Cross combinations between genetically diverse varieties, which are available in Indian cauliflower, are capable of giving high hybrid vigour. Snowball cauliflowers are genetically less diverse, therefore, the hybrid vigour in this type of cauliflower is not very much pronounced.

### Back-cross breeding

This method is practised when simply inherited one or two characters like disease resistance or curd quality are transferred from a variety to another well adapted variety. It can also be used to transfer male sterility from one variety to the other. After six repeated back crosses male sterility can be transferred to another variety.

### Anther culture

Anther culture has been successfully attempted in cauliflower. Diploid plants obtained from anther culture are of particular interest to the plant breeders. Such plants are used as parents for the production of $F_1$ hybrids or to develop open pollinated commercial cultivars.

### Somatic hybridization

Somatic hybrids were produced between radish and cauliflower by electrofusion of protoplasts from hypocotyls. Ogura-type cytoplasmic male sterility was transferred in cauliflower by donor recipient protoplast fusion and cytoplasmic male sterile plants were obtained.

### Genetic markers

Attempts have been made to develop transformed plants in cauliflower. Oncogenic strains of *Agrobacterium tumefaciens* harbouring $T_1$ plasmid have been used as vehicles of transformation for resistance against insects.

### Molecular markers

Molecular markers have been used to assign genes on different chromosomes and to identify genes for resistance and other economic traits. Molecular markers (RAPD) provide a quick and reliable alternative to identification of cauliflower cultivars.

## Salient breeding achievements

- Varieties have been evolved resistant to whiptail (Lecerf, M16/1), mosaic (Native Dwarf Erfurt, AH4-6) *Rhizoctonia solanii* (South Specific) and striped flea-beetle (Snowball A).

- Breeding for earliness has resulted in the selection of variety Comet from Erfurt Dwarf which takes only 60 days from transplanting to harvest.

- Takii 2 is suitable for cold storage.

- Neptune $F_1$ is recommended for canning.

## Future prospects

- Exploitation of heterosis, breeding for disease resistance and breeding for processing are areas where further emphasis has to be given.

- Breeding varieties for non-traditional areas is another aspect which needs to be strengthened.

- Tissue culture and androgenesis are certain potential techniques to be used for the maintenance of inbred lines.

## Cucurbitaceae

### Keys to the genera

Fruit fleshy many seeded pepo. Petals entire, calyx tube of male short, anther free or slightly cohering, usually exert. Stamens free, inserted on the tube of the calyx. Male flowers solitary or fasciculate. Calyx lobed subulate, entire, erect, pistillode glandular, seeds compressed, usually smooth. Connectives protruded, tendril 2-3 fid. Corolla companulate, 5 lobed to the middle or lower, tendrils usually simple : *Cucumis*.

## Cucumber

Cucumber (syn. gherkins), genus *Cucumis, C. Sativus* L. Warm season fruit vegetable. In developed countries it is grown as a glasshouse vegetable and in developing countries as an open-field vegetable. Cucumber forms an essential item of dietary in the West.

## Origin

According to De Candolle (1967) cucumber is an indigenous vegetable to India. Burma could be regarded as a secondary centre of origin of this crop.

## Floral biology

Opening and closing of the male flowers are mainly influenced by the sunrise and sunset, that is, by light and the time of the day. Anther in all varieties dehisces between temperature range of 20.5-21.5°C. Pollen fertility is considerable up to noon and by afternoon (2.00 pm), fertility is greatly reduced, and it is negligible by the evening. Stigmatic receptivity is of very short duration and pollination should be carried out within two hours after anthesis. Rise in temperature causes early drying of stigmatic secretion. Different floral abnormalities like mixed inflorescence, hermaphroditism, fusion, dimorphic female flowers, reduction and increase in the floral parts are also observed. A great tendency is observed in exotic collections towards the abnormalities.

## Breeding goals

- Early fruiting.

- High female to male sex ratio.

- Attractive green or dark green fruits with smooth surface and without prominent spines or prickles.

- Uniform long cylindrical shape without crook neck.

- Fruits free from carpel separation without hollow spots.

- Fruits free from bitterness.

- Less seeds at edible maturity.

- Resistance to powdery mildew, downy mildew, anthracnose, cucumber mosaic virus, insect pests and abiotic stresses.

## Cytogenetics

*C. sativus* has 2n = 14. The kayotypes of *C. sativus* are described by examining cells from shoots in metaphase association. It is suggested that cucumber is a secondary polyploid with x = 3. In cucumber, maximum number of nucleoli (6) corresponded exactly with the number of secondary constrictions plus satellites. This information confirms the theory of evolution of cucumber by fragmentation of chromosomes.

## Breeding methods

The breeding methods used for improvement of cucumber are introduction, selection, hybridization and heterosis breeding.

### Introduction

Varietal improvement programme in India started with introduction of varieties from foreign countries and these were recommended for commercial cultivation. These include Japanese Long Green and Straight Eight and Poinsette.

### Selection

Some varieties such as Sheetal, Khira 90 and Khira 75 are developed by employing selection procedure and released.

## Hybridization

Two varieties, Himangi (Poinsette x Kalyanpur Ageti) and Phule Shubhangi (Poinsette x Kalyanpur Ageti) are developed by hybridization followed by selection.

## Heterosis breeding

Heterosis breeding has brought about much improvement in cucumber. The open pollinated cultivars of cucumber are monoecious in sex form. A few attempts have been made to exploit monoecy in heterosis breeding in this crop. Examples are: Poona Khira x Japanese Long Green, Chaubatia Local x Solan Local, White Long Cucumber x Poinsette and Kalyanpur Ageti x Panvel. However, more of these hybrids are commercially exploited or grown by farmers.

Presence of gynoecious sex forms has made possible phenomenal progress in exploitation of $F_1$ hybrid under glasshouse conditions. The first $F_1$ hybrid, Pusa Sanyog developed in India in 1971 has a gynoecious female parent. It is suited to subtropical and cooler conditions also. This hybrid has shown heterosis for early yield, total yield, and number of fruits per plant.

## Resistance breeding

The Indian variety Banglora was the starting point for developing downy mildew resistance in cultivars in the United States. The cucumber cultivars, All Season, Improved Long Green, Korea Dutch, are less susceptible to red pumpkin beetle. Moderate resistance to *Meloidogyne incognita* has been observed in line GY 59 37-587.

## Salient breeding achievements

Lines resistant to parasitic and non-parasitic diseases have been evolved

| Diseases | Resistant varieties |
| --- | --- |
| **Non-parasitic** | |
| Low temperature | Azerbaijan. |
| Gummosis | Supert OE 48. |
| Drought | Hanski 264. |
| **Fungal** | |
| Downy mildew | Palmetto, Ashley, Chinese Long Stono. |
| Powdery mildew | Polaris, Ambra, Yamaki. |
| Scab | Wisconsin SR 12, Beleanto, Ashe Fletcher. |
| Anthracnose | Hybrid 517. |
| Fusarium wilt | Hybrids. |
| Leaf spot | Hivergreen, Biveryel. |
| **Bacterial** | |
| Wilt | Pl 200816, Pl 196477, Pl 200817. |
| Angular leaf spot | Poinsett, Dixie Gemin |
| **Viral** | |
| Mosaic | Kyoto-3-Feet, Table Green 65, Market More. |
| **Nematodal** | |
| Root knot nematode | West Indian Gherkins |

## Future prospects

Reduced number of nodes to first female flower, bushy plant type, self staked habit, higher female : male sex ratio, more flesh thickness, desirable flesh colour, fruit shape as per local preference, less or negligible cucurbitacin content, uniformity of fruit, less seediness, medium fruit size, better transportability, resistance to parasitic and non-parasitic disease are a few prospects in cucumber.

## Leguminosae

### Keys to the genera

Flowers zygomorphic, stamens definite, corolla papilionaceous, petals imbricate, the uppermost (standard the outer most) and the 4 others in 2 apposite pairs, stamens usually combined. Climbing or prostrate, rarely erect, herbs or shrubs, rarely trees, leaves pinnately trifoliate, rarely one or 5-7 foliate, stamens monodelphous or diadelphous, pod dehiscent not jointed, leaves not gland, dotted, leaflets stipulate. Style bearded below the stigma, stamens diadelphous, stigma oblique, keel not spiral: *Vigna*

## Cowpea

Cowpea (*Vigna unguiculata*) also called as *southern pea* and *blackeyed pea* belongs to family leguminosae. Less attention has been given to the vegetable cowpea. It is a nutritious leguminous crop low in anti-nutritional factors. It has a wide range of ecological adaptations and can be more widely grown. In fact, it probably has the greatest potential among all food legumes in the semi-arid to sub-humid tropical areas. The typical vegetable cowpea is characterized by a long pods (>30 cm), stringless pods, fleshy pod pericarp, thin and long seeds and higher monosaccharide: polysaccharide ratio. Pole and bushy types are available in vegetable cowpea.

# Origin

Africa is considered as the primary centre of origin. Of 170 species of genus *vigna*, 22 are found in India. The crop spread from Africa to Asia and Europe through Egypt. Now cowpea is widely distributed throughout the tropics and subtropics.

# Floral biology

The flower opens early in the morning, closes before noon and falls the same day. The extra floral nectaries at the base of the corolla attract ants, flies and bees but a heavy insect is required to depress the wing petal and expose; the stamen and stigma. Cowpea flower bloom between 6.30 a.m and 7.00 a.m. The process of opening of corolla takes 45-60 minutes. Anther matures and starts dehiscing about 78 hours before the time of opening of corolla. Pollen remains viable for 42 hours at room temperature and relative humidity 91%. Stigma becomes receptive from 12 hours before blooming to 6 hours after blooming.

# Breeding goals

- High green pod yield (vegetable type varieties).

- High seed yield (dry-seed type varieties).

- High fodder yield (fodder type varieties).

- Dual purpose (seed and vegetable type and seed and fodder type).

- Earliness.

- Appropriate plant type (erect, determinate for vegetable and seed type cultivars and spreading for fodder type cultivars).

- Wider adaptability.

- Photo-insensitive.

- Short tender pods for whole pod processing.

- Long tender and stringless pods for fresh consumption.

- Varieties suitable for inter cropping.

- Resistance to disease.

- Resistance to insects.

- Better seed quality.

# Cytogenetics

It is true diploid with 2n = 22. Sen and Bhowal (1960) made cytotaxonomic studies on *Vigna sinensis* (L.) Savi, *V. catjang* (Bur.M) Waqlp. and *V. sesquipedalis* (L.) Fruw, popularly known as *cowpea, catjang bean* and *asparagus bean*. The three cultivated species could be easily crossed among themselves. The hybrids among them were fertile like inter-varietal hybrids. Chromosome number of each of the three cultivated species were found 2n = 22. As mentioned elsewhere, all the above cultivated species are put under *Vigna unguiculata.*

# Breeding methods

## *Pedigree selection*

The pedigree system of breeding is the most common method used by cowpea breeders. This method has been successful in developing cowpea cultivars with new combinations of characteristics and resistance to diseases. Single plant selections are carried out within large $F_2/F_3$ populations. Individual plant progenies are planted in one or more rows, 4-6 m in length and 1.5 m apart.

## *Back-cross method*

The back-cross breeding procedure has been found efficient for transferring single-gene resistance to specific diseases into cowpea cultivars. For example, this procedure has been used to transfer Cls gene, which provides resistance to *cercospora* leaf spot, into the susceptible cultivar Colossus in USA.

### Heterosis breeding

Seven *Vigna unguiculata* var. *sesquipedalis* varieties were crossed in diallel. Of the 42 hybrids, only three exceeded the higher parent in seed protein content and pods/plant. The methods of $F_1$ hybrid production are hand emasculation and pollination, use of 'protruded stigma' types female, and use of male sterile line.

### Mutation breeding

Mutation breeding in cowpea has been utilized on a limited scale through irradiation (9-40 Krad) by gamma rays to isolate mutants with increased yield and earliness.

## Salient breeding achievements

- Varieties Pusa Phalguni suited for spring season, Pusa Barsati for rainy season and Pusa Dofasli for both the seasons have been evolved.

- K 1552 (Pusa Komal) recently evolved is early crop and resistant to bacterial blight.

- Varieties Grant resistant to *Fusarium* wilt, Chinese Red resistant to *Phytophthora* stem rot Mississipi 57-1 and Iron resistant to root-knot nematode have been evolved.

- Variety Goit, Dixies-Cream and Alabunch are resistant to cowpea mosaic.

## Future prospects

- Many valuable genes like male sterility, high protein and methionine content, resistance to thrips, aphids, pod-borers, root-knot nematodes, anthracnose and cowpea mosaic virus are available which need further utilization. Bush types of yard-long bean have been developed (Los Banos Bush Sitao No.1). There is a need to evolve high essential amino acid lines in cowpea. The extent of damage due to various diseases has not been estimated except for a few diseases.

## Malvaceae

### Keys to the genera

Plants destitute of peltate scales, leaves simple, entire, lobed or palmatified, carpals not separating, fruit capsular, bractioles neither spreading nor fimbriate, capsule 5-valved, epicalyx, calyx, corolla and androecioum fall as one unit after fruit set : **Abelmoschus**.

## Okra or Lady's finger or *Bhindi*

Bhindi, *Abelmoschus esculentus* (L.) Moench. Syn. Okra, *bhindi*, lady's finger or gumbo belongs to family malvaceae. It is a warm-season fruit vegetable in the tropical and subtropical countries of the world. The dehydrated *bhindi* is a processed product for preservation and export. *Bhindi* seeds form a nutritious ingredient of cattle feed and are a source of vegetable oil.

## Origin

The cultivated okra is of old world origin. According to Zeven and Zhukovsky (1975), Okra is believed to have originated in the Hindustani Centre of Origin, chiefly India, Pakistan and Burma. However, according to some other authors, *A. esculentus* originated in India (Masters, 1875), Ethiopia (De Candolle, 1883; Vavilov, 1951), West Africa (Chevalier, 1940; Murdock, 1959) and tropical Asia (Grubben, 1977).

## Floral biology

The flowering starts from below to upwards. Dehiscence usually occurs around 8-10 a.m., about 20 minutes after anthesis. Flowers remain open for shorter duration and wither in the afternoon. The stigma is receptive during anthesis, hence pollination is not very successful at bud stage. It is basically a self pollinated crop, however, cross-pollination through insects can be as high as 19% hence okra is classified as often-cross-pollinated vegetable crop.

## Breeding goals

- High pod yield.

- Dark green, tender, thin medium long, smooth, 4-5 ridged pods at marketable stage.

- Pods free from conspicuous hairs.

- Early and prolonged harvest.

- Short plant with more number of nodes/short internodes.

- Optimum seed setting ability.

- Pods suitable for processing industry and export market.

- Resistance to disease (Yellow vein mosaic virus, *Fusarium* wilt, *Cercospora* leaf spot, Fruit rot).

- Resistance/tolerance to insects (Fruit and shoot borer, Jassids and Whitefly).

- Tolerance to abiotic stresses (low temperature, excessive rains, saline and alkaline soils).

## Cytogenetics

The small size and large number of chromosomes makes exact chromosome count difficult in okra. Majority of the researchers have reported 2n = 130 in okra. However, other diploid chromosome numbers like 58, 60, 68, 72, 124, 182 and 194 have also been reported in *Abelmoschus*.

## Breeding methods

The common breeding methods applicable to any autogamous crop are used for improvement of okra. These are as follows:

### Introduction

A cultivar from Africa (Ghana) known as *Abelmoschus manihot* spp *manihot* introduced into India has been successfully used as a source of resistance to YVMV.

### Pure line selection

This is applicable to land races/cultivars collected from farmer's field, for example, Pusa Makhmali was bred from a material collected from West Bengal. Similarly, CO1 is a single plant selection from Red Wonder.

### Pedigree selection

This method is applicable for segregating generations after hybridization between donors. The individual plant selection starts in the $F_2$ generation and continues till $F_5$ or $F_6$. For example, Pusa Sawani was developed through this method in an inter varietal cross. Punjab Padmini, Prabhani Krani, P-7, Arka Anamika and Arka Abhaya are examples following interspecific hybridization.

### Mutation breeding

Ethyl methane sulfonate (EMS) has been found to be an effective mutagen to induce useful mutants carrying resistance to YVM and tolerance to fruit borer.

### Heterosis breeding

Heterosis in okra has been reported for various economic traits, *viz.*, early and late flowering, plant height, number, weight and size of pods, number of ridges, marketable and total yield. Using hand emasculation and pollination, commercial hybrids are developed. A few promising hybrids under private sector seed companies are H-7, H-8, Aroh1.

## Salient breeding achievements

- Variety Pusa Sawani, a selection from a cross Pusa Makhmali x IC 1542, was observed to be high yielding and highly stable over different locations.

- Variety Long Green Smooth has shown high resistance to nematode.

- High yielding varieties Lam Selection, Sel 2, Vaishali Vadhu, Lam Hybrid, R7, EMS8, Co1 and Beltes Five have been evolved.

## Future prospects

- The yellow-vein mosaic disease is the most serious disease in okra. There is need for continuous breeding programme to evolve resistant lines. Cultivation of okra during summer season has been problematic and less economic. There is a need to evolve varieties resistant to YVM disease and suitable for summer cultivation. Hybrid technology in okra needs to be popularized.

## Solanaceae

### *Keys to the genera*

Fruit indehiscent, a berry. Anthers connivent in a cone, longer than the filaments. Anther dehiscing introrsely by longitudinal slits, the tips empty, leaves pinnatesect : *Lycopersicon*.

## Tomato

Tomato (*Lycopersicon esculentum* Miller) is one of the most important warm-season fruit vegetable grown throughout the world. Among the vegetables, maximum attempts have been made to improve this crop. Short duration of crop, easiness in cultivation and a large number of seeds per fruit have made it an ideal crop for many research works.

### Origin

Tomato is a native of Peru in South America. The crop spread to North America primarily by migrating birds. The largest concentration of wild tomatoes is in Mexico.

### Floral biology

Dehiscence of anther occurs 1-2 days after the opening of corolla. It occurs from base to top and is longitudinal. It is mainly self-pollinated but certain percentage of cross-pollination also occurs. If the pollen is shed as the style grows up through the anther tube, self fertilization is the rule. Self-pollination also takes place when the style is short and the stigma is not extroverted beyond the connivent anther. A certain degree of cross-pollination happens when the stigma protrudes outside the level of anther (extrovert).

### Breeding goals

- Earliness.
- Increased fruit yield.
- Fruit quality.
- Indeterminate cultivars for green house production.
- Resistance to disease (Wilt, Blight, Anthracnose, Mosaic and Root knot nematode).
- Resistance to insects (Fruit borer, Whitefly).

- Resistance to abiotic stresses:

    (i) Cold set varieties.

    (ii) Hot set varieties.

    (iii) Drought tolerance.

    (iv) Salt tolerance.

    (v) Low temperature germination and growth.

    (vi) Chilling injury tolerance.

    (vii) Herbicide tolerance.

## Cytogenetics

Tomato is a true diplid with 2n=24. Haploids, tetraploids, trisomics and monosomics have also been produced in the tomato.

## Breeding methods

Tomato varieties have been developed by introduction, selection and hybridization procedures.

### Introduction

A few varieties like Sioux, Marglobe, Pest of All, Roma, Keckruth Ageti, La Bonita, etc. were introduced from America and adopted by growers at several places in India.

### Pedigree selection

Two parents are selected based on genetic divergence and desirability of characters. The $F_1$ and $F_2$ are developed. Selection starts from the $F_2$ generation onwards. This method is followed when character under improvement is governed both by non-additive and additive gene action. Pusa Ruby is such a selection from the cross between Improved Meeruti and Sioux.

## *Mutation breeding*

This method has been successfully utilised to evolve S-12 from Sioux, and Pusa Lal Meeruti from Improved Meeruti. Seeds are irradiated with 15-30 k gamma rays. Selection can be initiated from Mi generation. As considerable natural variability exists in tomato, utility of mutation breeding is limited as no methodology exists for directed mutagenesis.

## *Heterosis breeding*

Tomato is a classical example of self-pollinated vegetables where heterosis is being exploited on a commercial scale. Heterosis, being a function of specific combining ability of parental combinations, depends on the genetic divergence of the parents involved in the cross. This method has been successfully utilized to evolve Pusa Hybrid 2, Swarna 12, ARTH 3, ARTHA, NARF 101, Nath Amruth 501, Nath Amruth 601, FM2, KT4, MTH6, FM1.

## Salient breeding achievements

Varieties have been evolved resistant to diseases and pests.

| Disease and Pest | Resistant varieties |
|---|---|
| Fruit cracking | Pelican, Florida 10111 |
| Catface | K7, K10 |
| Puffiness | K7, K10 |
| Drought injury | *L. cheesmani* |
| Sunscald | Vivid |
| Late blight | Red Cherry |
| *Fusarium* wilt | Step 530, IRB 301-30 |
| Anthracnose | Pl 272636 |
| *Cercospora* leaf spot | Marglobe, Dwarf Stone, Floradel |
| Mosaic | Sonato, Estrelia |
| Root-knot nematode | Heralani, Nematex |

## Future prospects

● Twenty-six desirable characteristics have been ascribed to a tomato ideotype, but there is still a long way to go to achieve the objectives.

## Umbelliferae

### Keys to the genera

Umbels compound, fruit subterete, not winged, secondary ridges of the mericarps prominent, petals radiant. Fruit setose, involucral bracts prominent, pinnate : *Daucus*.

## Carrot

Carrot (*Daucus carota* L.) is a cool season root vegetable grown round the world in temperate climates during spring, summer and autumn seasons and in subtropical climates during winter. Carrot is an important source of carotene content (pro-vitamin A) and reportedly has anti-carcinogenic effect.

## Origin

Afghanistan is believed to be the primary centre of genetic diversity. There are evidences that purple carrot together with a yellow variant spread from Afghanistan to the Mediterranean region as early as the tenth or eleventh century. The white and orange carrots are probably mutations of the yellow form. The domestic carrot readily crosses with widely adapted wild carrot known as *Queen Anne's Lace*.

## Floral biology

The inflorescence of carrot is a compound umbel. Floral development is centripetal. Carrot is protandrous. This phenomenon is responsible for cross-pollination. The stigma becomes receptive on the fifth day after flowers open and remains active for 8 days but the better fruit sets are from pollination on 6th to 11th days after flower opening. Over 95 per cent of cross-pollination has been observed in carrot.

## Breeding goals

- High root yield.
- Scarlet/orange colour roots.
- High carotene content in roots.
- Uniformity in root shape and size.
- Thick flesh roots.
- Thin and self coloured core in roots.
- Early rooting.
- Cracking-free roots.
- High sugar and dry matter in roots.
- Slow bolting habbit.
- Smooth root surface.
- Resistance to *Alternaria* blight, *Cercospora* leaf blight.

## Cytogenetics

All the cultivated species of *Daucus* have 2n = 18. Neither polyploid nor structural changes in chromosomes seem to have played a role in the differentiation of the species.

## Breeding methods

### Introduction

Introduction is an important method of breeding in carrot. The varieties successfully introduced in India are Imperator, Danvers, Nantes and Perfection.

## Mass selection

Spontaneous mutation coupled with selection is mainly responsible for the development of cultivated carrots. Mass selection of root length resulted in the selection of high yielding lines.

## Bulk population

This method has been successfully used to evolve Pusa Kesar as a selection from a cross between Nantes and Local Asiatic.

## Mutation breeding

Chemical mutagens NEU (N-mitroso-N-ethyl urea) @ 0.025-0.6 per cent was successfully used to develop sterile line in carrot.

## Polyploidy breeding

Tetraploids (2n = 36) and Octaploids (2n = 72) have been developed in carrot. The polyploids have only limited utility in carrot.

## Heterosis breeding

Heterosis has been reported for earliness, tuber length, tuber yield, carotene content, top weight, core diameter and root diameter. In $F_2$, considerable loss in vigour has also been observed. Male sterility and self compatibility are the two plant phenomena which are being frequently used in heterosis breeding programme. Although over 95 per cent crossing occurred under field conditions, the commercial production of $F_1$ hybrid seed by natural cross-pollination is considered impracticable because of the large proportion of pollinator varieties that would have to be used.

| $F_1$ hybrid | Salient features |
| --- | --- |
| Chantaney x Genanda | Increased yield (25-33%) |
| Italian x Nantes | Increased yield (12%) |
| Virginia Savoy x Old Domain | Resistant to cucumber-mosaic virus |

## Salient breeding achievements

**Varieties resistant to disease and pest have been evolved**

| Disease/Pest | Resistant variety |
|---|---|
| *Cercospora carotae* | Inbred line WCRI |
| Motley virus | Sweet Crop, Top Weight |
| CMV | Virginia Savoy |
| *Pythium* spp. | Nantes, Danvers |
| *Heterodera carotae* (nematode) | Vilmorium 66 |
| Tip burn (physiological disorder) | Scarlet, Nantes |

## Future prospects

- Carrot is an important source of carotene. There is need to develop heat tolerant carrot lines rich in carotene. Breeding varieties suitable for non-traditional areas like warm-humid tropics would be an important and useful step. Breeding for resistance needs to be further intensified.

### Varietal attributes

Normally, vegetable varieties are selected for uniformity, better seed setting quality, resistance, tolerance to adverse weather conditions and reproductibility characteristics. These vegetable varieties may be self-pollinated, open-pollinated or hybrids, the seed of the former two groups is usually produced by the farmer himself after proper selection, reproduction and the basic seed is maintained for future commercial seed production, but in case of hybrids he will have to purchase seed every season for raising the crop for seed prduction. The elite or improved vegetable varieties are the result from systematic breeding and seed multiplication only by the well trained scientific personnel. However, seed quality of an improved variety depends on the skill of the vegetable breeder who develops them and the actual seed producers who multiply them commercially.

## Vegetable varieties and their charactersitcs

| Vegetable/Variety | Characteristics |
|---|---|
| **Amaranths** | |
| Japanese Yellow | A dwarf variety (52 cm tall), leaves are yellow green, softened smooth texture and very broad type. Medium to late in flowering, suitable for vegetable cultivation. Seed yield 12 q/ha. |
| Pusa Kiran | Medium-tall variety (90 cm) with 10-12 branch number per plant. The leaves are light green, smooth and broad. Seed colour black. Seed yield 11 q/ha. |
| Pusa Kirti | Tall variety (140 cm) with 12-14 branches/plant. Leaves are light green smooth, tender and broad. Seed colour balck. Seed yield 10-14 q/ha. |
| Choti Chauli | Suitable for sowing in spring and gives green leaves throughout summer. Dwarf variety with small leaves and rejuvenate quickly after each cutting. Gives six cutting. Leaves are ready after 20-25 days of seed sowing. Average seed yield is 8-10 q/ha. |
| Badi Chauli | Tall variety with thicker stems, tender and larger green leaves. Suitable for growing summer and rainy season. 3-4 cuttings are taken. Average seed yield is 11-12 q/ha. |

| Vegetable/Variety | Characteristics |
|---|---|
| Pusa Chauli | Plant height 90 cm. Stem green, medium thick and tender. Leaves are medium and green. Petioles green. Late bolter, inflorescence terminal and medium sized. Seed yield 12 q/ha. Other varieties CO-1, CO-2, CO-3. |

**Asparagus**

| | |
|---|---|
| Perfection | An early uniform very productive variety, delicious with high food value. The spears are large, green, succulent and light tipped. Average yield of spears is 80-100 q/ha. |
| Selection-841 | Bushy type, medium uniform plants and productive variety. The spears are 15-20 cm long, succulent, tender, green with better flavour suitable for soup preparation. Average yield of spears 90-110. |

**Beet Leaf**

| | |
|---|---|
| Banerjee Giant | Plant robust, leaves large and fleshy, double the size of ordinary spinach, suited to hills. It prefers rich soils. Seed yield 12-15 q/ha. |
| Long Standing | Leaves dark green, angular, thick and long. A slow bolting type, heavy yielder, adapted more particularly to high hills. Seed yield is 11-12 q/ha. |

| Vegetable/Variety | Characteristics |
|---|---|
| Pusa Harit | A cross between Sugarbeet x Local Palak, Plants upright, vigorously growing with uniform, thick, green slighly crinkled giant size leaves 3-4 or more cuttings. Heavy yielder with remarkable ability for rejuvenation. Late bolting habit and wide range of adaptability or varying climates. It is suitable for hills throughout the year. Seed yield 14-15 q/ha. Other varieites-All green, Pusa Palak, Pusa Jyoti, Jobner Green, HS-23, Palak No. 15-16. |

**Bhindi**

| | |
|---|---|
| Pusa Sawani | Intervarietal hybrid selection from a cross between I.C. 1542 and Pusa Makhmali, Stem and leaves are moderately hairy. Top leaves are palmately cut deeply lobed (3-5). First fruit appears on $3^{rd}$ to $5^{th}$ node. Fruits dark green, smooth, five ridges, 10-15 cm long ready in 50 days for market. It has lost its tolerance to YVMV.Suitable for spring and rainy season. Average seed yield 13-15 q/ha. |
| Harbhajan | A very prolific variety, fruits smooth, 15-20 cm long tender, dark green and five ridged. Field resistance to YVMV. Average seed yield is 12-13 q/ha. |

| Vegetable/Variety | Characteristics |
|---|---|
| P-7 | Plants medium to tall with splashes of pigmentation present on the stem. Leaves are deeply lobed upto the base of petioles and margins are less serrated. The basal portion of the petiole is deeply pigmented. Leaves, stems and petioles are sparsely hairy. Fruits are medium-long, green and five ridged. Fruit tip slightly furrowed and blunt. It carries high degree of resistance to YVMV. Seed yield 13-14 q/ha. |
| Parbani Kranti | Derived from *A. manihot* to Pusa Sawami through back crosses, resistant to YVMV developed by Maharashtra State Seed Committee. Plants are tall, single stemmed with dark green foliage. Fruit picking starts after 55 days after sowing. Fruits are slender. Seed yield 12-14 q/ha. Other varieties-P-8, Varsa Uphar, Arka Anamika, Arka Abhay, Punjab-Padmini, Hissar Unnat, A-4 (Pusa). |

**Bitter Gourd**

| Vegetable/Variety | Characteristics |
|---|---|
| Solan Hara | Fruits 20-25 cm long, 4-5 cm thick green, flesh white and spongy having high keeping quality. Ready for market in 80 days. Seed yield 4-5 q/ha. |
| Solan Safaid | Fruits 22-25 cm long, 4-6 cm thick, uniform, agreeable bitter taste. Flesh white and spongy. Rady for market in 80 days. Seed yield is 4.5 q/ha. |

| Vegetable/Variety | Characteristics |
|---|---|
| **Bottle Gourd** | |
| Pusa Summer Prolific Long | A local selection at IARI, prolific fruiting, 10-15 fruits per plant. trailer, flower white. ovaries long. Fruits long 15-18 cm long and 20-25 cm in diameter uniform, yellowish grccn, narrower at stalk and much thicker at distal end. Seed yield is 5 q/ha. |
| Pusa Summer Prolific Round | A local selection at IARI, prolific fruiting, trailer, flower white, ovaries round. Fruits round 15-18 cm in diameter when young, uniform, yellowish green. Heavy yielder. Seed yield 4-5.5 q/ha. |
| **Brinjal** | |
| Pusa Kranti | Medium tall growing, suitable for late sowing. Fruits are oblong 15-20 cm long, dark purple with shining green calyx and shy seeder. Fruits are attractive tender and borne singly. Seed yield 100-125 kg/ha. |
| Pusa Purple Long | Plant semi-erect habit, medium height, leaves and stems light green, spineless, leaves with cut edges. Fruits long (20-25 cm) bright purple, tender with thick skin. Take 100-110 days to maturity. Seed yield 150-200 kg/ha. |

| Vegetable/Variety | Characteristics |
|---|---|
| Pusa Purple Cluster | A medium early variety. Stem and leaves purple and spineless. Fruits are 10-12 cm long, deep purple in colour, attractive, tender and borne in cluster 4-6. Moderately reistant to bacterial wilt. Seed yield 125-175 kg/ha. |
| Hissar Syamal (H-8) | A prolific bearer, fruits 10-12 cm round, dark purple, tender with thick skin, borne singly. Plant height 50 cm, takes 70 days to maturity. Stem and leaves light purple in colour. Seed yield 150-190 kg/ha. |
| Pusa Anupam (Kt-4) | It has been developed at Katrain by a cross between PCC x Pusa Kranti. Fruits are longer than PPC and are borne in cluster. It is a suitable variety for pickles. Seed yield is 160-180 kg/ha. |
| Pusa Purple Round | Plant tall, erect, vigorous and sturdy. Leaves and stems dark green, spineless leaves thick, highly serrated with deep green in colour. Fruits round, purple, glossy, smooth and large. Each fruit weight 450-675 g. Plants bear 8-10 fruits. Resistant to little leaf disease and shoot borer. Seed yield 150-160 kg/ha. Other varieties-Pusa Bhairav, Surya, Arka, Nidhi, Arka, Keshav, Arka Neelkanth, ARV-2C. |

| Vegetable/Variety | Characteristics |
|---|---|
| **Brood Bean** | |
| Large Poded | Plants are tall, erect and pods long, borne in cluster. Pod yield varies 80-90 q/ha. Seed yield 15-18 q/ha. |
| Small Poded | Plants are small, erect, pods small fleshy, and borne in cluster (3-50). Seed yield is 10-15 q/ha. |
| **Brussel's Sprout** | |
| Hild's | Ideal plant height is 50-60 cm with 45-55 sprouts and 55 leaves/plant. The diameter of the sprout is 7-8 cm. Sprouts are compact with good flavour. Takes 115 days to mature and 3-4 pickings at 10 days intervals. Seed yield 4-5 q/ha. |
| Brussel's Dwarf | Plants dwarf 50 cm in height. Early cropper and suitable for areas of short growing season. Sprouts are of medium size, high yielding and single quality harvest. Seed yields is 3-4 q/ha. |
| **Bunching Onion** | |
| PBO-1 | Selection from exotic material. Perennial, leaves light green in colour, cylindrical, stem pure white and do not form bulb. Day neutral can tolerate frost and snow-fall. Grown in October-December and July-August. Average yield is 200-250 q/ha. |

| Vegetable/Variety | Characteristics |
|---|---|
| **Cabbage** | |
| Golden Acre | Selection from Copenhagan Market. Plants are short stalked with small frame. Early maturing variety with uniform, solid and round head, clear white interior, compact with a few wrapper leaves (4-5), which are cup shaped. Each head weighing 1.0 to 1.5 kg of excellent quality. Gets ready in 60-75 days. Average seed yield is 4-5 q/ha. |
| Pride of India | Selection from Compenhagan Market. Stem short, wrapper leaves 4-5, dark green in colour. Heads round, small to medium each weighing 1.5 to 2 kg, compact with better flavour. Get ready in 70-80 days. Suitable for offseason growing in hills. Seed yield 4-6 q/ha. |
| Pusa Drum Head | Selection made at Katrain from an introduction from Japan (EC6774). This is a late, main season variety with uniform heads, light green, flat, large, solid. Short stemmed variety with a wider frame and less outer green leaves have prominent veins. It possesses field resistance to black leg. Gets ready in 75-80 days. Average seed yield is 45 q/ha. |

| Vegetable/Variety | Characteristics |
| --- | --- |
| Large Late Drum Head | Late variety, head large drum shape. Ready after 100-120 days. Seed yield is 4-6 q/ha. |
| Pusa-Mukta (Sel-8) | Developed from an inter-varietal cross of EC-24855 and EC 10109, Pusa Mukta is an early variety with medium sized solid round heads. Slightly late than Golden Acre. It is resistant to black rot and suitable for disease prone areas. Seed yield 3-5 q/ha. Other varieties: Express, BRH-5, Copenhagen Market, September. |

**Carrot**

| | |
| --- | --- |
| Nantes | European cultivar, tops are small with green leaves. Roots are cylindrical with light tops, flesh orange, self coloured, half long, slim, well shaped, stumpy, abruptly ending in a small thin tail, deliciously flavoured. fine grained, tender and sweet with small self-coloured core. Get ready in 110-120 days. Seed yield is 6-7 /ha. |
| Chanteney | Roots tapering long with blunt end, flesh orange, self coloured core. Ready after 110-130 days. Excellent cultivar for canning and storage. Seed yield is 5-6 q/ha. |
| Pusa Yamadagni | Roots are 16-20 cm long, slightly tapering to semi-stumpy with medium tops, orange coloured with self-coloured core. Get ready in 80-120 days. Seed yield is 5-7 q/ha. |

| Vegetable/Variety | Characteristics |
|---|---|
| Pusa Kesar | Selection from a cross between Local Red and Nantes Half Long. Leaf top short, root red, long and tapering to a point, central core narrow and self coloured, less branching (forking). Contain high amount of carotene. Late bolting can tolerate high temperature. Seed yield is 6-8 q/ha. Other varieties- Zeno, Scarlet Nantes, Danver Half Long, A Plus, Desi carrot. |

**Cauliflower**

| | |
|---|---|
| Early Kanwari | Typical Indian type, adapted to hot and humid climate. Recommended for early sowing in May and transplanted to June. Developed by PAU. Plants are short stemmed, curd small and loose, pale white, ready for harvest in 100-120 days. Seed yield 2-3 q/ha. |
| Pusa Katki | Early variety have medium sized plants and bluish green leaves, medium sized curds tend to grow loose faster. Seed yield is 3-4 q/ha. |
| Pusa Deepali | Early variety plant height medium, uniform, leaves green erect with round tip. Curd small, white and compact. Seed yield 4.5 q/ha. |
| Improved Japanese | An introduction from Israel. Early, plants erect bluish green leaves, curd compact, white and medium |

| Vegetable/Variety | Characteristics |
| --- | --- |
| | to large sized. Seed is sown in August and crop is ready by late November to mid December. Seed yield is 4-5 q/ha. |
| Pusa Him Jyoti | A selection from MGS-2-4. Erect bluish green leaves with a waxy coating covering the curd tightly. Curds are solid, pure white, retain their colour even after exposure and with good shelf life. Seed yield 4-6 q/ha. |
| Late Dania | Variety developed at Kalimpong, successful in eastern hills. It has very sturdy plants with waxy leaves and medium deep curds. It is sensitive to fluctuating environments. Seed yield 3-5 q/ha. |
| Pusa Snowball-l | Developed by hybridization between EC-12013 and EC 12012. Self-blanching habit with inner leaves covering the curd tightly. Curds solid, medium sized, white with good shelf-life. Outer leaves large, straight and upright. Get ready in 100 days. Seed yield 4-5 q/ha. |
| Pusa SB-K-l | A selection from exotic material. Curd snow white, solid, slightly raised in the centre. Foliage light green with puckered margins and inner leaves cover the curd tightly. Possesses field resistance to black rot. Get ready in 115-120 days. Seed yield 3-4 /ha. |

| Vegetable/Variety | Characteristics |
|---|---|
| Pusa Snowball-16 | Late variety, curd medium, snow-white in colour. Seed yield 3-3.5 q/ha. |

**Capsicum**

| | |
|---|---|
| California wonder | Late variety, plant medium in height erect and sturdy. Fruits very thick, 3-4 lobes, bright green in colour. Fruit picking in 75 days. Seed yield 1-1.5 q/ha. |
| Yolo Wonder | Late variety with stocky and upright plants. Fruits are three lobed, smoth thick walled, pendent, blocky or square shaped, heavy and deep green turning bright crimson at maturity. The flesh is very thick, sweet and fine flavoured. First picking after 70 days. Seed yield 1.60 q/ha. |
| Bharat | Hybrid variety released by Indo-American Hybrid Seeds, Bangalore. Plants vigorous and productive, fruits uniform, smooth bright green in colour, thick walled, mostly 4 lobed, length and diameter 8-10 cm. First picking after 80 days. Duration of crop is 100-120 days. Resistant to TMV. Seed yield 1.5 q/ha. |
| Solan Hybrid | Early maturing hybrid, heavy yielder, resistant to fruit rot disease. Seed yield is 1.4-1.7 q/ha. |

| Vegetable/Variety | Characteristics |
| --- | --- |
| Solan Hybrid-2 | Plant tall, prolific bearer, early 3-4 lobed fruits, resistant to fruit rot and virus. Seed yield 1.0-1.0 q/ha. |
| Kt-1 | $F_1$ hybrid of Yolo Wonder and Russian Yellow. Plants bushy, vigorous, 55.56 cm tall with erect fruit bearing habit. Fruits available in 70-75 days. Fruits are light green, conical 8-11 cm long, 3-5 cm diameter with thick flesh. It can withstand long transportation. Tolerant to bacterial leaf spot and anthracnose. Average seed yield 5-1.8 q/ha. |

**Chicory**

| | |
| --- | --- |
| Kalpa Sel-1 | Selection from exotic germplasm at Kalpa. Roots thick, tapering, 30-35 cm long, flesh white. Leaves broad oblong, dark green crowded at crown. Plant erect, vigorous and quick growing. Free from bolters. Average seed yield-10 q/ha. |
| K-13 | Selection from exotic germplasm. Roots thick, non- tapering and do not break while uprooting. Flesh white suitable variety. Average seed yield 7-9 q/ha. |

**Chillies**

| | |
| --- | --- |
| Solan Yellow | Selected from local material, very pungent and bush type. Fruit 4-5 cm long erect bearing, oblong, turning yellow at maturity. Seed yield 1.5-1.8 q/ha. |

| Vegetable/Variety | Characteristics |
|---|---|
| Pachhad Yellow | Local selection. Fruit long, tapering pendent and turning yellow at maturity. Seed yield is 2.0 q/ha. |
| Hot Portugal | Fruit 12-15 cm long, pendent, good for pickling, green turns deep red at maturity. Seed yield 15 to 1.7 q/ha. |
| Hungarian Wax | Fruits thick, fleshy, light yellow, turning red at maturity. Mild in pungency. Fruits 10-16 cm long. Suitable for pickling. Seed yield 1.5-1.6 q/ha. |
| Surajmukhi | Plant erect 50-60 cm, branches (7-8) arises from base of the main stem. It bears 9-14 fruits in clusters. Fruits are erect and pure bright red on maturity and are highly pungent. Fruit remain green for a longer time on the plant, leaves are much broader than other chilli cultivars. Resistant to bacterial wilt. For green picking gets ready in 60-75 days. Seed yield 1.8-2.0 q/ha. |
| Sweet Banana | Fruit thick, fleshy, light yellow, turning red at maturity, 18-20 cm long, broad, round at pedicle end, tapering and non-pungent. Good for table and pickling purposes. Seed yield 1.2-1.5 q/ha. |
| Punjab Lal | Plants semi-dwarf. Fruits upright, small in size, immature fruit green in colour, turns deep red at |

| Vegetable/Variety | Characteristics |
|---|---|
| | maturity, extremely pungent. Resistant to many viral diseases. Seed yield 1.3-1.7 q/ha. |
| **Chinese Cabbage** | |
| Palampur Green | Leaves green, tender (30x25 cm size), stem creamy, late in flowering. First cutting after 25-30 days and subsequent cuttings at 15 days interval and 5-6 cuttings. Seed yield 8-9 q/ha. |
| Solan Selection | Leaves are smooth, light green with fleshy petioles. Seed yield 6- 7 q/ha. |
| Pusa Sag | A cross from Wongbok (Suttons) x Turnip. Taste like local sarson. Seed yield 10-11 q/ha. |
| **Celery** | |
| Utah-52-70 | High yielding, tall, medium, late variety, producing large white stalks of fine quality. Seed yield 6-7 q/ha. |
| **Coriander** | |
| Indian Type | Plant medium to tall, leaves green in colour. Stem, leaves and seeds have aromatic flavour and taste due to essential oils ranging from 0.4 to 0.8 per cent. Have wider adaptability. Average seed yield is 9-10 q/ha. |
| European Type | Plant small to medium in height and have more aromatic flavour and |

| Vegetable/Variety | Characteristics |
|---|---|
| | taste. The essential oil content contained is 1.4 to 2.0 per cent, suitable for cooler climates. Average seed yield is 7-8 q/ha. |
| **Cucumber** | |
| Khira-75 | Local selection, very prolific fruits, well filled to end, smooth, light green, cylindrical, 11-15 cm long. First picking after 75 days. Average seed yield 3-5 q/ha. |
| Khira-90 | Local selection, very prolific, well filled upto end, smooth, light green, cylindrical, 15-20 cm long, first picking after 90 days. Average seed yield 4-5 q/ha. |
| Poinsette | An introduction from U.S.A. Fruits dark green, 25-30 cm long, with white spines, resistant to downy mildew, powdery mildew, anthracnose, angular leaf spot and fruit fly. Gets ready in 60 days. Seed yield 3-4 q/ha. |
| Pusa Sanyog | $F_1$ Hybrid developed at Katrain, high yielding and early in maturity. Vines are not very long. Fruits 25-30 cm long cylinderical, light green, stripes, flesh crisp and attractive. Average seed yield is 25-3 q/ha. |
| Japanese Long Green | Early variety matures in 45 days and prolific bearer. Fruits are green |

| Vegetable/Variety | Characteristics |
|---|---|
| | and 30-40 cm long with whitish green surface. The flesh is light green and quite crisp. It essentially requires staking if straight fruits are desired. Average seed yield is 3-4 q/ha. |
| Local Khira | Improved cultivar from local germplasm. Fruits light green cylinderical 12-15 cm long, tender, smooth, skin thick, juicy flesh, greenish white, sweet, better taste and flavour. Resistant to fruit fly, can be transported to long distances. Seed yield is 4-6 q/ha. |
| Kh-1 | Early hybrid variety, fruits shining 10-12 cm long, light-green colour, ready in 65 days. Seed yield 3-4 q/ha. |
| **French Bean** | |
| Contender | An introduction from U.S.A., bush type with pink flowers. An early maturing (50-55 days) variety. Pods 12-14 cm long, green large with slightly curved tip, stringless and meaty. Mature seeds are light brown. This variety is tolerant to mosaic and powdery mildew. Seed yield is 10-12 q/ha. |
| Arka Komal | A dwarf variety, heavy yielder and resistant to anthracnose disease, suitable for low and mid-hills. Average seed yields 12-14 q/ha. |

| Vegetable/Variety | Characteristics |
|---|---|
| Premier | A dwarf variety. Pods tender, dark green, flattened and 13 cm long. An early variety. Gets ready in 55 days. Seed yield 10-12 q/ha. |
| VL-Boni-1 | Dwarf variety, pods light green, stringless, round, little curved and gets ready in 50-55 days. Seed yield 8-12 q/ha. |
| Pusa Parvati | Dwarf vareity, green pods, well filled, stringless, 15-18 cm long pods, gets ready in 50 days, seed light-brown colour, resistant to powdery mildew and mosaic. Seed yield is 9-12 q/ha. |
| Kentucky Wonder | A pole type and late maturing variety, very prolific bearer, three pods per cluster, pods are green, fleshy, large, curved, round, thick, meaty and become constricted and stringy at later stage. Mature seeds are light brown. Gets ready in 65 days. Seed yield 12-14 q/ha. |
| SVM-1 | A pole variety. Fruits 13-14 cm long, green, stringless, round, 8-10 brown shining seeds per pod. Resistant to angular leaf spot. Gets ready in 65 days. Seed yield 10-15 q/ha. |
| Lakshmi (P-37) | A pole variety, prolific bearer, 3 pods per cluster. Pods 10-12 cm long. Stringless, round, green and attractive. Gets ready in 60-70 days. Seed white with light yellow scar. Seed yield 9-12 q/ha. |

| Vegetable/Variety | Characteristics |
| --- | --- |

## Garden Beet

Crimson Globe

Tips are medium to tall, leaves elliptical, large and bright green with maroon shade, veination is prominently coloured in young and older ones. Edible roots are globular to flattened globe, medium red with small shoulders where flesh is medium dark red within distinct zones and non- corrosive in taste when taken raw. Gets ready in 60-65 days and suitable for salad. Seed yield 8-12 q/ha.

Detroit Dark Red

Tops are small, the leaves glossy, dark green, tinged with maroon (more prominent on older ones), where veination is prominent, mid-rib is thin from dorsal side and appears to be wider. Roots perfectly round, smooth, uniform, attractive, skin deep red, small tap root, flesh very dark blood with light red zoning, fine grained, tender and corrosive when eaten raw. Gets ready in 80 to 100 days. Seed yield is 9-12 q/ha.

## Garden Pea

Arkel

A French introduction. Plants are dwarf, early, vigorous and may grow up to 50 cm, flowers in 35-40 days, from the sixth node onwards. Pods dark green, long contain 8-10

| Vegetable/Variety | Characteristics |
|---|---|
| | seeds. Seeds are wrinkled, green sweet and larger in size. Gets ready in 55-60 days. Suitable catch crop. Highly susceptible to powdery mildew. Seed yield 12-14 q/ha. |
| Mater Ageta | A variety developed at PAU, Ludhiana. Plant very dwarf 15-20 cm in height. Light green, can tolerate higher temperature. Earliest variety, gets ready in 25-30 days, short duration variety. Pods green 6-7 immature seed (ovules) per pod, sweet and attractive. Seed yield 10-12 q/ha. |
| Palam Priya (DPP-68) | A variety developed at HPKV, Palampur, through hybridization. Mid-season, tolerant to powdery mildew, pods are green, 6-8 immature seeds (ovules) per pod, attractive and sweet. Seed yield 9-12 q/ha. |
| Bonneville | Exotic variety from USA, dwarf attaining a height of 60-70 cm. Gets ready in 85 days. Peduncles usually bear two white flowers. Pods are 8 cm long with 8-9 ovules/pod. Pods well filled, light green and sweet. Seeds wrinkled, and green. Susceptible to powdery mildew. Seed yield 10-12/ha. |
| Lincoln | Plants are medium-tall and double podded. Pods dark green, |

| Vegetable/Variety | Characteristics |
|---|---|
| | attractive, curved 9-10 cm long, well filled, with 8-9 sweet green ovules. The seed wrinkled. Gets ready in 85-90 days. Prolific bearer in the hills. Retains green colour in transit. Seed yield 12-15 q/ha. |
| VL-3 | Medium-tall, suitable for main-season, pods 7-8 cm long, green in colour containing 7-8 ovules. Seed wrinkled. Gets ready in 100-110 days. Seed yield 10-12 q/ha. |
| Early Giant | Well suited to hill cultivation. Early pole type with dark green pods of excellent quality having sweet, creamy-green, bold and wrinkled seeds. It continues to give green pods for a longer time. Seed yield 12-15 q/ha. |
| VL.-7 | Early, matures in 60-65 days, pods dwarf, wrinkled seed, light green in colour with 6-8 ovules/pod. Seed yield 9-12 q/ha. |
| Solan Nirog (Sel. 8-1) | Mid-season variety, Pods 8-10 cm long, dark green, 8-9 seeds/pod. Resistant to powdery mildew. Gets ready in 90-95 days. Seed yield 10-15 q/ha. |
| Kinnauri | Climbing, needs staking, pods thick, light green, round seed, 5-6 seeds per pod. Ready for picking after 110-120 days. Seed yield 14-15 q/ha. |

| Vegetable/Variety | Characteristics |
|---|---|
| **Garlic** | |
| Selection-1 | Small segmented, compact bulbs with 12-15 cloves. Better flavour, cloves white, medium, uniform size and attractive. Average yield 150-190 q/ha. |
| GHC-1 | Selection from the local material. Plants are 30-35 cm tall. Leaves are light green colour. Medium segments, bulbs compact white 10-15 cloves. Good flavour, attractive bulb and cloves are peeled with ease. Average yield 180-225 q/ha. Suitable for pickles. |
| Large segmented | A selection from local material at Solan. Heavy yielder 2-5 cloves per bulbs, each clove large and handy, easy to peel. Aroma not as strong as in ordinary small segmented garlic. Average yield 190-250 q/ha. |
| **Ginger** | |
| Him Giri | Selection from local clone Dhalja, rhizomes are of medium size attractive. Variety high yielding tolerant to rhizome rot. Average yield of rhizomes is 125-150 q/ha. |
| Maron | Plant height about 30 cm, rhizome length 14-15 cm with 10-12 daughter rhizomes. No. of tillers produced 3-10 per plant. Rhizome rot disease, reaction is 7-8%. Average rhizome yield is 100-120 q/ha. |

| Vegetable/Variety | Characteristics |
|---|---|
| Kerala Local | Plant height about 30 cm with 10-12 tillers per plant. Leaves are long, narrow, green in colour. Rhizomes are 14-15 cm long, 10-12 daughter rhizomes. Disease reaction with rhizome rot is 3-4%. Average yield is 120-140 q/ha. |
| Him. Sel (Local) | Plant height is 35-40 cm. Plants vigorous, green foliage with 16-18 cm pseudostems (tillers). Rhizome length 14-15 cm with 14-15 daughter rhizomes. The disease reaction was 8-10%. Average yield is 120-140 q/ha. |

**Knol-khol**

| | |
|---|---|
| White Vienna | Early, dwarf variety with medium green leaves and stems. Knobs are oval, globular with distinct small tops, medium sized, light green or nearly white, smooth and flesh is creamy white and tender with delicate flavour. Gets ready in 50-60 days. Seed yield 5-6 q/ha. |
| Purple Vienna | Purple-blue leaves, knobs are globular round, large in size with bluish-purple tinge, flesh light green. Knob formation in 55-65 days. Seed yield 4-6 q/ha. |
| Large Green | Green round knobs, early with small tops, large sized, tender, delicate and flavoured. Flesh white, gets ready in 70 days. Average seed yield is 5-6 q/ha. |

| Vegetable/Variety | Characteristics |
| --- | --- |
| DPKK-1 | A selection from exotic germplasm. Early variety, a week earlier to the existing variety *viz*. White Vienna. Knobs are green tender, good flavouring type. Seed yield 4-5 q/ha. (Pre-released variety). |

**Leek**

| | |
| --- | --- |
| PPL-1 | Selection from exotic germplasm. The hollow plant is consumed. The leaves are light green. Do not form bulb, swollen stem. Ready in 150-160 days. Good replacement for green onion. Seed yield 5-6 q/ha. |

**Lettuce**

| | |
| --- | --- |
| Great Lake | Heading type variety with large firm heads. Foliage dark green and the outer leaves are blistered. Seeds are white. Resistant to tip bum. Seed yield 5-6 q/ha. |
| Alamo-1 | An open-pollinated heading variety. Leaves are crisp, curly and dark green, heads compact. Average head weight 500-800 g. Gets ready in 90-100 days. Seed yield 5-7 q/ha. |
| Chinese Yellow | Early variety and good cropper. Leaves are loose, light green, crisp and tender, flowers are yellow and seeds white. Seeds yield 4-6 q/ha. |
| Simpson Black Seeded | Leaves large light green becomes bitter when mature. Leaves open. Seeds black. Seed yield 5-6 q/ha. |

| Vegetable/Variety | Characteristics |
|---|---|
| PH Lettuce | Selection from exotic germplasm. Early heading type. Light green colour, head compact, matures in 100-120 days. Seed yield 4-5 q/ha (Pre-release variety). |
| **Methi** | |
| IC- 74 | A very high yielding, Indigenous selection, tender leaves with good flavour, grows well on fertile soils. Gets ready in 50 days. Seeds used as condiment. Seed yield 12-15 /ha. |
| Pusa Kasuri | A good yielder, plant spreading gives three cuttings of tender leaves. Highly flavouring and tasty variety. Gets ready in 60 days. Suitable at tender stage for drying in off-season consumption. Takes 155 days for seed maturity. Rosette habit, late flowering. Flower yellow, pods sickle shaped, flat, small green. Seed yield 6-7 q/ha. |
| Pusa Early Bunching | Shoots upright, vigorous, height 40-45 cm. Flower axillary, white. Pods long, drawn out to a beak, flat, moderately green. Large seeded. Good yielder. Average seed yield 12-15 q/ha. Other varieties CO-1, Methi No-14 and 47, EC-4911. |
| **Onion** | |
| Nasik Red | Bulbs red round with high keeping quality. Seed yield 5-6 q/ha. |

| Vegetable/Variety | Characteristics |
| --- | --- |
| Pusa Red | A very popular short to intermediate day-long variety. Bulbs medium, flat to globular, red average weight 70-90 g and less pungent. It has less bolting. Gets ready in 125-140 days. Seed yield 4-6 q/ha. |
| Nafad-53 (N-53) | Developed by HAU, Hissar suitable for *kharif* season. Bulbs medium in size, scarlet red in colour, round and less pungent. Average bulb weight is 70-100 g. Gets ready in 150-165 days. Seed yield 5-7 q/ha. |
| Agri Found Dark Red | Selection made by Assoc. Agr. Dev. Found. from local germplasm, bulbs round globular in shape with tight skin, moderately pungent and dark red with bulb diameter of 5-8 cm. Takes 150-160 days for maturity. Better keeping quality and storing capability. Seed yield 4-7 q/ha. |
| Brown Spanish | An introduction, long day type. Bulbs are attractive, round to slightly oblong with brown coloured scales. Skin tight and more pungent. Maturity at a time when (September) no fresh onion available in the market. Very long shelf life and storability. Seed yield 4-5 q/ha. Other varieties-Arka Kalyan, Patna Red, Punjab Red Round. |

| Vegetable/Variety | Characteristics |
|---|---|
| **Parsley** | |
| Mosscurled Champion | Plants are dwarf, compact, bushy, and productive which can also be used for garnishing an decoration purpose. Its leaves are very dark green, exceedingly fine cut, serrated and deeply curled. Seed yield 6-7 /ha. |
| **Potato** | |
| Kufri Chandermukhi | Early variety, wider adaptability. Matures in 110-130 days. Tubers are large, smooth surface, oval and flattened shape, white skin colour, flat eyes and dull white flesh. Because of its regular shape, its amenability to post-harvest treatment long transit and temporary storage in cool sheds, low ability to form sugars at low temperatures meet all the quality requisites for export purpose as seed. It has good keeping quality and shows resistance under field conditions to leaf roll and virus 'Y'. Average yield is 150-200 q/ha in mid- hills and 250 q/ha in dry temperate zone. |
| Kufri Jyoti | A medium maturing variety and takes about 130-150 days to mature. It is highly resistant to late blight, both on foliage and tubers. It is high yielding variety, also |

| Vegetable/Variety | Characteristics |
|---|---|
| | suitable for plants and highly suitable for Himachal Pradesh. Average yield 150-175 q/ha. This variety should not be allowed to grow, beyond its maturity time as tubers will grow larger in size and bound to show some cracking which is not a desirable quality. |
| **Pumpkin** | |
| Solan Badami | Vines vigorous, fruits orange in colour, medium sized, 2-4 kg in weight, fruit size 24x17.5 cm. Body globular with lobes, flesh thick and attractive. Yellow green stem with good keeping quality. Early and prolific. First fruiting in 100-120 days. Seed yield 4-6 q/ha. |
| **Radish** | |
| **(a) Asiatic types** | |
| Japanese White | An introduction from Japan with medium large top and deeply cut leaves. Roots are pure white 20-30 cm long cylinderical, thick, smooth stumpy with blunt end. The flesh is snow white, crisp, solid, sweet and mildly pungent. Late bolting habit cultivated in the main season of hills and plains, it matures in 60-65 days. Seed yield 9-10 q/ha. |
| Chinese Pink | Root pink, cylindrical with blunt end, flesh white, leaves long with |

| Vegetable/Variety | Characteristics |
|---|---|
| | red midrib, ready after 45 days. Seed yield 8-9 q/ha. Other varieties- Pusa Desi, Pusa Rashmi, Pusa Chetki, Jaunpuri etc. |
| (b) European types | |
| White Icicle | Top medium short. Roots pure white without - green shoulder, thin, tender, solid, icicle-shaped, straight tapered. Becomes pithy in delayed harvesting. The flesh is icy white, crispy and juicy with milk and delightfully sweet flavour with just enough pungency to appeal to the appetite. A very good early variety. Ready in 30 days. Suitable for kitchen gardens and high hills. Average seed yield 5-6 q/ha. |
| Rapid Red White Tipped | Temperate variety, table type and very early with short tops. Roots are small, round, smooth, attractive shape and fine quality having bright red skin with white tip. The flesh is pure white, crisp, snappy and pungent in flavour. Gets ready in 25 days. Suitable for kitchen garden and high hills. Seed yield 4-6 q/ha. |
| Pusa Himani | Developed from a cross of Black x Japanese White. This variety produces white, long (30x25 cm) medium, thick tapering with white green shoulder, attractive roots. |

| Vegetable/Variety | Characteristics |
|---|---|
| | Flesh pure white, crisp and sweet flavoured with mild pungency. Ready in 55-60 days. Most suitable for sowing throughout the year in hills. Seed yield 6-8 q/ha. |
| Sel. 9 | A new line, developed at Katrain to improve White Icicle, a European type radish for its pithiness by crossing with EC 9005. Roots long, tapering with green stem end. Seed yield is 6- 7 q/ha. Other varieties-Scarlet Globe. |
| **Red Cabbage** | |
| Red Acre | Leaves are deep purplish-red colour, which is of great attraction in Kitchen garden. It is a novelty to five star hotels. Yields are usually low. It has a distinct coating of wax. Produces heads of 1-2 kg. Matures in 90-100 days. Shape of head is some what round. Average seed yield is 7-8 q/ha. Other Varieties-Mammoth Rock Red, Large Red, Red Danish. |
| **Sardha Melon** | |
| Sel. 1 | Early, small fruit size, weighing 700-1000 g, sweet, better flavour. Matures in 100 days. Total soluble solids 13 per cent. Seed yield is 4-6 q/ha. |

| Vegetable/Variety | Characteristics |
| --- | --- |
| Sel. 9 | Matures in 110 days. Fruit weight is more than 1 kg. Total soluble solids above 13 per cent. Seed yield 4-5 q/ha. |

### Spinach

| | |
| --- | --- |
| Verginia Savoy | Growth vigorous and upright having blistered, dark green leaves with thick tips crumpled and tender. Seeds are smooth 5-6 cuttings at 15-20 days interval. Late bolting variety. Seed yield 7-9 q/ha. |
| Early Smooth Leaf | Smooth round seed, smaller light green smooth leaves with a pointed apex. Seed yield 7-8 q/ha. |
| Shennihon (AG-41) | Prickly seeded variety. Leaves are smooth, green with pointed apex. 4-6 cuttings. Seed yield 8-9 q/ha. |
| SH-1 | $F_1$ hybrid of Verginia Savoy and Japanese. Green semi-savoyed with blunt pointed apex. Seed yield 8-10. |

### Sprouting Broccoli

| | |
| --- | --- |
| DPGB-1 (Palam Samridhi) | Selection from exotic germplasm. Sprouting type. Terminal head, sprouts also appear from the axil of leaves which supplement additional yield at later stage. Sprouts are green, tender, nutritious with better flavour. Each terminal head weights 400 g. Leaves are large green. Seed yield 3-4 q/ha. (Pre-released variety). |

| Vegetable/Variety | Characteristics |
|---|---|
| **Summer Squash** | |
| Australian Green | Bush type, very early and prolific variety. Fruits are dark green with longitudinal stripes of whitish colour all over, 25 to 30 cm in size, very tender and delicious without crooked neck. Seed yield 5-6 q/ha. |
| **Sugarbeet** | |
| Ramonaskaya-06 | Variety introduced from USSR. Root is tapering, long (25-27 cm) smooth, top medium to heavy. Flesh white, juicy containing 16-18 per cent sugar. Ready in 140-150 days to form good roots. Suitable for seed production in dry temperate hills. Seed yield 15 q/ha. |
| Erosetype-E | It was introduced from West Germany. Roots long 20-25 cm, tops medium. Sugar content 14-17 per cent. Suitable for seed production in dry temperate zone. Seed yield 14 q/ha. |
| **Tomato** | |
| **Inderminate** | |
| Solan Gola | Fruits medium, smooth round, turning red at maturity, skin thick and withstands long transport. Needs staking. First picking after 75 days. Seed yield 1.8 to 2 q/ha. |

| Vegetable/Variety | Characteristics |
|---|---|
| Solar Bajr | Selection from Solan Gola. High yielding and resistant variety. Fruit heart shaped, solid, thick skin, weighs 70 g, gets ready in 70-75 days and tolerates distant transport. Seed yield 2-3 q/ha. |
| Yashwant (A-2) | Fruits thick skinned, slightly flattened, uniform in colour. First picking after 70 days, resistant to buck eye rot. Seed yield 1.6-2.2 q/ha. |
| Sioux | Early and prolific bearer. Fruits are medium large, whitish green when unripe, red colour on ripening, roundish, smooth and uniform. Seed yield 1.2 to 2.4 q/ha. |
| Marglobe | Tall and mid-season variety. Grows large, erect and vigorously; fruits are large, globular, smooth, thick walled, juicy, mildly acidic and fair keeper with green shoulder. Seed yield 2-2.5 q/ha. |
| Naveen | Hybrid variety. Average height 75-125 cm depending on training system. Fruits medium sized round, thick skin, with stands long transport, smooth with deep red colour when ripe. Average fruit weight 65-70 g. Gets ready in 80-90 days .Seed yield 1.5 to 2.0 q/ha. |

| Vegetable/Variety | Characteristics |
|---|---|
| PTOM : 9301 | A indeterminate variety developed through hybridization from exotic germplasm. Resistant to bacterial wilt and suitable in rainy season. Fruits are medium in size with 40-50 g weight. Seed yield 1.5-2.0 q/ha. (Pre-released variety). |
| Solan Shagun | Hybrid developed by UHF, Solan for mid-hills. Leaves are deep green colour. Fruit slightly oval deep red, average fruit weight 65 g. Total soluble solids 6 Brix. Gets ready in 70-75 days. Tolerant to early blight and buck-eye rot. Seed yield 2-2.5 q/ha. Suitable for long transportation. |

**Determinate**

| | |
|---|---|
| Roma (EC 13513) | Plants are semi-dwarf, good foliage cover, excellent bearing habit, pear shaped, red colored, borne in clusters and thick walled. Suitable for long distant transportation and best suited for processing *viz.* tomato ketchup and puree. Seed yield 2.0-2.5 q/ha. |
| Rupali | Hybrid variety matures in 80-85 days, 60 cm tall, fruits round and uniform red colour. Packing can be done in baskets. Seed yield 2.0-2.5 q/ha. |
| MTH-15 | Hybrid variety, fruits round, uniform red colour. Gets ready in |

| Vegetable/Variety | Characteristics |
|---|---|
| | 80-85 days. Plant height about 70 cm. Seed yield 1.5-2.0 q/ha. |
| PTOM-18 | A variety developed through hybridization from exotic germplasm. A determinate variety resistant to bacterial wilt and prolific bearer. Fruits are medium in size, oblong in shape. Suitable for hilly areas. Prone to bacterial wilt. Seed yield 1.5-2.5 q/ha (Pre-released variety). Other varieties-NT- 3, Sel- 18, HS-101, Pusa Ruby, Flora-Dada, Punjab Chhuhara, Hissar Arun, Sonali, Ajanta, Swama, Century and Sakthi, Arka Alok. |
| **Turnip** | |
| Purple Top White Globe | Top dark green, erect and leaves are deeply cut, edible roots are large, nearly round, smooth, bright purplish red in the upper part extending above the surface and white in lower portion. Flesh is creamy white, firm, crisp and sweet flavoured. It matures in 60-65 days. Seed yield 12-15 q/ha. |
| Golden Ball | Tops small, erect, leaves cut/lobed. Roots are perfectly globe shaped, medium sized and smooth having bright creamy yellow skin with appetizing, pale amber coloured, refreshing flesh of fine texture and flavour. It matures in 70-75 days. Seed yield 10-14 q/ha. |

| Vegetable/Variety | Characteristics |
|---|---|
| Snow Ball | Much more like purple Top White Globe but roots smaller, round and completely creamy white. Ready in 60 days. Seed yield 10-14 /ha. |
| Pusa Chandrima | Selection-4 from a inter-varietal cross between Snowball x Japanese White. Better performance both in hills and plains. Large sized, globular to flattened globe, smooth with pure white skin. Tops medium and not so deeply cut. Flesh fine grained, sweet and tender. Gets ready in 60 days. Earlier cropper and suitable for October sowing in hills. Seed yield 9-12 q/ha. |
| Pusa Swarnima | Selection-3 from a inter-varietal cross between Golden Ball x Japanese White: Early, tops medium and deeply cut. Flatish round roots with creamy yellow skin. Flesh fine textured good flavour, pale and ambered, coloured. Gets ready in 70 days. Seed yield 10-12 q/ha. |
| Pusa Sweti | Asiatic cultivar suitable for sowing in October. Roots pure white attractive medium in size. Gets ready in 4-45 days after sowing. Seed yield 9-14 q/ha. |
| Vegetable Cowpea | Birsa Sweta, Red Prince, Yard Long Bean, Check Barbati, Pusa Dofasli, Improved Pusa Dofasli and Local varieties. |

# FIFTEEN

## Vegetable Seed Technology

Seed is a basic ingradient for existence, survival, perpetuation and multiplication. Seed comprises one of the most important inputs for crop production. In broad sense, any plant part which is used for commercial multiplication of crop is called *seed*. In strict sense, seed is the product of fertilized ovule that consists of embryo, seed coat and cotyledon(s).

Quality seed plays a vital role in maximizing the production and productivity of crops. Vegetable growers recognize quality seed of improved varieties as the most strategic resource for higher and better vegetable yields. India rank 8th in seed trade the world over.

### Seed quality traits

A quality seed constitutes the kingpin in crop production. The performance of a crop depends upon the quality of the seed used. A good quality seed should essentially embody the following traits :

- It should be genetically pure or true to type.

- A good quality seed must be viable, that is, capable of germination giving rise to a healthy seedling.

- It should have an all-round good health and should essentially be of uniform size, shape, colour, texture, weight, appearance and plumpness.

- A quality seed should free from all kinds of inert materials including the seeds of other crops as well as weed seeds.

- A good quality seed should be free from seed borne disease(s), external insect damage and/or insect eggs, borne externally or internally.

- A good quality should not have been blended or bleached either during development stage or during post-maturity stage.

- It should contain only the prescribed moisture content.

- A broken, peeled off, partially rotten, and a dampness affected seed has poor quality.

- It should have been produced, processed and truthfully labelled *as per* prescribed procedures and standards by an approved seed grower/seller.

## Seed classification

The seed classes denote the different stages in seed production wherein a generation system of multiplication is followed. There are four following classes of seed :

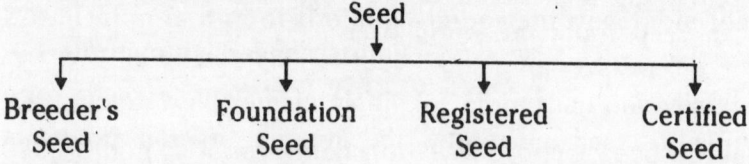

Seed

Breeder's Seed    Foundation Seed    Registered Seed    Certified Seed

## Breeder's seed

Breeder's seed or nucleus seed or propagating material is initially available only in small quantity. It is under the direct control of the breeder concerned. If found suitable for release as a cultivar superior to the existing one on testing at different levels of inputs and management practices, the initial seed is to be multiplied in a specified quantity. The multiplication is done by the breeder himself/herself or it is done under the direct supervision of the breeder concerned. The sponsoring breeder or the institution ensures adequate and timely supply of the seed in required quality both for initial and recurring multiplication of the foundation seed.

The International Convention for the protection of New Varieties of Plants Cordinated by UPOV (1978) with the headquarters in Geneva regulates the grant of right in the office of the World Intellectual Property Organisation (WIPO). Its lebelling tag is white in colour.

## Foundation seed

This seed is also known as *elite* seed in Canada. It represents the second stage (next to breeder's seed) or the propagation material available in limited quantity. It is also so multiplied that the specific genetic identity and purity are maintained. All the same, foundation seed may very slightly degenerate during course of multiplication. The sponsoring institution either identifies the grower or increases this seed by itself and distributes through the pre-identified representatives. It is an important class of seed as it comprises the sole source of all the other classes of seed for further multiplication. Hence it has been rightly called *mother seed*. It also carries white tag.

## Registered seed

It is usually a non-commercial class of seed available as a planting stock for the certified seed production. It represents the progeny of either foundation seed or registered seed so multiplied that the satisfactory level of genetic identity and purity are maintained in accordance with the prescribed standards. The objective is to increase the foundation seed for an additional generation before the production of certified seed. It is essentially to tube proved and certified by certifying agency. This class of seed is eliminated in a number of cross pollinated crops particularly with the species where seed is produced outside the area of adaptation. In a number of cases, no registered seed has produced certified seed. Its multiplication may be done under the special contract between a grower and the certifying agency. This class of seed is lebelled with purple tag.

## Certified seed

This seed represents the final product of a seed production programme. It is the progeny of either foundation seed or registered seed or even certified seed itself. The multiplication is done by the approved growers under the supervision of seed producing agency through seed technicians. A satisfactory level of genetic identity and purity has to be maintained to meet the prescribed standards. It is available in large quantities for commercial crop production. The certification is done by the certifying agency. The seed is processed according to the procedures and tested before sale. There are two classes of certified seed *viz.*, $F_1$ and $F_2$. However, seed from $F_1$ generation is not allowed for recertification.

## History of Vegetable Seed Production

Before World War-II, no one used to produce vegetable seeds on commercial scale.

| | |
|---|---|
| 1719 | Vegetable seeds were being exported to America from Europe. World War-I had a global effect on seed industry and seed supplies from European countries were cut-off and USA and Canada met their needs through indigenous seed production. |
| 1800-1900 | Most of the European vegetables introduced in India by Portuguese and English. |
| 1860 | Vegetable seed mostly supplied by M/S Sutton and Sons, Reading, London. |
| 1867 | Potato Breeder Seed Scheme was initiated at CPRI, Shimla. |
| 1876 | A Hand Book of Seed Testing was published |

and seed testing station was established. The World's first seed testing station was established by Prof. F. Nobbe inTharandt Saxony, Germany.

1908    Association of Official Seed Analysts for standardization of seed testing methods was proposed.

1914    International Crop Improvement Association started the process of joint inspection and certification in USA.

1916    M/S Sutton and Sons established office at Calcutta to supply seed in India.

1922    The Society of Commercial Seed Technologists (SCST) and the Commercial Seed Analysis Association of Canada were organised.

1924    International Seed Testing Association (ISTA) was borne in Norway. Royal commission made recommendations for production and distribution of superior seed material in India.

1925    The first analysis of seed production needs and problems in India was done by the Royal Commission on Agriculture.

1928    On the basis of report issued by the Royal Commission on Agriculture, there should be a complete organisation for the supply of quality seed to the cultivators. Also the private sector of proved integrity should be encouraged.

1939    Association of Official Seed Analysts for standardization of seed testing method was finally established.

1939-45 Temperate vegetable seeds were being imported from abroad.

1942    Seed production of temperate vegetable

varieties was started at Quetta, (Pakistan) as the seed supplies were cut-off due to World War-II.

The topographic or tetrazolium test was developed by Lakon in Germany to know seed viability quickly.

1942-43    Seed production program also started at Katrain (H.P.) and Kashmir valley. Vegetable seed industry made a rapid progress.

1945    M/s Sutton, Pocha's, Palekars etc. developed temperate vegetable seed production facilities in Quetta and Kashmir Valley.

1946    The All India Seed Growers, Merchants and Nurserymen's Association was established.

1947    Supplies of vegetable seeds were cut-off from Quetta after partition of the country.

1948    Vegetable seed industry in Kashmir Valley was considerably affected by disturbances.

1949    Seed production program was started at Central Vegetable Breeding Station at Katrain, Kullu valley by the Govt. of India.

Central Potato Research Institute was established at Shimla-research on development of varieties and production technology on potato.

The first seed testing laboratory was established in Saxony, Germany.

1951    First five year plan started with a program primarily to distribute and multiply seeds.

1955    Central Vegetable Breeding Station, Katrain was transferred to the Indian Agricultural Research

Institute, New Delhi with a view to intensify the improvement work on temperate vegetables and renamed as IARI, Regional Station.

| | |
|---|---|
| 1956 | Second Five Year Plan started with an idea to establish 25 acre farm in each Extension Service Block Setting up Seed Testing Stations to ensure vegetable seed quality standards production of nucleus into foundation seed at block level and distribution among farmers. |
| 1958 | The first Vegetable Seed Testing Station was established in IARI, New Delhi. |
| 1960 | A suitable organisation for nucleus seed production and legislation for quality control was set up. |
| | A Central Seed Corporation was proposed by the State Ministers of Agriculture in August. |
| 1961 | The proposed Central Seed Corporation considered by the State Ministers of Agriculture was approved by the Union Cabinet. |
| | Rock-fellor Society equipped the Vegetable Seed Testing Laboratory (IARI) and was designated as Central Seed Testing Laboratory. |
| | Declared and celebrated as World Seed Year of FAO of United Nations. |
| | Systematic research work on temperate vegetables/varieties, sugar beet and chicory was initiated at Kalpa and Solan (H.P.) |
| 1963 | National Seeds Corporation was established to make all efforts to develop Indian Seed Industry. |
| | Harrington demonstrated the value of moisture proof containers in increasing seed longevity under high humidity conditions. |

|  | Hand Book of Tolerances and Measures of Precision for Seed Testing was published by ISTA. |
|---|---|
| 1963-64 | NSC was made responsible for foundation seed of other crops including vegetables. |
| 1966 | Indian Seed Act was passed by Govt. of India with a view to control quality of seeds on 29th December. |
| 1967 | Seed Plot Technique in potato was developed for raising healthy seed stocks. |
|  | Govt. appointed the Seed Review Team to study in depth the seed industry in India. |
| 1968 | The Seed Rules were framed in India in consultation with ISTA. |
|  | NSC came into full existence. |
|  | NSC established its own Quality Control Seed Testing Laboratory. |
| 1969 | The Seed Act came into force throughout the country on 2nd October. |
| 1970 | Indian Society of Seed Technology (ISST) served as an educational link among Seed Technologists. |
|  | Started publishing Biennial Seed Research and Quarterly Seed Tech. News. |
|  | All India Co-ordinated Vegetable Improvement Project was established at IARI, New Delhi. |
|  | A Center under AICVIP was stated at IARI, Regional Station, Katrain. |
| 1971 | The Central Seed Committee was framed by the Govt. of India to fix genetic purity seed standards. |

First Indian vegetable hybrid Pusa Meghdoot in Bottle gourd was developed and released by IARI, New Delhi.

| | |
|---|---|
| 1972 | Roberts developed formulae to describe the relationship between temperature, seed moisture and period of viability. |
| 1974 | National Seed Project was launched by Govt. of India with the assistance of World Bank. |
| 1974-75 | NSC produced record of 8000 tonnes of vegetable seeds of 29 kinds and 60 varieties. Same quantity of seed was produced by private seed trade. |
| 1976 | International Seed Testing Rules were developed. |
| | National Commission on Agriculture submitted the report reviewing all aspects of seed industry including teaching, training and research. |
| 1976 | Maharashtra State Seed Corporation Ltd. was incorporated under Companies Act with registered Head Office at Akola. |
| 1977 | A National Seed Program financed by World Bank was launched to strengthen breeder, foundation and certified seed production. |
| | NSC started acting as Co-ordinating for planning and advisory organisation for seed production, processing and marketing, creating new and strengthening the existing facilities for seed testing, certification and seed technology research. |
| 1978 | ICAR organised the Eleventh Workshop of Vegetable Scientists at Solan and concluded that H.P. and J & K could meet the entire demand of |

the country for quality seed of temperate vegetable/varieties, sugar-beet and chicory

| | |
|---|---|
| 1982 | Govt. formed an inter-agency apex body, the National Biotechnology Board. |
| 1983 | Enforcement of the Seeds (Control) Order governing quality of seeds through Seeds Act and Seed Rules. Seeds have been declared as an essential commodity. |
| 1985 | Seed Testing Procedures in India were followed according to the rules prescribed by ISTA. |
| 1986 | Elevation of status of AICVIP to the level of Project Directorate of Vegetable Research. |
| | ISST organized the All India Seed Seminar on Indian Seed Industry: Retrospects and prospects on Nov. 11-13th at New Delhi. |
| 1988 | Indian Minimum Seed Certification Standards were reviewed. |
| | Announcement of New Seed Policy called New Liberalized Seed Policy by Govt. of India on seed development on 16th September. |
| | A specially designed machine with axial flow vegetable seed extraction was developed at PAU, Ludhiana. |
| | Indian Minimum Seed Certification Standards published by the Central Seed Certification Board; Department of Agri. and Co-op., Mini. of Agric., Govt. of India, New Delhi in July. |
| 1989 | There was 22.27% increase in vegetable production in India over 1979–81. |
| | Seed Industry sought further incentives/concessions in a meeting with the Secretary A & C, Govt. of India. |

| | |
|---|---|
| 1991 | Navdana (Nine seeds), the seed keepers was established by the Research Foundation for Science and Technology and National Resource Policy (NGO). |
| 1992 | Delinking of PDVR form IARI and shifted to Varanasi (U.P.) at its new Headquarters. |
| 1994 | Drafting of legislation on plant variety protection law was made public in India on 2nd Feb. |
| 1996 | ICAR launched a special project promotion of hybrid research in vegetables at 4 ICAR Institutes and 11 SAU'S. |
| 1997 | A Designated Member Laboratory of ISTA and also a notified Seed Testing Laboratory under Seeds Act, 1996 was shifted to the main building of the NSC Headquarters at Pusa Complex, New Delhi. |
| 1998 | Applied Training Course Agro Forestry Seeds at Ranchi organised by NSC Ltd. 3-10th March. |
| | The American Seed Trade Association and the Canadian Seed Trade Association 115th Annual conference 21-25th June at Toronto Ontario, Canada. Asian Seed 98 Conference at Manila Philippines, 23-25th Sept. |
| 1999 | 6th International Workshop on Seeds at Merida, Mexico Jan. 24-28th. |
| | World Seed Conference, Cambridge GB-United Kingdom September 6-8th. |
| | The World Seed Conference co-organised by ISTA, FIS/ASSINSEL, OECD and UPOV held 6-8th Sept., 1999 in Cambridge, U.K. where ISTA was founded in 1924. |

Retooling the Seed Industry for a Global Market. The 21st Annual Seed Technology Conference organized by the Seed Science Center of Iowa State University, Y2K, Feb., 15-17th, 1999.

Seed Testing, Seed Certification and Field Plot Testing organized by The Danish Plan Directorate from Feb. to June, 1999.

International Course on Seed Production and Seed Technology, organized by IAC, in co-operation with Wageningen Agricultural University on 11-15 July, 1999.

## Seed production organisation

To strengthen the vegetable seed industry in the country, two types of government/public sector organisations responsible for seed production and certification in India. The first type of organisations, are represented by the National Seeds Corporation (NSC), which has responsibilities for the entire country. The second type of organisations are State Seeds Corporations (SSCs) and State Seed Certification Agencies (SSCAs) that have state wise responsibilities.

### National Seeds Corporation

The National Seeds Corporation (NSC) was initiated in 1961 under the Indian Council of Agricultural Research. Later, on 7 March, 1963, it was registered as a limited company in the public sector. The headquarters of NSC is in Pusa Campus, New Delhi. The NSC was established to serve two main objectives :

1. To promote the development of a seed industry in India, and

2. To produce and supply the foundation seeds of various crops.

NSC produced 12 tonnes of seeds of four vegetables in 1974-75. It produced 8000 tonnes of vegetable seeds of 60 varieties of 28 different kinds. Some of the functions of NSC have now been delegated to the other two agencies, *viz.*, State Seeds Corporations, and State Seed Certification Agencies.

## State Seeds Corporations

The National Seed Project was launched in 1976 to establish State Seeds Corporations. The States Seeds Corporations are chiefly concerned with the production and supply of certified seed, and within the state marketing of certified seed. State Seeds Corporations have been recently established in order to reduce the work load of NSC. These corporations were established in view of the great success of and the impact made by the Tarai Development Corporation (TDC), Pantnagar (established on February 27, 1969), which had gained a virtual strength hold on the seed market of U.P. almost to the extent of exclusion of NSC. It is hoped that the State Seeds Corporations would able to function more efficiently and would be able to stimulate a faster growth of the vegetable seed industry.

## State Seed Certification Agencies

The State Seed Certification Agencies (SSCAs) are responsible for seed certification in the concerned states. The inspectors of SSCAs make required number of field inspections and conduct purity and germination test for certification purpose. SSCA also organizes short term training courses for seed growers, keeps check on the source of seed used for raising seed crop etc.

Each state has a State Seed Certification Board which supervises the activities of its SSCA.

Persons involved in seed processing and distribution, including businessmen dealing with production, processing and marketing of seeds, and scientists for agricultural universities, are the members of this board.

In addition, there is a Central Seed Certification Board, which advises the state governments and their SSCAs on the matters of seed certification; the chairman of this board is nominated by the central government. The members of this board are drawn from among the officials of the different state departments of agriculture scientists from the agricultural universities, and persons from seed industry. The board may also appoint committees for specific task.

## Private Seed Companies

There are several private companies that take up the work of quality seed production. The names of some private seed companies are :

- Maharashtra Hybrid Corporation, Jalna.
- Pioneer Seed Company, Hydrabad.
- Nath Seed Company Pvt. Ltd., Aurangabad.
- Hindustan Lever Ltd., Bombay.
- Proagro Seed Company, New Delhi.
- Ankur Seed Company, Nagpur.
- Vijay Seed Company, Jalna.
- Mahendra Seed Company, Jalna.
- Indo-American Hybrid Seeds, Bangalore.

These seed companies are engaged in production of hybrid seeds of various crops including vegetable crops. According to an estimate, the private sector produces about 60% of the seed sold in the country.

## Seed certification

Seed certification is a scientific and systematically designed, legally sanctioned process for quality control of seeds through verification of seed sources, field inspection of seed crop, supervision at post-harvest stages including processing and packaging, seed sampling and analysis, grant of certificate, certification tags, labels and sealing.

## Objectives

To ensure genuineness and quality of seed to the purchaser, provision of a continuous supply of comparable material by careful maintenance is a must.

With the promulgation of the Seeds Act, 1966, certification system is accepted for notified varieties of crops. For implementing the provisions of section 8 of the Act, Seed Certification Agencies have been established under the Societies Act, 1886 at State level and a Central Seed Certification Board has been established under section 8(a) of the Act, amended in 1972, primarily to render advisory services on scientific and operational matters to the Central Government and State Seed Certification Agencies.

## Certification standards

To determine the quality of the seed, prescribing the minimum standards under the certification programme is essential. These are formulated by the officially constituted committees or agencies. These standards are required for the verification of varietal purity, isolation, seed-borne diseases, weed seeds and other crop seeds, germination, seed health, moisture content at field inspection of the standing crop and during the tests conducted in the laboratory. Under the Indian Seeds Act, 1966, seed certification standards have been developed and are continuously reviewed for making necessary amendments according to the needs and based on scientific findings over the years.

The Indian Minimum Seed Certification Standards should be applicable uniformly throughout the country. These certification standards are in two parts – general and specific. Only those varieties are eligible for certification which are notified by the Central Seed Committee.

## Seed testing

Seed testing is very important for distinguishing good quality seed from the seed below the prescribed standards to ensure the planting value of the seed. The International Seed Testing Association (ISTA), established in 1924, has developed uniform rules and regulations for seed testing. These rules prescribe testing techniques based on scientific evidence, which are accurate within the stated statistical limits and practicable for everyday operations. The international rules and regulations for seed testing should be followed.

The trained manpower and a good laboratory with standard equipments are the prerequisites for seed testing programme. The physical infrastructure and facilities should be planned on the basis of average expected workload during the peak season for efficient handling without undue delay. The seed testing process can be subdivided into:

### Seed sampling

A sample from a seed lot is obtained by taking small portions (primary sample) at random from different portions in the lot and combining them (composite sample). From this sample, smaller samples (submitted samples) are obtained for sending to the laboratory. The samples are obtained preferably with the help of seed drier from the seed lot. A part of the submitted sample is known as *working sample* in the laboratory which is subjected to quality tests.

### Moisture determination

The object of moisture determination is to find out the moisture content of seed at the time it was sampled.

To ensure this the sample must be submitted in a moisture proof container and analysed without any delay. The moisture can be determined by hot air oven method. This is the most standard method used commonly. Its principle is elimination of water from the seed by heating under precisely controlled conditions. The moisture content is expressed as percentage of the original weight. The quick method should be caliberated against the standard air oven method.

## Purity analysis

The working sample is separated into 3 component parts- pure seed, other seeds and inert matter. The percentage of each part is determined by weight. The seed quality is considered superior if pure seed percentage is above 98. The percentage of seeds of other species and inert matter should be nil or negligible. The purity test is done with the object of determining the composition by weight of the sample being tested, and by inference, the composition of the seed lot. The result of purity analysis should be given to one decimal place and the percentage of all components must be 100. Components of less than 0.05% shall be reported as "Trace".

## Evaluation for germination

It is necessary to precisely understand germination, percentage of germination, normal seedlings, abnormal seedlings, hard seed, fresh ungerminated seeds and dead seeds.

The paper, sand and soil are permissible as substrata. The test may be conducted either in cabinet germinators or specially assigned rooms for the purpose as walk in germinators. The rules specify that at least 400 seeds should be tested for germination in replications of 100, 50 or 25 seeds. The seeds from the pure seed fraction of purity test are used for the germination test. Every seedling is evaluated in accordance with the general principles prescribed in the ISTA rules. For evaluation, the stage of development of the essential structures must be sufficient to permit detection of any abnormal seedlings. At the end of normal germination test, the percentage of ungerminated seeds may be counted.

### Bio-chemical test for seed viability

These tests are conducted to determine the viability of seeds of certain species which germinate slowly or show a high degree of dormancy. When the germination test reveals a high percentage of dormant seeds in a sample, the sample is subjected to a bio-chemical test to determine the viability of a working sample or of individual seeds.

### Verification of genetic purity

This is required to determine the extent to which the submitted sample confirms to the variety claimed for it. The characters to be compared may be morphological, physiological, cytological or chemical. The test is carried out on seeds, seedlings, more mature plants grown in glass house or growth chamber or field plots. The results are expressed as percentages of the number of seedlings examined.

## Seed packaging

Seeds should be packed in smaller units to avoid risks of physical gradients, particularly vapour pressure which arise in large bulks. Packaging in smaller units makes identification, transportation, handling and marketing easier. The choice of packaging materials and amount of seeds to be packed will depend on several factors such as the kind of seed, duration of storage, storage environment, moisture content. Packaging materials can be classified into 3 types:

**Moisture-vapour permeable:** Jute bag, cloth bag, paper bag etc.

**Moisture-vapour resistant:** Jute bag laminated with thin polythene film.

**Moisture-vapour proof:** Tin cans, polythene bags of 700 gauge or more, aluminium foil pouch etc.

### Packaging operations

● Filling of seed bags to an exact weight.

● Placing leaflets in the seed bags regarding improved cultivation practices, if needed.

● Attaching labels, certification tags on the seed bags and sewing of the bags.

● Storage/shipment of seed bags.

The baggage weigher, bag closer and platform weighing scales are the equipments used for seed packaging. Paper bag, cardboard box, aluminium foil pouches, polythene bags etc., are used for packaging vegetable seeds. If moisture proof packings are generally used, the moisture content must not exceed beyond a critical level. Adequate facilities, therefore should be available for seed drying.

### Seed storage

The storage of seeds may be short term (3-9 months but occasionally up to 18 months), medium term (18 months to 5 or 6 years) and long term (up to 10 years or more). The purpose of seed storage is to maintain the seeds in good physical and physiological condition from the time they are harvested till they are planted.

### Factors affecting seed longevity in storage

● Kind of seed

● Initial seed quality

● Moisture content

● Temperature and relative humidity during storage

## Principles of seed storage

- Storage in a cool and dry place
- Effective storage pest control
- Proper sanitation in seed store
- Drying seeds to safe moisture limits before storage
- Controlling storage conditions depending upon length of storage period and prevailing climatic conditions
- Storing high quality seeds only

## Quality control in storage

- Stacking
- Germination and vigour tests
- Temperature control (ventilation, insulation, refrigeration)
- Moisture control (ventilation, moisture proofing, dehumidification, sealed containers, desiccants)

## Pre-storage preventive measures

(i) Before arrival of new produce, all processing and storage structures should be thoroughly cleaned and disinfected with residual sprays of insecticides, e.g. Malathion 50 EC (one part in 25 parts of water) @ 5 litres/100 m³ area.

(ii) The seed moisture content should be reduced to specified levels. Most species of insects don't breed at such a low moisture content.

(iii) Delay in processing and subsequent storage damages the seed. In case of insect infestation, seed should be fumigated with Aluminium phosphide @ 2 tablets of 3 g each/tonne with an exposure period of 3-5 days. Moisture content of seeds to be fumigated should not be more than 12%.

(iv) Processed seed may be treated with malathion 5% dust @ 0.5 g/kg seed.

(v) For seed storage, preferably new bags should be used to avoid insect infestation and mechanical mixture and processed seed should not be stored in an area where unprocessed or carry-over seeds are kept.

During storage, different types of seed should be stored separately and inspected at fortnightly intervals. If required, the fumigation may be done and after fumigation the godown should be cleaned and all dead insects removed. To prevent reinfestation, surface treatment with 5% Malathion dust @ 3-4 kg/100 m² should be given.

## Processes of seed formation and development

A true seed is defined as a fertilized mature ovule consisting of embryo, stored material and protective coats. Actually seed is a ripened ovule containing embryo or unit of reproduction.

Fruit is a ripened ovary containing seeds. The purpose of a flower is the production of seed and the purpose of seed is the production of flower.

**Important steps**

1. Microsporogenesis,
2. Megasporogenesis,
3. Pollen germination,
4. Fertilization,
5. Embryo and seed development and
6. Seed maturity.

## Objectives

1. To study the fertilization process.

2. To study the embryo and ovule development by cell division.

3. To acquaint with accumulation of reserve food materials.

4. To study seed maturity.

5. To study loss of moisture content in seed.

## Microsporogenesis

Process occuring within the anther which finally results in the formation of pollen grains. The microspore mother cell undergoes two nuclear divisions (called *meiosis*) which result in the formation of four cells called *microspores* which upon equational division, become the pollen grains. The number of chromosomes within any of the four cells is one half of that found in the original micro-spore mother cell and is known as the *genetic number*. The pollen nuclei further divides mitotically into two each called *sperm* with 'n' number sperm cells (gametes).

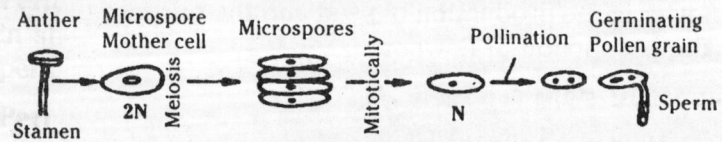

*Microsporogenesis*

## Megasporogenesis

The process occuring within the ovule which finally result to in the formation of the embryo sac on the female side. The corresponding development occurs several days after the formation of male element. The ovule at first arises as a tiny protuberance from the placenta in the cavity of the

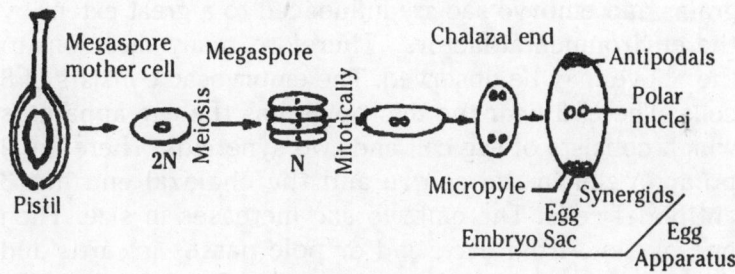

*Megasporogenesis*

ovary. In it at a very early stage a cell i.e. MMC (megaspore mother cell) of the embryosac becomes evident in the nucleus. This MMC increase in size and divides twice. Following the formation of the megaspore. mother cell (megasporocyte), the meiotic division occur, giving rise to à linear tetrad of 4 spores (megaspores), 3 of which normally degenerate. This megaspore which will develop into the embryosac, enlarges in size and its nucleus divides the two resulting nuclei give rise to 4 and by a third division the 8 nuclei of the embryo sac (mega gametophyte) are formed. Both thése development (micro and megasporogenesis) leading to the formation of pollen

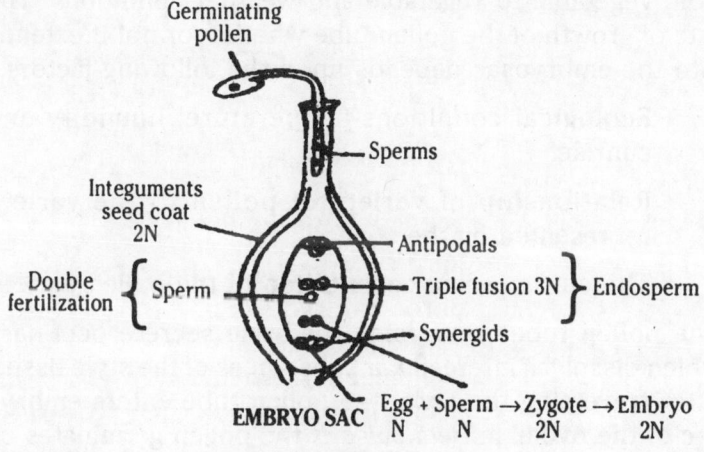

*Result of fertilization process*

grains and embryo sac are influenced to a great extent by the environmental factors. Therefore, many variations to the above may be observed. The embryosac consists of 8 cells. The end near the micropyle has the egg apparatus which consists of egg cell and two synergids. There are 2 polar nuclei in the centre and the chalazal end has 3 antipodal cells. The embryo sac increases in size. Then one nucleus from each end or pole passes inwards and the two polar nuclei fuse together somewhere in the middle, forming the definitive nucleus. Synergids are more or less pear-shaped and the egg cell which is enlarged lie below them.

## Pollen germination

The pollen grain after landing on the stigma germinates and long slender pollen tube grows through the style. The surface of the stigma secretes substances which may provide optimum conditions for pollen germination and subsequent pollen tube growth. Pollen germination is dependent upon the receptivity of the stigma, a condition assumed to be associated with the secretion of the stigmatic fluid. Period and duration of receptivity of stigma varies from vegetable to vegetable and weather conditions. The rate of growth of the pollen tube whether or not it extends into the embryosac depends upon the following factors:

- Ecological conditions-temperature, humidity and sunrise.

- Relationship of variety of pollen to the variety represented by the style.

- The chromosome constitution of the pollen variety.

The pollen tubes traversing the style secrete pectinase which dissolves intercellular substances of the style tissue. After traversing the style, the pollen tube enters embryo sac of the ovule. In *Beta vulgaris* the pollen germinates on

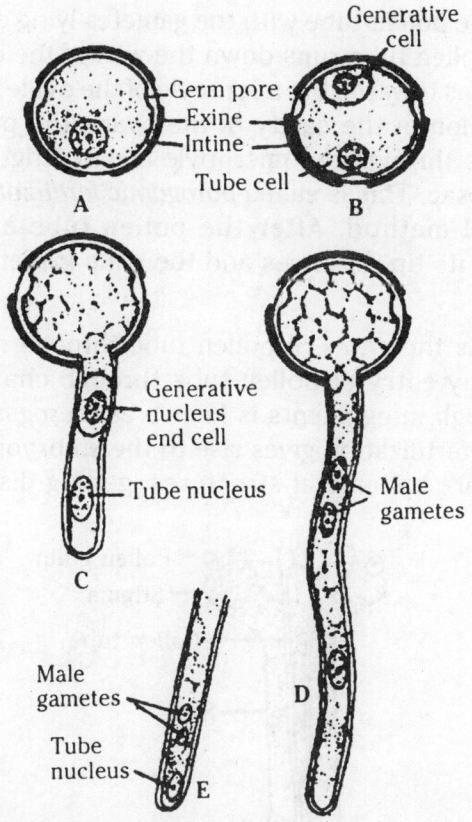

*Pollen and its germination. Changes inside stigma, style and ovary*

the stigma in two hours. After pollination, the pollen grains reach to stigma, the intine of each grows out into a tube called *pollen tube*. The growth of pollen tube is stimulated by the sugary substances secreted by the stigma. The tube penetrates carrying with it the tube nucleus and generative nucleus. The generative nucleus soon divides forming two male gametes, which the tube nucleus gets disorganised sooner or later. Sometimes the generative nucleus divides before pollination. A mass of cytoplasm accumulates at the tip of the pollen tube with the gametes lying embedded in it. The pollen tube runs down the wall of the ovary and finally it turns towards the micropyle of the ovule, whatever be its position in the cavity of the ovary, the pollen-tube then passes through the micropyle and at length reaches the embryosac. This is called *porogamic fertilization* and is the normal method. After the pollen tube enters the embryosac its tip dissolves and the male gametes are set free.

Porogamy is the entry of pollen tube through micropyle, chalazogamy entry of pollen tube through chalaza while entry through integuments is known as *mesogamous*. The egg-cell on fertilization gives rise to the embryo; while the synergids are ephemeral structures, getting disorganized

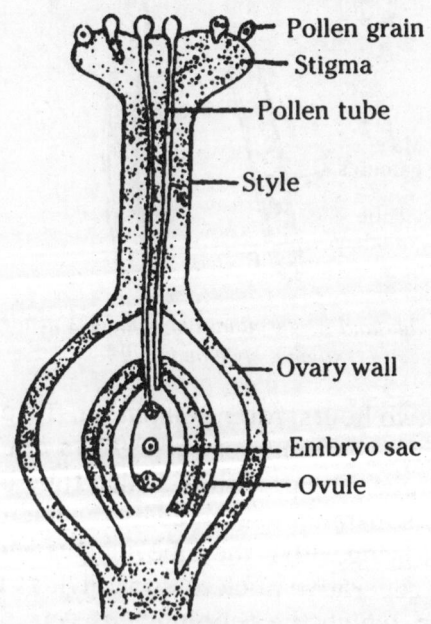

*L.S. of ovary and ovule-germination of pollen and course of pollen tube*

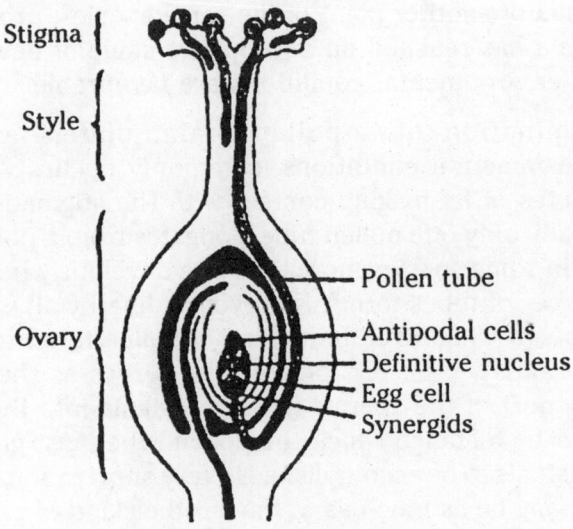

Stigma

Style

Ovary

Pollen tube
Antipodal cells
Definitive nucleus
Egg cell
Synergids

*L.S. showing process of fertilization*

soon after fertilization or even before or during it, the antipodal cells appear to have no definite function; so sooner or later, they also get disorganised. They may however be nutritive in function. They possibly represent vegetative cells of the extremely reduced female gametophyte. The definitive nucleus on fertilization (now called the *endosperm nucleus*) gives rise to the endosperm. Numerous pollen grains may alight on or be deposited on a single stigma. Under conditions of abundant pollination, from 600 to 800 pollen grains may be present on a single stigma. Even foreign pollen grains sometimes germinate on a stigma, but growth of the pollen tube usually fails to occur or takes place very slowly. If a pollen grain from one plant falls on a stigma of the same plant, germination usually occurs, although not always. Pollen from a given plant often fails to germinate on the same individual plant; or if germination occurs, the pollen tube grows sluggishly. This is one cause of self-sterility in plants. On the other hand, pollen from a given plant usually germinates on the stigma of another plant of the same species, provided the stigma has reached on appropriate stage of development and environmental conditions are favourable.

Germination of a pollen grain, under favourable environmental conditions, commonly occurs within few minutes of its making contact with the stigmatic surface. Usually only one pollen tube elongates from a pollen grain, but in some species more than one develops when a plural number of tubes forms, however, ordinarily all except one cease growing. Even branching of pollen tubes may occur in some species. The pollen tube serves as the mode of transport of the sperms from the stigma into the embryo sac. The distance which the pollen tube must grow if this transfer is to be accomplished is very short in some species, but may be as much as 30 cm in other kind of plants such as sweet corn, with long styles.

The time interval between germination of a pollen and fertilization differs greatly from one kind of vegetable to another, but for many lies, within the range of 12 to 48 hours. In a few vegetables however, this time interval is less than an hour and in some it may be months. The absolute rate of elongation of pollen tubes ranges up to about 34 mm per hour. The rate of growth of pollen tubes is markedly influenced by environmental conditions, especially temperature. In tomato, for example, their maximum rate of growth occurs at about 20°C, being less at higher and lower temperature. Apart from environmental conditions the rate of pollen tube growth is markedly influenced by the degree of physiological compatibility between the pollen tube and the tissue of the pistil. Incompatibility usually exists between the pollen and pistillary tissues in many species when self-pollination occurs. Although, germination of the pollen grains may occur, the pollen tubes grow very slowly, and fertilization seldom results. When cross-pollination occurs in the same species, however, rapid growth of the pollen tube is the usual occurrence. The length of time for which pollen grains of different vegetable species remain viable under ordinary air dry conditions ranges from few hours to several days. The life duration of many vegetable varieties of pollen can be extended to a desired period by using a various techniques when it is desired to cross two varieties/ species which bloom at different seasons.

## Fertilization

Union of the male and female nuclei. Fertilization involves the union of both sperm nuclei from the pollen tube with the egg nucleus and the polar nuclei of the embryosac. One sperm nucleus unites with the egg nucleus to form the diploid zygote (embryo) while the other fuses with the polar nuclei or with a single nucleus, in case the polar nuclei have previously fused, to produce a triploid nucleus which form the primary endosperm nucleus. This union of the two sperm nuclei with the nuclei in the embryo sac is called double fertilization. That failure to set fruit is occasionally due to the fact that *double fertilization* did not occur seems probable. Fertilization depends upon the chromosome constitution of the nuclei involved in the process. In certain varieties of many species, fruits may develop without fertilization and such fruits are termed *parthenocarpic*. There can be many varieties of the noble/normal phenomena.

## Triple fusion

Fusion of male gamete with polar nuclei produce triploid nucleus which is the primary endosperm nucleus. Sometimes there is syngamy or triple fusion. Syngamy is the fusion of the male gamete with the egg. The time involved between pollination and fertilization varies a good deal in different plants. Generally, the time taken is from a few hours to a few days, but in some cases from a few to several months.

## Seed development

The fertilized egg develops into a rudimentary plant, the embryo of the seed, the starting point of the next plant germination. The fertilized polar nuclei develops into a tissue called the *endosperm* which surrounds and nourishes the growing embryo. The endosperm in most vegetable seeds is absorbed completely by the time the seed matures. Seeds of bean, watermelon, garden pea and pumpkin contain no endosperm. After fertilization, the embryo, which starts as a single cell, grows rapidly and the ovules expand to accommodate the enlarging structures within. The embryo is a mass of undifferentiated cells in its early stages. As enlargement continues, three well defined structures are formed.

(i)   The epicotyle or young shoot.

(ii)  The hypocotyle or young root.

(iii) The one or two cotyledons or seed leaves.

Usually the cotyledons of the embryo become thickened to allow storage of food materials, such as starch, sugar, oil or protein. This process is indicative of the seed maturity. The period of filling of the embryo or endosperm, is one of stress on the mother plant, because large amounts of organic food materials need to be manufactured by the leaves and transported to the developing seed. During the course of seed development, reserve food materials are accumulated in the endosperm from the adjacent tissues. Finally, enlargement of the embryo ceases, the parts become dry, the seed becomes a dormant living organism capable of withstanding adverse conditions. Structurally, the seed is a resting embryo plant, which is surrounded by a seed coat and may have an endosperm. The embryo has one or more cotyledons, which in many cases serve as foliage leaves after the germination of seed. One of the wonders of the seed is that the entire above ground part of the plant develops from the tiny epicotyl, and the elaborate root system originates in the small hypocotyl. Since lot of energy or food supplied are needed for the completion of these processes, the seed is important as a storage organ.

## Seed coat development

Integuments of the ovule undergo marked reorganization and histological, changes during maturation to form seed-coats in bite genic ovules (which have two integuments), the seed coat may be derived from both the integuments or from the outer integument only; the inner integument may disintegrate. The seed coat or testa is developed from one or two integuments of the ovules, which form a covering to protect:

- the embryo against drying out

- the mechanical injury

- attacks by insects, fungi and bacteria

Usually the outer coat is hard dry and durable (exine) and the inner (intine) one is thin and membraneous. Often the two seems to be fused into one layer. The seed coat and fruit wall may develop appendages or special structures that help the seed to certain modes of dissemination and protection. The development of seeds is usually an outcome of the process of pollination, it is by no means an invariable one. Failure of seeds to form a usually cross-pollinated species is self-pollinated or when cross-pollination occurs between different species or varieties of the same species. The principal causes of a failure of seeds to develop are:

- Pollen fails to germinate after pollination.

- Pollen tubes may grow too slowly for fertilization to be accomplished or the tubes may burst before reaching the embryo sac.

- Fertilization fails to take place.

- Fertilization occurs, but abortion follows before more than a few divisions of the fertilized egg occur.

- Fertilization occurs and embryo growth proceeds but is arrested at a later stage at its development.

Failure of the embryo to develop to maturity appears to result from physiological conditions inherent in the young embryo or in the surrounding endosperm.

The first four conditions listed above commonly lead to the development of empty seeds, the last one usually results in the formation of shrunken and usually non-viable seeds. The seed usually matures at the same time the fruit ripens.

## Embryo development

The first division of the zygote is transverse and it results in a small apical cell and a large basal cell. The apical cell divides vertically forming two just aposed cells and basal cells undergoes a transverse divisions forming two superimpose cells. This results in a T-shaped 4-celled proembryo. The first superimpose cell divides transversely giving rise to n & n. These two cells divide further resulting in a row of 3 or 4 cells, forming suspensor. The second superimpose cell and its derivatives undergo vertical divisions forming a group of 4 to 6 cells. This group divides by oblique-periclinal walls forming a set of inner cells and a row of outer cells. The inner cells form the initials of the root apex and the outer cells form the root cap. The 2 cells formed as a result of the division of apical cell again divide vertically forming quadrants. Each cell of the quadrant divides transversely and thus an octant containing two tiers of cells are formed. The cells of the octant undergo vertical division resulting in a globular proembryo. Periclinal divisions occur in the peripheral cells of the globular proembryo which delimit an outer layer, the dermatogen.

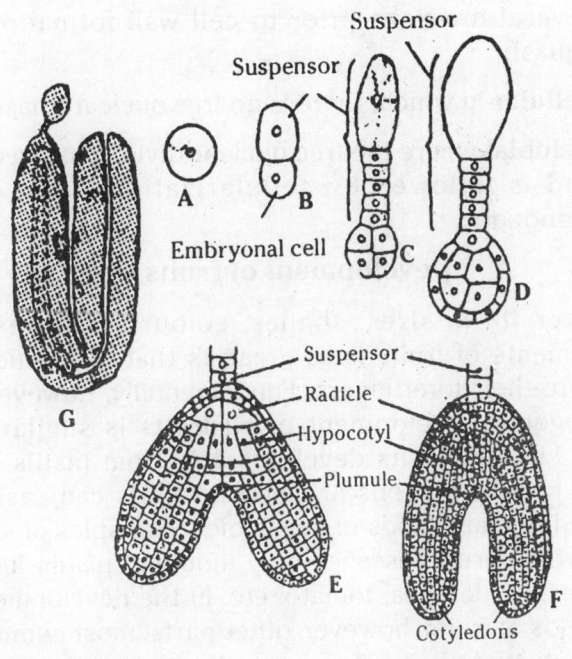

*Development of embryo A-Fertilized egg. B,C and
D-Development of suspensor and young embryo. E, F-Differentiation of
embryo. G-Mature dicotyledonous embryo*

The tier gives rise to cotyledons and shoot apex and other
forms hypocotyle-radicle axis. In monocotyledons, the
apical cell remains undivided and develops into a
haustorial cell of the suspensor. Apical cell divides into two
by a transverse division. The terminal cell of these two by
repeated divisions in different planes gives rise to a single
cotyledon. The other cell by a number of divisions gives
rise to stem apex, hypocotyl and root tip.

## Endosperm development

There are three type of endosperm development :

(i)   **Nuclear**-where the endosperm nucleus undergoes

several divisions prior to cell wall formation e.g. squash

(ii) **Cellular**-in which, there is no free nuclear phase, and

(iii) **Helobial**-where the free nuclear division is preceded and is followed by cellularization, as in some monocots.

## Development of fruits

The diversity in sizes, shapes, colours, textures and arrangements of fruits is as great as that of the flowers which are their forerunners. Fundamentally, however, the morphogenic development of all fruits is similar. The simpler kinds of fruits develop solely from pistils. Such gradual transformations of pistils to fruits can easily be observed in many kinds of vegetables. Examples of simple fruits, which are in essence only modified pistils include the beans, garden pea, tomato, etc. In the development of some kinds of fruits, however, other parts, most commonly the receptacle or floral cup, ripen along with the pistil and are incorporated into the final structure of the fruit e.g. cucumber.

In general, development of a fruit and its enclosed seed or seeds occurs concomitantly and in a reciprocally coordinated fashion. During the transformation of some pistils or pistils plus adjacent parts, into fruits a considerable enlargement may occur, as much as several hundred fold in some vegetables. The resulting mature fruit tissues may be soft and fleshy as in tomato or hard and dry as in gourds. In other vegetables development of floral organs in fruits may occur with little enlargement of the tissues, the resulting fruits more frequently being dry and hard than soft and fleshy.

The processes of pollination and fertilization exert marked influence on the development of most fruits. Failure of pollination to occur usually results in obsession of the pistil and a consequent failure of fruit to form. The influence of pollination upon the development of fruits is; probably mediated at least in part through auxins. Pollen grains and tubes are known to contain auxins in considerable quantities. It seems unlikely however, that the quantity of auxins furnished by the pollen tubes is sufficient to be the sole cause of fruit development. Ovularies contain very little active auxin but considerable quantities of bound auxin. A primary effect of its increased auxin content is the prevention of abscission of the pistil. Soon after pollination, but before fertilization, the ovules and ovularies start to grow. Enlargement of these organs, at least during the early stages of fruit development is presumably induced by auxin set free. A number of vegetables are now known in which formation of parthenocarpic fruits can be induced by application of suitable chemicals in landing to the style of young flowers in the absence of pollination e.g. water melon, tomato, egg plant and pumpkin. The principal compounds known to be effective in the induction of partheocarpic fruit formation are indolepropionic, indolebutyric, phebylacetic, $\alpha$-naphthalene acetic, $\alpha$-naphthoxyacetic and phenoxyacetic acids. Most of the compounds are auxins and the effective concentrations ranges 50 to 3000 ppm.

To identify seeds and fruits correctly, developments of these parts to maturity needs to be followed, the botanical definition of a fruit is much broader than the popular meaning of the word. For instance, the mature bean pod is the fruit of the bean plants and bean seeds are ripened ovules. The bean pod originates in the bean flower as a minute ovary which carries minute ovules. Since fruit and seeds are present in from miniature form in the flower as ovary and ovules, hence the importance; of the flower in the development of the seed is well established.

In sweet corn many ovaries are arranged together on a common receptacle, which later becomes the cob of the ear of corn. A grain of sweet corn is a familiar example of a single seeded fruits commonly called a *seed*. (sweet corn produces 20-25 m pollens/plants). Lettuce, carrot etc. also produce one seeded fruit. But the ovary of the watermelon flower possess many ovules which produce the many seeded watermelon fruit.

The embryo after its complete differentiation becomes dormant.

## Fate of endosperm and seed formation

As the embryo is developing, the endosperm also goes on growing at the expense of nucellus. The endosperm ultimately fills the whole ovules and replaces the nucellus completely. In some plants, the nucellus may persist outside the endosperm in the form of a thick layer called *perisperm*. The cells of the perisperm are stored with reserve food material and provide nourishment to the young embryo at the time of germination.

## Seed producing regions

The growth of plant and quality seed production are strongly influenced by genetic factors and environmental conditions. These factors include temperature, rainfall, wind velocity, soil condition and insect activity and their relationship with varietal adaptability in any given locality. These factors are natural and many of them can be modified to a great extent to the optimum level for the production of quality seed. Thus it becomes easier to classify various agroclimatic zones successfully to plan production programmes of different kinds of vegetable crops. Production technology for seeds varies from location to location and from crop-to-crop. However, a broad general recommendation can be adopted which could be suitably modified on the basis of individual

## Seed-production regions

| Vegetable crops | Regions |
| --- | --- |
| Deccan Plateau of North Andhra Pradesh (Telengana region), North Karnataka, Eastern Maharashtra including Marathwada and Vidharbha | Brinjal, chilli, cucumber, gourds, okra, onion and tomato |
| Coastal districts of Andhra Pradesh, Tamil Nadu (small areas only), Rajasthan (Kota) Nimar Valley and Indore region of Madhya Pradesh | Bottle gourd, bitter gourd, pea. cauliflower (mid-season), okra, onion and watermelon |
| Gujarat and Rajasthan (Sriganganagar) | Bitter gourd, cauliflower, cucumber and pea |
| Uttar Pradesh: Districts of Etawah, Kanpur, Aligarh, Mainpuri and Bareilly | Bitter gourd, cucumber and mid-season cauliflower |
| Hajipur region of North Bihar and Saproon Valley (Solan), Mandi District of Himachal Pradesh and Jammu and Kashmir | Bitter gourd, cucumber and early cauliflower |
| Himachal Pradesh (Lower hills) and Kashmir Valley | Beet, cabbage, capsicum, carrot, knol-khol and radish (temperate) |
| Punjab and Haryana (small areas) | Asiatic radish, carrot, cauliflower, okra and turnip |
| West Bengal | Leafy vegetables, pumpkin and tomato |
| North Bengal (Kalimpong) | Capsicum. cauliflower, French bean, leafy vegetables. pumpkin and tomato |
| Bomdila district of Arunachal Pradesh | Cauliflower (snowball), French beans, radish (temperate) |

vegetables. Timely operations not only ensure a rich harvest but also guarantee varietal purity and freedom from undesirable weeds, diseases and pests. Emphasis should always be laid on those factors, which contribute to and affect seed quality e.g., seed source, method of sowing, roguing, harvesting and postharvest operations.

## Self-fertilized solanaceous vegetables

### Solanaceous vegetables

### Tomato

*Determinate types:* Arka Meghali, CO 3, Hissar Anmol, Hissar Lalima, Hisar Lalit, HS 101, HS 102, HS 110, KS 2, La-Bonita, Pusa Early Dwarf, Punjab Chhuhara, Pusa Gaurav, Punjab Kesri, Pusa Sheetal, Roma, S-12, Selection 7.

*Indeterminate types:* Arka Saurabh, Arka Vikas, Best of All, Marglobe, Pant Bahar, Pusa Ruby, Pant T-1, Pant T-3, Sel-120, Sioux.

### Brinjal

*Long type:* Arka Keshav, Azad Kranti, Arka Nidhi, Arka Sheel, BE 26, Hissar-Jamuni, PPC, PPL, Pusa Anupam, Pusa Kranti, Pant Samrat, Punjab Barsati.

*Round type:* Arka Navneet, Aruna, Hissar Shyamal, Jamuni Gol, KS 224, Punjab Bahar, PPR, Pant Rituraj, T-3.

### Green brinjal

Arka Kusumakar, Arka Shirish, Green Long, Rajendra Baigan, Ramnagar Giant

### Chilli

Andhra Jyoti, Bhagya Lakshmi, Kalyanpur Type 1, Kalyanpur Type 2, NP 46 A, Pant C I, Punjab Lal, Pusa Jwala, Pusa Sadabahar, CA 960, HC 28, JM 218, K I, K 2, MDU 1, X 235

### Capsicum

Arka Basant, Arka Gaurav, Arka Mohini, California Wonder, Yolo wonder.

The method of cultivation for seed production is nearly the same as that of cultivation for fruit production. Individual plants with good fruiting are marked and ripe fruits collected for seed purpose. Roguing of seed crop throughout the crop period is a must to maintain the true-to-type plants. The specified isolation distance must be adhered to strictly for maintaining the purity of a particular variety.

In tomato, the extraction of seed from ripe fruit is done by fermenting the crushed fruits for 1-2 days and then putting it in water, so that the seeds settle down and pulp and skin float which are easily separated. Seed separation can also be done using commercial HCl. It takes only about half an hour's time, after which the seeds are cleaned up and dried. Sometimes alkali method is also used to separate out the seeds from pulp. In this method, the pulp containing seeds is mixed with 10% washing soda in equal quantities and kept in an earthen vessel for about 12 hrs. Thereafter, seeds are washed with water and dried. The quantity of fruit required to produce 1 kg tomato seed varies from 160-210 kg, depending on the variety. On an average 100-150 kg/ha tomato seed can be obtained.

In brinjal, the ripe and yellow fruits are crushed and stored for overnight and the seeds thereafter are washed, sieved and dried. About 200-300 kg seeds can be obtained from a hectare of brinjal crop. Acid method can also be used for seed extraction in brinjal.

In chilli and capsicum, the ripe fruits are harvested and dried. Drying can be done by spreading the fruits under the sun which may take 10-15 days time, depending on the light intensity or the fruits can be dried in hot-air oven at about 54.5°C in 2-3 days. The seeds are extracted by breaking open the dried fruits by hand. An axial flow vegetable seed extracting machine can alternatively be used for seed extraction from tomato, brinjal and chilli fruits. The machine can extract tomato seeds @1.25 kg/hr, brinjal@ 1.8 kg/hr and chillies @ about 3.0 kg/hr. On an average about 200-300 kg/ha of seeds in chilli and about

# Prescribed field standards for production of vegetable seeds

| Crop | Minimum isolation distance | | Minimum number of field inspections and stages | Off-type (%) Maximum permitted | |
|---|---|---|---|---|---|
| | F | C | | F | C |
| Cowpea | 10 | 5 | 2: From flowering to fruiting | 0.10 | 0.20 |
| Garden pea | 10 | 5 | 3: First before flowering, second and third at edible pod stage | | |
| Chilli and Capsicum | 400 | 200 | 3: First before flowering, second at flowering and third at the mature fruit stage | 0.10* | 0.20* |
| Cauliflower | 1,600 | 1,000 | 4: First before the marketable stage, second at start of curd head formation, third when most plants have formed curd and fourth at flowering stage | 0.10* | 0.20** |
| Lettuce | 50 | 25 | 3: First before full grown stage in non-heading types, second full grown stage in non-heading types and third at flowering | 0.10* | 0.20* |
| Carrot | 1,000 | 800 | 4: First 20-30 days after sowing, second at lifting and replanting, third at flowering and fourth at maturity | 0.10* | 0.20** |
| Onion | 1,000 | 500 | 4: When seed crop is raised by the transplanting method: first early (20-30 days after sowing), second when bulbs are lifted, third when bulbs are replanted and fourth at flowering | 0.10** | 0.20** |
| | | | 3: When seed crop is raised by seed to seed method: first 20-30 days after sowing, second when bulbs are formed and third at flowering | | |
| Radish and Turnip | 1,600 | 1,000 | 3: First 20-30 days after sowing, second when lifted and replanted and third at flowering | 0.10** | 0.20** |
| Okra | 400 | 200 | 3: First before flowering, second at full flowering and fruiting and third at mature fruit stage | 0.10** | 0.20** |
| Tomato | 50 | 25 | 3: First before flowering, second during flowering and the immature fruit stage and third at mature fruit stage | 0.10* | 0.20* |
| Cucurbits | 1,000 | 500 | 3: First before flowering, second during flowering and the immature fruit stage and third during mature fruit stage | 0.10** | 0.20** |

*Maximum permitted at the final inspection; **maximum permitted at end after flowering; F, foundation seed; C, certified seed

100-150 kg/ha of seeds in capsicum can be obtained.

### Peas and beans

### Peas

*Early varieties*: Ageta 6, Arkel, Jawahar Matar 3, Jawahar Matar 4, Hisar Harit, Pant Sabzi Matar, PM 2, VL 7.

*Mid season varieties*: AP 1, AP 3, Bonneville, Jawahar Matar 1, JM 2, JP 71, JP 83, Lincoln, P 88, Pant Uphar, VL 3.

### French bean

*Pole types*: Kentucky Wonder, Pusa Himalata, SVM 1, VL 12, VL 17

*Bush types* : Arka Komal, Contender, Giant Stringless, Pant Anupama, Pant Bean 2, Premier, Pusa Parvati, VL Boni 1.

### Cowpea

Arka Garima, Pusa Do Fasali, Pusa Komal, Pusa Phalguni, Pusa Rituraj, S 263.

### Dolichos bean

CO1, CO 2, Jawahar Sem 53, Jawahar Sem 85, Kalyanpur Type 2, Konkan Bhushan, Prolific, Pusa Early Rajani.

### Cluster bean

Pusa Mausami, Pusa Nav Bahar, Pusa Sadabahar.

The cultivation practices for seed crop of peas and beans are more or less the same as for green pods. Although a self-pollinated crop, pea, is well known for producing off-type plants. Hence, rigorous roguing must be undertaken at flowering and fruiting stage. When almost 90% pods on the plants mature and turn dry, the whole plants are uprooted and collected. After about a week, the seeds are separated out from the pods by threshing and winnowing. The ripe and dry pods can also be picked up by hand and threshed. Usually the moisture content of seeds at this time is higher. Therefore, the drying must be resorted to maintain the specified moisture; content of these crops. On an average, from a hectare of seed crop about 1,500 kg pea, 1,000 kg French bean, 1,000kg cowpea, 700kg dolichos bean and 700kg clusterbean seeds can be obtained.

## Okra

Pusa Sawani, Selection 2, Harbhajan, Punjab Padmini, P 7, CO 1, MDU 1, Parbhani Kranti, Varsha Uphar, Arka Abhay and Arka Anamika are important varieties for seed production. There are no specific practices to be followed for the crop raised for seed production. Since it is an often cross-pollinated crop, maintaining proper isolation distance is a must. The YVMV affected plants should be rogued out in addition to roguing off-type plants.

Its fruits are harvested when they become dry on plants. The seeds are taken out from the pods. The seeds are dried to specified moisture level, cleaned, treated and stored. For one year storage 21°C temperature and 12% humidity should be maintained. An hectare of okra crop gives about 1,000-1,200kg seeds.

## Cross-fertilized cole crops

### Cole crops

### Cauliflower

*Early varieties* : Early Kunwari, Improved Japanese, Pusa Deepali, Pusa Early Synthetic, Pant Gobhi 3, Pusa Katki.

*Mid-season varieties:* Pant Shubhra, Pusa Aghani, Pant Gobhi 4, Pusa Himjyoti, Pusa Synthetic, Pusa Shubhra.

*Late varieties:* Pusa Snowball, Pusa Snowball K I, Snowball 16.

## Seed standards for vegetables

| Crop | Class of seed | Germination (%) (min.) | Pure seed (%) (min.) | Inert matter (%) (max.) | Other crops seed (max.) (no./kg) | Weed seed (max.) (no./kg) | Ordinary pack | Vapour proof pack |
|------|------|------|------|------|------|------|------|------|
| Tomato | FS | 70 | 98 | 2 | 5 | None | 8 | 6 |
|  | CS | 70 | 98 | 2 | 10 | None | 8 | 6 |
| Brinjal | FS | 70 | 98 | 2 | None | None | 8 | 6 |
|  | CS | 70 | 98 | 2 | None | None | 8 | 6 |
| Chilli | FS | 60 | 98 | 2 | 5 | 5 | 8 | 6 |
| Capsicum | CS | 60 | 98 | 2 | 10 | 10 | 8 | 6 |
| Okra | FS | 65 | 99 | 1 | None | None | 10 | 8 |
|  | CS | 65 | 99 | 1 | 5 | None | 10 | 8 |
| Cabbage | FS | 70 | 98 | 2 | 5 | 5 | 7 | 5 |
|  | CS | 70 | 98 | 2 | 10 | 10 | 7 | 5 |
| Cauliflower | FS | 65 | 98 | 2 | 5 | 5 | 7 | 5 |
| Knol-khol | CS | 65 | 98 | 2 | 10 | 10 | 7 | 5 |
| Radish | FS | 70 | 98 | 2 | 5 | 10 | 6 | 5 |
|  | CS | 70 | 98 | 2 | 10 | 20 | 6 | 5 |
| Carrot | FS | 60 | 95 | 5 | 5 | 5 | 8 | 7 |
|  | CS | 60 | 95 | 5 | 10 | 10 | 8 | 7 |
| Beet | FS | 60 | 96 | 4 | 5 | 5 | 9 | 8 |
|  | CS | 60 | 96 | 4 | 10 | 10 | 9 | 8 |
| Spinach | FS | 60 | 96 | 4 | 5 | 5 | 9 | 8 |
|  | CS | 60 | 96 | 4 | 10 | 10 | 9 | 8 |
| Amaranthus | FS | 70 | 95 | 5 | 5 | 10 | 8 | 6 |
|  | CS | 70 | 95 | 5 | 10 | 20 | 8 | 6 |
| Lettuce | FS | 70 | 98 | 2 | None | 5 | 8 | 7 |
|  | CS | 70 | 98 | 2 | None | 10 | 8 | 7 |
| Onion | FS | 70 | 98 | 2 | 5 | 5 | 8 | 6 |
|  | CS | 70 | 98 | 2 | 10 | 10 | 8 | 6 |
| Garden pea | FS | 75 | 98 | 2 | None | None | 9 | 8 |
|  | CS | 75 | 98 | 2 | 5 | None | 9 | 8 |
| Cowpea | FS | 75 | 98 | 2 | 5 | 5 | 9 | 8 |
|  | CS | 75 | 98 | 2 | 10 | 10 | 9 | 8 |
| Frenchbean | FS | 75 | 98 | 2 | None | None | 9 | 7 |
|  | CS | 75 | 98 | 2 | None | 10 | 9 | 7 |
| Cucurbits | FS | 60 | 98 | - | - | - | 7 | 6 |
|  | CS | 60 | 98 | 2 | - | - | 7 | 6 |

## Cabbage

Copenhagen Market, Golden Acre, Late K I, Pride of India, Pusa Drum Head, Pusa Mukta.

## Knol-khol

King of North, Large Green, Purple Vienna, White Vienna.

The method of cultivation of seed crop of cole crops is nearly the same as that of cultivation for fruit production. Individual plants with good fruiting should be marked and ripe fruits be collected for seed purpose. Roguing of seed crop throughout the crop period is essential to maintain true-to-the type plants. The specified isolation distance must be adhered to strictly for maintaining the purity of seed. Since cole crops are basically cool season crops, they require low temperatures for flowering and seed setting. Hence their seed production is restricted mainly to hilly areas. Early and mid-group season varieties of cauliflower, however, can flower and produce the seeds in different parts of the plains. The methods of seed production in cole crops are:

### Seed-to-seed method

The plants with good curd are left in the field, where they flower, and produce seeds. But this method occupies a lot of area because selected plants for seed production are left scattered in the field. However, in early cauliflower, this method yields relatively better quantity and quality of seeds. The seeds ripen from March to May.

In cabbage, the transplanting is done in last week of September and seeds are ready by May-June next year. Cabbage is biennial in nature. In the first year, the heads are formed and the next year seeds are produced. It requires chilling temperature for about 30-60 days for seed production. Because of it, the cabbage seed cannot be produced in plains.

### Head-to-seed method

The selected heads of cabbage or curds of cauliflower are uprooted carefully and replanted in a compact block for seed production. While transplanting, a spacing of 75 cm x 75 cm may be maintained. In cauliflower, at the time of transplanting, a scooping or incision in the middle of the curd is given to facilitate the growth of side stalks which permit better quality seed production. In cabbage, the transplanting can be done by the following methods.

## Stump method

The heads are cut just below the base by a sharp knife, keeping the stem with the outer leaves intact. The beheaded portion is called *stump*. In the following season when the dormancy is broken, the buds sprout from the axis of all the leaves and leaf scars. In this method though the seed yield is more, the flower shoots require heavy, staking otherwise they breakdown easily during interculture operations.

## Stump with central core intact method

The heads are chopped on all sides with downward perpendicular cuts in such a way that the central core is not damaged. During the last week of February and till 15 March when the heads start busting, two vertical cross cuts are given to the head taking care that the central growing point is not injured.

## Head intact method

If crop matures in the first season, the heads are examined for trueness to the type. The plants with off type heads are removed. The head is kept intact and only a cross cut is given to facilitate the emergence of a stalk. On an average, 300-400 kg/ha seed yield from cauliflower and 400-500 kg/ha seed yield from cabbage can be obtained.

## Root Crops

## Radish

*Asiatic/tropical types:* Arka Nishant, Chinese Pink, CO 1, Japanese White, Kalyanpur No.1, Pusa Chetki, Pusa Reshmi.

*European temperate types:* White Icicle. Pusa Himani, Rapid Red White Tip, Scarlet Globe.

## Carrot

*Asiatic types:* Pusa Kesar, Pusa Meghali, Selection 223.

*European types:* Chantaney, Early Nantes, Jeno, Pusa Yamdagni.

## Turnip

*Asiatic types:* Punjab Safed, Pusa Kanchan, Pusa Sweti.

*European types:* Golden Bold, Pusa Swarnima, Pusa Chandrima.

## Beet

*European types:* Detroit Dark Red, Crimson Globe, Crosby Egyptian, Early Wonders.

The method of cultivation is nearly the same as for fresh roots. However, the seed is produced by seed-to-seed method or transplanting root-to-seed method. Transplanted root-to-seed method is better since it gives an opportunity to rogue out off-type roots at the time of transplanting, maintaining only true-to-type plants for seed production. In plains, it is sown from September to January and in hilly areas from March to August. Its seed crop should be harvested when most of the pods turn yellow. After that it is dried and seeds are threshed out from the pods. A hectare of radish seed crop yields 600-1,000kg seed.

The carrot is sown from August-November in plains and from March-July in hilly areas. In carrot, the seeds are formed in umbels. The first and largest umbel is formed on main flowering stalk and is known as *primary* or *king umbel*. Secondary umbels are formed at the terminus of branches from the main flowering stem and flower in a sequence from the top to the bottom of the inflorescence. Tertiary umbels originate on secondary umbel stem. The seeds from primary umbels are heavier, more mature and of high quality. Since in carrot all the umbels do not mature together, harvesting is done 2-3 times. However, the seed crop can be harvested when all the secondary umbels mature and tertiary umbels turn yellow. A hectare of carrot seed crop yields 450-500 kg seeds.

In turnip, the asiatic types are sown from July-September and European types from October-December in the plains and from March-May in hilly area. The harvesting is done during summer when pods mature becoming brownish-red. After threshing, the seeds are separated out. About 500-600kg seeds/ha can be obtained from turnip.

## Cucurbits

### Bottle gourd

Arka Bahar, PSPL, PSPR, Pusa Naveen, Punjab Round, Punjab Komal.

### Bitter gourd

Arka Harit, Kalyanpur Baramasi, MDU 1, Priya, Punjab 14, Pusa Do Mausami, Pusa Vishesh.

### Sponge gourd

Kalyanpur Chikni, Pusa Chikni.

### Ridge gourd

Pusa Nasdar, Kalyanpur Dharidar, PKM 1, CO1, Arka Sujat.

### Pumpkin

Arka Chandan, CM 14, CO 1, CO 2, Pusa Vikas, Pusa Vishwas.

### Cucumber

Japanese Long Green, Kalyanpur Green, Khira 90, Pinsette, Poona Khira, Sheetal.

## Muskmelon

Arka Rajhans, Durgapura Madhu, Hara Madhu, Hissar Madhur, Pusa Madhuras, Punjab Rasila, Pusa Rasraj, Pusa Sharbati, Punjab Sunehri.

## Watermelon

Arka Manik, Asahi Yamato, Durgapura Kesar, Durgapura Meetha, Sugar Baby.

## Longmelon

Lucknow Early, Arka Sheetal.

## Round melon

Arka Tinda, S 48.

The method of cultivation for seed crop is the same as that of fresh fruits. However, the harvesting is done when fruits are fully mature and become dry. Roguing of seed crop throughout the crop period is a must to maintain true-to-type plants. Generally, seeds are extracted by cutting open the fruits longitudinally. In watermelon, pumpkin etc. seeds are embedded in the pulp. Therefore, different methods are applied for extraction of seeds.

In mechanical methods, axial flow vegetable seed extractor is used to separate out the pulp from seeds, whereas in chemical methods commercial grade HCl is used to separate the pulp from seeds within 15-20 minutes. Thereafter, the seeds are washed in water and dried to prescribed moisture levels. From a hectare of seed crop, 300-500 kg bottlegourd seeds, 100-300 kg bitter gourd seeds, 300-400 kg luffa, pumpkin, cucumber and round melon seeds, 200-300 kg muskmelon seeds and 400-500 kg watermelon seeds can be obtained.

## Onion

The important varieties are Arka Kalyan, Arka Niketan, Agri

Found Dark Red, Agri Found Light Red, B 780, N 2-4-1, N 53, Pusa Red and Pusa Ratnar. Onion seeds are largely produced in Gujarat, Maharashtra, Madhya Pradesh and Rajasthan. In Uttar Pradesh, Haryana, Punjab and Bihar it suffers from Purple blotch resulting in a low seed yield which is uneconomic.

Onion is a biennial crop taking two full seasons to produce seeds. In the first year, bulbs are produced and in second year seed stalks are produced. Onion is a long day plant but seed production is day neutral. It requires cool condition during early development of the bulb crop and again prior to and during early growth of seed stalk. Varieties bolt readily between 10 and 15°C. During harvesting and curing of the seed, fairly high temperature and low humidity are desirable.

Onion is largely cross-pollinated crop with up to 93% natural cross-pollination (NCP), but some self-pollination does occur. It is chiefly pollinated by honeybees. For seed production the field should be isolated from fields of other varieties of onion or fields of the same variety not conforming to varietal standards, at least by 1,000 m for FS and 500 m for CS.

Mostly bulb-to-seed method is used for seed production. Seed-to-seed method is used if bulbs have lower keeping quality.

Well-matured bulbs should be harvested, and topped, leaving an 1.27 cm mark. Before storage, a thorough section and curing of bulbs should be done. The time required for curing depends on weather conditions and may take 3-4 weeks. The mature bulbs should be stored in well ventilated, cool stores (0-4.5°C) until 3-4 weeks prior to planting.

The best time for planting of bulbs is the second fortnight of October. The roguing carried out in the field when bulbs are not harvested. After harvesting, true-to type bulbs are selected. Seed is ready for harvesting when first formed seed in the heads get blackened. A total of 2-3 pickings may be necessary to harvest the heads. Seed heads are cut, snapped off, keeping a small portion of the stalk attached. Seed heads after harvesting should be thoroughly dried. Seeds from heads are removed. After that seeds are cleaned by putting them in water and dried under the sun or by drier and stored. The moisture content should not be more than 6-8%. Seed yield is 850-1,000 kg/ha.

# ACRONYMS

| | |
|---|---|
| AA | Appellate Authority |
| BT | Bio-Technology |
| CPRI | Central Potato Research Institute |
| CSC | Central Seed Committee |
| CSCB | Central Seed Certification Board |
| CSCO | Chief Seed Certification Officer |
| CSTL | Central Seed Testing Laboratory |
| DIA | Designated Inspection Agency |
| DUS | Distinctiveness, Uniformity and Stability |
| FVOS | Flower, Vegetable and Ornamental Seeds |
| GOT | Grow Out Test |
| IARI | Indian Agricultural Research Institute |
| ICAR | Indian Council of Agricultural Research |
| IMSCS | Indian Minimum Seed Certification Standards |
| IP | Import-Permit |
| ISST | Inter-State Seed Certification |
| ISTA | International Seed Testing Association |
| NBPGR | National Bureau of Plant Genetic Resources |
| NPQAC | National Plant Quarantine Advisory Committee |
| NSC | National Seeds Corporation |
| NSIP | New Seeds Import Policy |
| NSTL | Notified Seed Testing Laboratory |
| OGL | Open General Licence |

| | |
|---|---|
| PDVR | Project Directorate of Vegetable Research |
| PEQ | Post-Entry Quarantine |
| PER | Pre-Entry Requirement |
| PPA | Plant Protection Authority/Advisor |
| PQC | Plant Quarantine Counter |
| PQF | Plant Quarantine and Fumigation Stations |
| PQO | Plant Quarantine Organisation |
| PSCE | Phytosanitary Certificate Existing |
| RL | Research Laboratories |
| RSCO | Regional Seed Certification Officer |
| SA | Seed Act, 1966 |
| SAU | State Agricultural University |
| SCA | Seed Certification Agency |
| SCI | Seed Certification Inspector |
| SCO | Seed Certification Officer/Order |
| SLEA | Seed Law Enforcement Agency |
| SSC | State Seeds Corporation |
| SSTL | State Seed Testing Laboratory |
| STM | Seed Testing Manual |
| TPS | True Potato Seed |
| VSAR | Vegetable Seed Analysis Report |

# SIXTEEN

## Vegetable Crop Queries

### General

★ Vegetables being short duration crop, can give six to eight times more yield than any cereal crops.

★ Vegetables are rich reservoir of nutrients particularly vitamins and minerals.

★ China is the largest producer of vegetables in the world followed by India.

★ India occupies first position in cauliflower, second in onion and third in cabbage in the world.

★ The area and production-wise largest vegetable growing states are West Bengal, Orissa and Uttar Pradesh.

★ India shares 13.38 per cent of world production of vegetables.

★ Vegetable crops in India occupy only 2.8 per cent of the total cultivated land.

★ The home or kitchen garden is the most ancient type of garden.

★ Garden for vegetable forcing is most intensive method of vegetable cultivation.

★ Truck gardening is an extensive method of vegetable cultivation.

### Nutritive value

★ Drumstick leaves are the richest source of fat and vitamin C (ascorbic acid).

★ Agathi (*Sesbania*) is the richest source of protein and calcium.

★ Rajgira leaves are the richest source of Vitamin A followed by colocasia leaves.

★ Radish leaves are the richest source of riboflavin followed by fenugreek leaves.

★ Giant chillies are the richest source of thiamine followed by peas.

★ Tapioca is the richest source of carbohydrate and calories followed by sweet potato.

## Solanaceous vegetables

### Tomato

★ Tomato, universally treated as Protecive Food, is being extensively grown as annual plant all over the world.

★ The specific requirements of tomato for export in West Asia are round, medium size and red colour, while cherry tomato is preferred for export to European countries.

★ The export quality tomato producing areas are Nasik and Pune in Maharashtra and Bangalore in Karnataka.

★ Processed products of tomato especially 'Tomato Puree' and 'Tomato Paste' have great demand for export.

★ The first known record of tomato dates back to the year 1554 in South America.

★ The Italians might have been the first Europeans to use tomato.

★ Tomato (*L. esculentum*) has five forms, *cerasifome;* small fruited type or cherry tomato; *pyrifomae*, pear tomato, *commune*, common tomato, *grandifolium*, potato leaf type and *validium*, upright tomato.

★ Muller (1940) divided the genus *Lycopersicon* into sub-genus *Eulycopersicon* and *Eriopersicon*. *Eulycopersicon* is a red fruited type, where as *Eriopersicon* is green fruited type.

★ Tomato has tap root system.

★ Growth habit in tomato is of two types : (1) *determinate type*—plant are dwarf and growth is restricted with the appearance of terminal flower; (2) *indeterminate type*— growth is continued and there is less initiation of flower and fruit on the stem.

★ Lycopene, which is responsible for red colour in tomato, is highest at 21°C - 24°C while the production of this pigment drops rapidly above 27°C.

★ For early crop of tomato, a sandy loam is the best. However, for higher yield, heavy soils rich in organic matter are preferred.

★ Tomato is moderately tolerant to acid soil (pH 5.5).

★ Varieties like Pusa Early Dwarf, Pusa Ruby, Arka Vikas, Arka Saurabh, Punjab Kesari and Pant Bahara are suitable for fresh market.

★ Varieties like Pusa Gaurav, Roma, Punjab Chhuhara, Pusa Uphar, Arka Saurabh are suitable for processing.

☆ Pusa Sheetal is suitable for low temperature regions.

☆ Pusa Hybrid 1 is suitable for high temperature regions.

☆ Pusa Sadabahar is suitable for high and low temperature regions.

☆ Normally, 125-175 g hybrid seed is required for the planting of one hectare land.

☆ Seedlings of tomato soaked for 24 hours in NAA at 0.1 ppm showed higher fruit set and increased early and total yield.

☆ The foliar application of Parachlorophenoxy Acetic Acid (PCPA) 50-100 ppm at the flowering stage increase the fruit set at low and high temperature.

☆ Seed treatment of tomato with 2, 4-D @ 2-5 ppm increases fruit set earliness, parthenocarpy.

☆ Pruning and training is generally followed in indeterminate tomatoes.

☆ Tomato is climactric fruit.

☆ In order to enhance the ripening of tomato fruits, ethrel (1,000 ppm) can be sprayed on the plants at the time of initiation of ripening.

☆ In indeterminate cultivars, fruits can normally be harvested 70-100 days after planting, while determinate cultivar may begin to fruit at 70 days depending on the environmental condition.

☆ Sel-120 is the first Root-knot nematode resistant variety of tomato.

☆ In 1992, Greshof and Doy produced haploid plants of tomato by anther culture.

## Brinjal

☆ Brinjal is also known as egg plant or aubergine.

☆ Brinjal fruits are good source of vitamin B.

☆ Pigmented, dark-purple brinjal has more vitamin C than those with white skin.

☆ Dry fruit of brinjal is reported to contain goitrogenic principles.

☆ Brinjal flowers are pentamerous, hermaphrodite and solitary.

☆ In brinjal heterostyly is common feature except in bunchy type cultivars.

☆ Brinjal fruit is a berry.

☆ Brinjal is a self-pollinated crop but cross-pollination also occurs in it. This is because it has pronounced heterostyly which farvours cross-pollination.

☆ Brinjal is a day-neutral plant.

☆ The optimum temperature for normal growth and development of brinjal plant and fruits is 21-30°C or below. If temperature rises above this the growth stops completely.

☆ In brinjal, opening of flowers takes place at 6- 7 : 30 hr in summer which is delayed to 11:15 hr in winter.

☆ High humidity and high temperature in the morning hours hasten the opening of flowers and dehiscence of anthers in brinjal.

☆ Brinjal flowers are divided into four types, namely (i) long styled, (ii) medium styled, (iii) pseudo-short styled and (iv) short styled.

★ The maximum fruit setting in different cultivars is in 'long styled flowers' which ranges from 70-80%. In medium styled, it ranges between 12-55%. There is no fruit setting in pseudo-short and short styled flowers.

★ Brinjal cultivars belong to three varieties, these are (a) *esculentum,* which includes round and egg-shaped cultivars; (b) *serpentinum*; long, slender types; and (c) *depressum*, which includes all early and dwarf brinjal cultivars.

★ It is stated that *S. melongena* is more closely related to *S. incanum* than to any other species of *Solanum.*

★ Pusa Purple Long is an extra early bearing maturity cultivar of brinjal.

★ Annamalai is aphids resistant cultivar of brinjal.

★ Arka Shirish is green coloured cultivar of brinjal.

★ KKM-1 is milky white coloured cultivar of brinjal.

★ Pusa Bhairav and Pusa Anupam are resistant to *Phomopsis* fruit rot.

★ Pusa Anupam is a cross between PPC and Pusa Kranti.

★ Pb. Neelam is a cross between Jamuni Gole and Pant Rituraj.

★ Vaishali is cross between Arka Kusumakar and Manjiri Gota.

★ Pragati is a cross between Vaishali and Manjari Gota.

★ For brinjal cultivation, sandy soils are good for an early crop production while silt loam or clay loam are good for heavy production.

★ Approximately 30,000-45,000 seedlings are sufficient to cover one hectare of land.

★ Application of copper and manganese increases the number of flowers and fruits while zinc improves the weight of fruits.

★ *Orobanchae* sp. is one of the serious weed of brinjal crop. It is a root parasite.

★ It is estimated that 100-110 cm irrigation is required for normal brinjal crop.

★ Application of 2, 4-D (2 ppm) at flowering induces parthenocarpy, increases fruit-set, advances fruit maturation and significantly increases the total yield of brinjal.

★ In brinjal, the highest yield was obtained from plants whose roots were dipped in $GA_3$ + ascorbic acid each at 250 ppm solution.

★ Brinjal green fruits have the longest shelf life of 4 weeks.

## Chilli and Capsicum

★ Andhra Pradesh has been leading both in area and production of chilli.

★ Chilli is also known as 'hot pepper' and capsicum as 'bell pepper'.

★ In food and beverage industries, chilli is being used in the form of oleoresin which permits better distribution of colour and flavour in food.

★ Capsicin is the pungent principle found in chilli.

★ The genus *Capsicum* originated in the new world tropic and subtropics.

☆ The Portuguese brought *Capsicum* from Brazil to India during 1584.

☆ *Capsicum baccatum* has very good yield potential and useful genes which are resistant to powdery mildew and anthracnose fruit rot disease.

☆ *C. annuum* and *C. frutesceus* have 'white flowers' while *C. pubesceus* have 'purple flowers'.

☆ Chilli and capsicum can be grouped in often-cross-pollinated crops. Cross pollination is reported upto 63%.

☆ Majority of the flowers open at 5 am.

☆ MDU 1 is a mutant (irradiation) variety of chilli.

☆ Sindhur is mild pungent variety of chilli.

☆ Jwalamukhi is a variety of chilli suitable for high-density planting and low pungency and is resistant to virus.

☆ Punjab Lal is resistant to TMV, CMV, Leaf curl virus, fruit rot and die back.

☆ Arka Abir is a paprika variety suitable for colour extraction.

☆ CH1 is $F_1$ hybrid (using GMS) of chilli.

☆ Pusa Deepti is $F_1$ hybrid of capsicum.

☆ Most extensive cultivation of chilli can be seen on vertisol covering the states of Andra Pradesh, Karnataka, Maharashtra and Tamil Nadu.

☆ Approximately 50,000-60,000 seedlings will be sufficient to cover one hectare of field.

☆ Compared to chilli, bell peppers are shy rooters and as such dipping the seedlings in 1.5% superphosphate solution before transplanting is advisable.

☆ Fruit set in chilli can be improved by application of growth regulators like GA (10-100 ppm) NAA (20-200 ppm) and CCC (1,000 ppm).

☆ Among the well known international institutes is the Asian Vegetables Research and Development Centre, Taiwan which does a lot of work on peppers.

## Cole crops

### Cauliflower

☆ The first illustration and description of cauliflower was presented by the herbalist Dodens (1544).

☆ Cauliflower has descended through mutation and selection from a common ancestor, the wild cabbage *Brassica oleracea* L. var. *sylvestris* L. Dr. Jemson.

☆ Cauliflower was introduced in India in 1822 by a botanist Dr. Jemson from Kew Garden, London.

☆ Pusa Hybrid 2 is first $F_1$ hybrid released by a public sector organisation.

☆ Pusa Himjyoti is the only variety which can be grown from April to July in the hills.

☆ Light soils are preferred for early crops while loamy and clay loam soils are suitable for mid and late maturity types.

☆ Cauliflower is a thermosensitive crop, i.e. temperature plays an important role, influencing vegetative, curdling and reproductive phases of plant.

☆ When tropical cultivars of cauliflower are grown in lower temperature, they form small sized curds. This is commonly termed as 'buttoning'.

☆ Generally, closer spacing is kept for early and mid season crop and wider spacing for late maturing cultivars of cauliflowers.

☆ NAA (10 ppm) treatment of cauliflower seedlings as starter solution has been found effective in respect of plant stand in the field and the vegetative growth.

☆ Application of $GA_4$ + $GA_7$ at the rate of 80 ppm shortens the period from transplanting to the harvest.

☆ If the harvests of cauliflower are late, the curds start loosening because of flower stalk emergence. They may become leafy, ricy and fuzzy.

## Cabbage

☆ Cabbage is distinguished by its swollen heads which is formed by thickening of edible bud with thick, tightly packed overlapping leaves manifesting a large head.

☆ Cabbage has an anti-cancer property. It protects against bowel cancer due to the presence of indole-3- carbinol.

☆ Cabbage belongs to the cole group of vegetables (cole is a plant of genus *Brassica*) which originated from a single wild ancestor *Brassica oleracea* L. var. *sylvestris* L. commonly known as wild cabbage, cliff cabbage or colewort.

☆ Cabbage seed production is taken up in hilly regions of India which fulfil its chilling requirement to enter into reproductive phase after a period of vegetative phase (6 weeks approx.).

* Cabbage bears seeds in a special kind of bicarpellary pod called the 'siliqua'.

* The cultivated cabbage is a biennial crop but its progenitor wild cabbage is herbaceous, usually perennial.

* Nieuwhof (1969) recognized the three cultivated forms of cabbage : *B. oleracea* var. *capitata* L. f. *alba* 1DC (white cabbage), *B. oleracea.* var. *capitata* L. f. *rubra* (L) Thell (red cabbage) and *B. oleracea.* var. *sabauda* L. (savoy cabbage).

* Self incompatibility is found in cabbage.

* The flowers of cabbage are protogynous.

* The early group of cabbage takes 55-70 days to mature the late group matures in 85-130 days while cultivars maturing in between fall under mid-season group.

* Pusa Drum Head possesses field resistance to black leg caused by *Phoma lingum*.

* Pusa Ageti is the first tropical variety developed for cultivation under high temperature conditions.

* Pusa Ageti can produce seeds under subtropical conditions of North Indian plain.

* Pusa Sambandh is a synthetic variety of cabbage.

* Cabbage grown In saline soils are more susceptible to disease like black leg.

* Cabbage grown on heavy soils (clay loam or silt loam) have good keeping quality due to slow growth and compactness of heads.

* Cabbage is more hardy than cauliflower and can withstand frost and extreme cold weather.

☆ Cabbage produces seed in the temperate areas only.

☆ Cabbage can be grown almost throughout the year in mid-hills and southern hills.

☆ Cabbage seedlings will become 'lanky' due to over application of nitrogen.

☆ Cabbage seedlings are transplanted at 4-6 true leaf stage.

☆ A spray of CCC or SADH (2,500-5,000 ppm) increases the low temperature resistance in cabbage.

☆ Maintenance of self-compatibility lines of cabbage is done by the use of 2-5% $CO_2$ gas and 5% sodium chloride solution.

☆ Pusa Mukta is resistant to black rot disease.

☆ Sauerkraut is an ancient preparation from cabbage and it was popular with seamen who used it as a remedy for scurvy. Storage varieties of white cabbage with compact heads, fine leaves with thin veins are preferred for sauerkraut.

## Knol-khol

☆ The inflorescence in knol-khol is racemose.

☆ Knol-khol have sporophytic self-incompatibility.

☆ In India, knol-khol is more popular in Kashmir.

☆ The edible part of knol-khol is the swollen stem called tuber or knob which arises from the thickening of stem tissue just above the ground.

☆ In Kashmir, knol-khol leaves are also used as greens.

☆ Early varieties of knol-khol are more prone to premature bolting.

★ Premature bolting in knol-khol may cause development of elongated or flattish round knobs. Even in case of cultivars, round globular or flattish-round knobs are formed.

★ Purple varieties are more susceptible to premature bolting.

## Chinese cabbage

★ Chinese cabbage (*Brassica campestris* L.) is the most important crop grown in China.

★ Chinese cabbage is grown as pot herb and also as salad crop.

★ Chinese cabbage might have originated from hybridization between turnip (*B. campestris* ssp. *rapifera*) from north China and pak-choi (*B. campestris* L. ssp. *chinensis* (Rupr). Olsson) from south China.

★ Chinese Sarson No.1 is a non-heading type, resistant to alternaria leaf spot variety of Chinese cabbage.

★ To raise one hectare crop of Chinese cabbage, 500 g seed is required through transplanting method and 2.5 kg for direct seedling.

## Sprouting broccoli

★ The inflorescence in broccoli is cymose.

★ Genic male sterility has been reported in broccoli by Cole (1959).

★ Broccoli have sporophytic self-incompatibility.

★ U.S.A. is the largest producer of sprouting broccoli in the world.

★ Broccoli is an Italian word derived from Latin 'Brachium' meaning 'an arm or branch'.

★ Morphologically, broccoli resembles cauliflower, but the plant forms head rather than curds consisting of green buds and thick fleshy floral stalk.

★ Sprouting broccoli has about 130 times more vitamin A contents than cauliflower and 22 times more than cabbage.

★ Sprouting broccoli is a rich source of sulphoraphane, a compound associated with reducing the risk of cancer.

★ Purple type cultivars are more hardy than white type and can withstand lower temperature.

★ Green type cultivars are more commonly cultivated than white type.

★ Purple Sicilian broccoli is also known as purple cauliflower.

★ Palam Samridhi is mainly recommended for subtropical conditions.

★ Calabrese type broccoli developed from Italian green sprouting broccoli are becoming more popular with the growers as well as consumers for both fresh market and processing.

★ Broccoli is sensitive to higher temperature as bud clusters of head become loose quickly.

★ In India, mainly early and mid types of broccoli are grown and these are susceptible to frost.

★ Early and mid types of broccoli are annual whereas late types are biennial in nature.

☆ The late type cultivar of broccoli requires vernalization for the development of head and initiation of flowering stalk.

## Brussel's sprouts

☆ The inflorescence in Brussel's sprouts is racemose.

☆ Genic male sterility has been reported in Brussel's sprouts by Johnson (1958).

☆ Brussel's sprouts have sporophytic self-incompatibility.

☆ The edible part of Brussel's sprouts is 'swollen axillary bud' known as *'sprouts'* or *buttons'* or *'mini cabbages'*.

☆ Brussel's sprouts is a cool and moisture loving crop which needs a long growing season.

☆ Brussel's sprouts is frost resistant and late cultivar can with-stand temperature as low as -10°C.

☆ Excessive application of nitrogen results in quick growing of plant and development of loose sprouts; so it should be avoided.

☆ Excessive application of potash imparts bitter taste to the sprouts.

## Kale

☆ Kale (*Brassica oleracea* var. *uccephala*) is a minor cole crop.

☆ Kale is propagated by seed.

☆ Kale is the hardiest crop and can withstand temperature as low as $-10^0$ to $-15^0$C.

☆ Dwarf Green Curled Scotch, Dwarf Moss Curled (dwarf) and Moss Curled, Hamburger Market (medium tall) are important dwarf type varieties of kale.

☆ Karam Sag is one of the medium tall variety mostly grown in Jammu & Kashmir.

☆ Scotish and Siberian varieties are more suited for temperate regions.

## Root crops

### Radish

☆ Radish is a root-cum-leafy vegetable suitable for tropical and temperate climate.

☆ The edible portion of radish root develops from both primary root and the hypocotyl.

☆ Radish is a cross-pollinated vegetable due to sporophytic system of self-incompatibility.

☆ Pusa Chetki is suitable for growing in hotter month, i.e. middle of March to middle of August when no other varieties can be grown successfully.

☆ Arka Nishant is resistant to pithiness, pre-mature bolting, root branching and forking.

☆ Pusa Himani is hybridization between Black and Japanese White.

☆ Pusa Himani is the only variety which can be grown throughout the year.

☆ A long day as well as temperature will result in bolting before proper root development. The roots also become hard, pithy and pungent during hot weather.

☆ Application of fresh undecomposed manure will result in forking of roots in radish.

☆ Akashin is a physiological disorder of radish caused due to boron deficiency.

☆ Excess application of NPK and soil moisture stress result in pithiness of radish.

☆ Wart is a physiological disorder which is a white protrusion of white inner root tissue mainly due to soil moisture deficit.

☆ Hollow root is a physiological disorder which is due to high temperature during 16-30 days of sowing.

## Carrot

☆ Carrot is an annul herb for root production and biennial for flowering and seed set.

☆ Carrot flower is protandrous. The inflorescence is a compound umbel.

☆ The temperate types form roots both under tropical and temperate conditions but set seed only under temperate conditions as they need low temperature of 5-8°C for 40-60 days before throwing out the seed stalks.

☆ The Asiatic types are high yielding and produce seed under tropical condition. But they are poor in carotene content.

☆ The wild forms of carrot are annual but the cultivated types believed to have been derived from the wild types are biennial.

☆ Pusa Kesar is a selection from a cross of Local Red and Nantes Half Long.

✫ Pusa Meghali is a selection from a cross between Pusa Kesar and Nantes.

✫ Chantaney is an excellent cultivar for canning and storage.

✫ Frequent irrigation encourages the growth of main root and prevents secondary root development.

✫ Cavity spot is a physiological disorder of carrot due to calcium deficiency.

✫ Forking results when the hard soil does not allow straight growth.

✫ Splitting of carrot root occurs when there is a sudden change in the soil-moisture status and also in the case of heavy application of nitrogen.

**Beetroot**

✫ The probable origin of beetroot is Mediterranean region for *Beta vulgaris* L. ssp. *maritimat,* possibly by hybridization with *B. patula.*

✫ Morphologically, the upper region root develops from the hypocotyl and lower portion from the tap root, from which the secondary roots arise.

✫ The colour of root is due to the presence of red-violet pigment, ß-cyanins and yellow pigment, ß-xanthins.

✫ The inflorescence is large spike, which normally develops in the second year.

✫ The seed ball contains more than one seeds, which are viable for 5-6 years under normal storage conditions.

✫ Under high warm weather, zoning is caused which is marked by the appearance of alternating light and dark concentric circles in the root.

☆ A temperature of 4.5°C–10°C for 15 days or more may induce premature bolting in beetroot.

☆ The seed of between root usually contain one or more seeds and are called seed balls.

☆ If the temperature is high during sowing, the roots produced would be coarse with woody flesh and dull colour.

☆ Viral diseases like beet mosaic yellow are transmitted by aphids and curly top by hoppers.

☆ Internal black spot or brown heart or heart rot of beetroot is caused due to boron deficiency.

☆ The beets are harvested when they attain a diameter of 3-5 cm.

## Turnip

☆ Turnip is an important root vegetable grown as a summer crop in temperate climate and as a winter vegetable in subtropical places where the winter is not severe.

☆ Turnip can be grown at an elevation of 1,500 m mean-sea-level or above but it is not suitable for growing in low lands of wet tropics.

☆ The fleshy thickened underground portion of turnip is actually the hypocotyl.

☆ A distinct tap root and secondary roots of turnip arise from the lower part of the swollen hypocotyl.

☆ Thickening of turnip begins in the central part of the hypocotyl, followed by the upper and lower parts.

☆ Normally the roots of turnip attain edible maturity in 40-80 days.

☆ The inflorescence is a terminal raceme on the main stem.

☆ Turnip has sporophytic system of self-incompatibility.

☆ Sodium chloride and carbon dioxide can be used for overcoming self-incompatibility.

☆ Pusa Swarnima was developed by hybridization and selection by involving Asiatic (Japanese white) and European (Golden Ball) types.

☆ Pusa Kanchan is a selection from the cross between Local Red Round (Asiatic) and Golden Ball (European).

☆ Pusa Kanchan sets seeds in the plains.

☆ Early Milan Red Top is an extra early and high yielding cultivar reaching maturity in 45 days.

☆ High temperature adversely affects quality, the roots become woody, tough and bitter in taste in hot weather.

☆ Temperature below 10°C induces flower in turnip.

☆ Turnip yellow mosaic virus disease is transmitted through flea beetle. Cabbage may serve as an alternate host for this virus.

## Bulb crops

### Onion

☆ India ranks second in area and production in world (after China) and third in export (after the Netherlands and Spain).

☆ In India, Maharashtra is the leading state with 23.4% area and 27.50% production followed by Karnataka (19.90% area, 10.80% production).

★ In India, the per hectare yield is highest in Gujarat followed by Punjab.

★ The Netherlands is the leading exporter of onion, contributing about 21% of world export.

★ Mostly, the Indians prefer red and pungent type varieties, while in Japan, Europe and America, the yellow coloured varieties are preferred.

★ The inflorescence is an umbel.

★ *Allium* belongs to class Monocotyledon.

★ The common onion is the most widely grown of all the cultivated *Alliums*.

★ The shallot is a perennial onion which rarely produces seed and which is perpetuated each year by replanting some of the bulbs which form in clusters on the surface of the soil.

★ The Egyptian ground onion is noted for its hardiness and early ripening than the common onion.

★ The tree onion or Egyptian tree onion is a viviparous plant that grows as a perennial bulb.

★ The tree onion is noted for its resistance to virtually all known pests and diseases of onions.

★ The flowers of onion are white or bluish.

★ Anther dehiscence usually occurs between 5 a.m. and 9 a.m.

★ Onion is highly cross-pollinated crop due to its protandry.

★ The onion is pollinated chiefly by honey bees. Blow flies are excellent pollinators, as they go from flower to flower.

☆    In onion, male-sterile lines have been isolated in Pusa Red at IARI.

☆    Pusa Red is very good in storage and wider adaptability.

☆    Arka Kalyan, Agrifound Dark Red, Baswant 780, N 53, and N2-4-1 are suitable under *kharif* season.

☆    Arka Bindu and Agrifound Rose are suitable for export, particularly to Malaysia and Singapore.

☆    Punjab 48 and Udaipur 102 are white coloured varieties of onion.

☆    Early Grano and Brown Spanish are yellow coloured varieties of onion.

☆    Brown Spanish is long-day type variety.

☆    Agrifound Dark Red is multiplier onion variety.

☆    Almost all cultivars grown in plains in India are short-day cultivars.

☆    Long-day varieties of onion do not bulb under short-day conditions, whereas short-day varieties, if planted under long-day condition, will develop early bulbs.

☆    In onion, temperature is more important than the day length for seed production, while photoperiod is more important than temperature for bulb formation.

☆    The usual method of propagation of common onion is by seed.

☆    Egyptian or tree onion produce topsets or bulbil in flower cluster.

☆    The multiplier or potato onion seldom produces flowers and seeds.

☆ If young aged seedlings are transplanted, the establishment will be poor. If average seedlings are transplanted there will be a problem of pre-mature bolting.

☆ Small pickling onion is planted 3 times in a year in Karnataka, i.e. March-April, June-July and September-October.

☆ For getting quality bulbs, 1.5-2 cm size sets are considered to be the best, while 2-2.5 cm size sets have been recommended for getting good yield.

☆ Onion is usually grown in winter season (*rabi*) in northern India, but it is grown both in *rabi* and rainy season (*kharif*) in Maharashtra, Gujarat, Andhra Pradesh, Karnataka and Tamil Nadu.

☆ The most favourable temperature for purple blotch (*Alternaria porri*) of onion is 28-30°C. with 80-90% relative humidity.

☆ Black mould (*Aspergillus niger*) is a very common storage disease of onion. Under field conditions, coloured varieties seem to be more susceptible than white ones.

☆ Thrips (*Thrips tabaci*) is considered to be the most harmful pest worldwide. The onion thrips probably cause more damage to onions and their allies than do all the other animal pests combined together.

☆ The consumption of green onion or spring onion is almost equal to that of mature onion (dry onion) in the world.

☆ Three varieties, Early Grano, Pusa White Flat and Pusa White Round are recommended for growing as green onion.

- ★ Brown Spanish, an introduction, has been found to be the best and most suited for growing in areas 1,000 m above mean-sea-level.

- ★ The best time to harvest *rabi* onion is one week after 50-70% neck fall.

- ★ In *kharif* season, since tops do not fall, soon after the colour of leaves changes to slightly yellow and tops start drying, the bulbs are harvested.

- ★ The average yield of multiplier onion is 15-18 tonnes/ hectare.

- ★ The purpose of drying to remove excess moisture and curing is an additional process which aids in the development of skin colour and removing field heat before the bulbs are stored.

- ★ A temperature of 0ºC with 65-70% relative humidity will keep onions dormant and reasonably free from decay for about 6-8 months.

- ★ For the *kharif* onion, spraying of maleic hydrazide 2,500 ppm at 75-90 days after transplanting checks the sprouting.

- ★ At the Bhabha Atomic Research Centre, Trombay, bulbs irradiated with gamma rays at very low dose level (4,000-9,000 krads) offered an effective physical method to control sprouting of onions.

- ★ Lasalgaon in Maharashtra is the biggest onion market in India.

- ★ The Netherlands is the leading exporter of onion, followed by Spain and India, who are at the second and third place respectively.

- ★ Indian export is about 12% of world's onion demand.

★ Pusa White Flat, Pusa White Round, Punjab 48(S-4-8), N 257-9-1 and Udaipur 102 have been identified for dehydration purposes.

★ About 90% export is of big onion varieties, viz. N-53, N-2-4-1, Bellary Red, Patna Red, Agrifound Dark Red, Agrifound Light Red, Pusa Red, Pusa Madhvi and Arka Niketan.

★ The export of small onion varieties, viz. Agrifound Rose, Arka Bindu, Banglore Rose and multiplier onion varieties viz. CO 3, CO 4 and Agrifound Red is 10%.

★ Cytoplasmic genetic male sterility was first found in onion.

★ Anthesis starts in *Allium cepa.* at 7.00 am.

## Garlic

★ China ranks first in area and production of garlic.

★ India ranks second in area and third in production of garlic.

★ In garlic productivity Egypt tops the list followed USA.

★ The per capita availability of garlic is highest in Korean Republic, i.e. 11.14 kg/year, whereas in India it is only 0.340 kg/year.

★ In India, Madhya Pradesh ranks first in area while Gujarat is first in production.

★ In India, the yield/hectare is highest in Punjab followed by Haryana.

★ Garlic is used as a spice or condiment throughout India.

☆ Garlic has higher nutritive value than other bulb crops.

☆ A colourless, odourless, water soluble amino acid known as allin is present in uninjured garlic.

☆ Allicin is the anti-bacterial substance of garlic and has the typical odour of fresh garlic.

☆ Diallyl disulfide is said to possess true garlic odour.

☆ *Allium longicuspis* Rgl. is believed to be garlic's wild ancestor.

☆ Garlic is a sexually sterile diploid.

☆ Garlic differs from onion in that the leaf base does not store food, but metric as dry scales enclosing the cloves.

☆ Yamuna Safed is recommended for cultivation all over India.

☆ G 282 is suitable for export purposes.

☆ Agrifound Parvati is a long-day type variety suitable for cultivation in hills of northern states.

☆ Garlic is a frost-hardy plant.

☆ In India, mostly short-day type varieties are grown.

☆ The critical day length for bulbing in garlic is 12 hours.

☆ Exposure of dormant cloves or young plants to around 20°C or lower temperature, depending upon the varieties, for 1-2 months hastens subsequent bulbing.

☆ Garlic is propagated vegetatively by cloves. Bulbils are also used as planting material.

★ The garlic is ready for harvesting when tops turn yellowish or brownish and show signs of drying up and bend over.

★ Garlic is produced only in one season, i.e. winter (*rabi*).

★ Curing is an additional process of drying to remove the excess moisture and to allow the bulbs to become compact and go into dormant stage.

★ Storage life of garlic is prolonged and loss in weight is reduced by spraying maleic hydrazide (3,000 ppm), three weeks before harvest.

★ Irradiation with 6 krad of cobalt 60 gamma rays has been recommended for successful storage of garlic.

★ Garlic is a condiment crop.

## Leek

★ Leek is a non-bulb forming member of the onion family and is grown for its blanched stem and leaves.

★ Leek belongs to the family Amaryllidaceae, genus *Allium* and species *porrum*.

★ The most common varieties of leek are London Flag and American Flag.

★ The leek is a biennial crop and the seed is produced in India at higher altitudes in the hills.

## Peas and Beans

### Pea

★ *Pisum elatius*, a wild species, is considered as the ancestor of pea (*Pisum sativum*).

★ Peas are diploid with a chromosome number of n = 7 (2n=14).

☆ There are two sub-species of peas, namely *Pisum arvense* known as field pea having coloured flowers, and *Pisum sativum,* which is white-flowered and is known as horticultural or garden pea or sweet pea.

☆ Pea is a herbaceous winter annual.

☆ Arka Ajit is resistant to powdery mildew and rust.

☆ UN 53 (6) is a snap pea (whole pod edible) line developed by IIHR, Bangalore.

☆ Ooty 1 is resistant to white fly.

☆ Arkel is the most popular exotic pea introduced from England.

☆ Bonneville is a popular variety in India introduced from USA.

☆ Peas sown earlier or later suffer from wilt and mildew attack, respectively.

☆ Soaking of pea seeds in $GA_3$ (10 ppm) for 12 hrs gives the highest germination.

☆ Pea gives good response to phosphorous application. It favours nitrogen fixation by increasing nodule formation.

☆ Foliar application of 0.1% ammonium molybdate at the time of flowering has shown favourable effect on yield and quality of peas.

☆ Flowering, fruit set and grain filling periods are the critical stages for irrigation in peas.

☆ Training or staking is an important operation especially in tall pea varieties. Training or staking should be done along the direction of the wind.

★ Spraying 15 ml of 10M solution of the growth regulator CCC at the five-node stage of development has favourable effect on the growth and yield of crop.

★ In the processing industry, the maturity of pea is tested with the help of tenderometer.

★ If the harvesting is delayed, the pod surface of pea becomes coarse which brings down the market value of the produce and also due to the conversion of sugar into starch.

## Cowpea

★ Cowpea is grown for its tender pods and also for its dry seeds which are used as pulse.

★ Cowpea pods are rich in vitamin B.

★ Cowpea is used as fodder, green manure and vegetable crop.

★ The most probable progenitor of cowpea is var. *mensensii*.

★ Cowpea belongs to the family Leguminosae, subfamily Fabaceae.

★ There are 5 distinct sub species of cowpea out of which two are wild, viz. *V. dekinotiana* and *V. mensensis* and three are cultivated in India, viz. *V. unguilulata*, subsp. *unguiculata*, subsp. cylindrical and subsp. *sesquipedulis*.

★ Cowpea is a vigorously growing annual with strong tap root system.

★ Pusa Dofali, Arka Garima and Arka Suman are photo-insensitive varieties, suitable for both summer and rainy seasons.

★ Cowpea is a drought hardy crop and comes up well under rainfed conditions tolerating moderate dry spells.

★ Cowpea is grown as a catch crop.

★ Cowpea is grown as a cover crop in basins of fruit orchards.

★ Flowering and pod development periods are the critical stages for water management in cowpea.

## French bean

★ The French bean is also known as kidney bean, haricot bean, snap bean, navy bean, and is one of the most important leguminous vegetables.

★ There are 3 types of French bean: Bush type with short internodes, semi-pole type with longer internodes than those in bush-type and pole-type having longer internodes than that of semi-pole-type.

★ *Phaseolus aborigineus* is the progenitor of French bean.

★ There are four cultivated species of *Phaseolus*, viz. *P. vulgaris, P. coccineus, P. lanatus* and *P. aconitifolius*. All the species are self-pollinated, except *P. coccineus* which is generally cross-pollinated.

★ Pusa Parvati is resistant to mosaic and powdery mildew.

★ Kentucky Wonder and Contender are introduced for USA.

★ Through pure-line selection, Arka Komal, Pant Anupama and IIHR 220 have been developed.

★ IIHR 909 was developed by hybridization between Contender and Blue Crop.

★ Variety IIHR 220 is resistant to rust.

★ Through mutation breeding, **Pusa Parvati has been** developed and released from IARI, Katrain.

★ Most of the French bean varieties are day-neutrals except some semi-pole varieties which are short-day types.

★ French bean is a shallow rooted crop.

★ Flowering and pod development periods are the critical stages for irrigation in French bean.

★ Staking is an important operation for pole beans.

★ Plant regulators like PCPA at 2 ppm, L-naphthyl acetamide or B-naphthal acetic acid at 5-25 ppm have shown favourable effect on fruit set.

★ $GA_3$ sprayed at 50-200 ppm proved effective in improving the crop growth.

## Cluster bean

★ Cluster bean is grown for green pods used as vegetable and dry seeds.

★ *C. senegalensis* might be the immediate ancestor of *guar*.

★ Variety, Pusa Sadabahar is non-branching and suitable for summer and rainy seasons.

★ Cluster bean is a hardy and drought tolerant crop.

★ Cluster bean is grown as a sole crop, mixed crop and a border crop around a main crop.

★ Flowering and pod development periods are the critical stages for irrigation in cluster bean.

### Indian bean

☆ Dolichos bean or Indian bean or Hyacinth bean is grown for its whole pod.

☆ Indian bean is a herbaceous perennial, but usually grown from seed as an annual.

☆ Two types of dolichos bean are recognized, *purpureus* var. *typicus* is a garden-type with soft edible pods as the pod walls have less fibre, and var. *lignosus* is field bean grown for fresh dry seeds as pulse.

☆ Pusa Early Prolific is a pole-type bean, suitable for autum and spring seasons.

☆ Arka Jay and Arka Vijay are bush type beans, photo insensitive, tolerant to heat and drought.

☆ Hebbal Avare is a bush type, pods are not edible but it can be used in hybridization programme for incorporation of dwarf plant characteristics.

☆ Arka Vijay and Konkan Bhusan are dual type varieties of Indian bean.

☆ Indian bean is a cool season and drought tolerant crop.

☆ In bush varieties, the crop is ready for harvest two months after sowing and in pole-types it takes three months for first harvest.

### Lima bean

☆ Lima bean is also known as double bean or butter bean.

☆ Lima bean, though of high nutritional value, is not commonly grown in India.

☆ King of the Garden, Karolina Butter, Challenger, Florida Butter are pole-type varieties of lima bean.

☆ Hopi and Wilbur are semi-pole type varieties of lima bean.

☆ Handerson Bush, Burpee Bush, Ford Hook 242, Baby Potato and Baby Ford Hook are bush type varieties of lima bean.

**Broad bean**

☆ The broad bean is also known as faba bean or horse bean.

☆ Broad bean is widely cultivated in Latin America.

☆ The stem of broad bean is square, erect and usually does not branch.

☆ Some people are allergic to the pollen of broad bean and also to the green pods.

☆ Aquadule Claudin, Imperial White Long Pod, Masterpiece Green Long Pod, Imperial Green Long Pod, Red Epicure are long pod type varieties of broad bean.

☆ Imperial White Windsor, Giant Four-Seeded Green Windsor, Imperial Green Windsor are Windsor type varieties of broad bean.

☆ Broad bean tolerates salinity.

☆ Broad bean is a hardy plant and is the only bean which is grown as an autumn and a winter crop as it can withstand cold temperatures as low as 4°C.

☆ Broad bean is mainly grown at higher elevation.

## Winged bean

☆  Winged bean also known as four-angled bean or Goa bean.

☆  Winged bean is a herbaceous, perennial, tropical legume of high value.

☆  Winged bean is exceptional among all food plants because practically all parts, including green pods, immature seeds, young leaves, flowers and tubers, are edible and are produced at every stage of the life cycle.

☆  Winged bean is photo sensitive and requires short days for normal flower induction.

☆  Winged bean cultivars available in India are IIHR selections, 20, 60, 71 and WBC 2.

☆  Winged bean is free from major pests and diseases. Recently minor diseases like false rust and leaf spot and pest like aphids were observed.

## Okra or Lady's finger or *Bhindi*

☆  India is the largest producer of okra in the world.

☆  Okra is more remunerative than the leafy vegetables.

☆  The root and stem of okra are useful for cleaning cane juice in preparation of jaggery.

☆  The species *Abelmoschus manihot* is also used as leafy vegetable.

☆  Okra has tremendous export potential as fresh vegetable. It accounts for 70% of the 30% foreign exchange earnings, other than onion, from export of vegetables.

☆ For fresh fruit export, okra fruits should be green, tender, 6-9 cm long.

☆ Among suitable varieties, Pusa Sawani, Prabhani Kranti, Varsha Uphar and Pusa A-4 have established in the areas producing the crop mainly for export.

☆ *Bhindi* flowers are self-fertile but usually up to 10% cross-pollination occurs by insects, thus it is classified as an often-cross-pollinated crop.

☆ Dr. Harbhajan Singh initiated systemic research work on improvement of *bhindi*.

☆ Pusa A-4 has excellent shelf life. It is resistant to YVMV and tolerant to jassids and shoot and fruit borers.

☆ MDU 1 is an induced mutant from Pusa Sawani by using gamma rays.

☆ Arka Anamika variety has been developed as a hybrid derivative selection from inter-specific cross between *A. esculentus* and *A. tetraphyllus*.

☆ Punjab Padmini variety has been developed as a hybrid derivative selection from inter-specific cross between *A.esculentus* and *A. manihot* ssp *manihot*.

☆ Punjab 7 has been developed as a hybrid derivative selection of a cross between *A. esculentus* cv. Pusa Sawani and *A. manihot* ssp. *manihot*.

☆ Punjab 8 is an induced mutant derived from Pusa Sawani by treating seed with EMS (1%).

☆ Prabhani Kranti, a YVMV resistant variety, was developed from interspecific cross.

☆ The variety, Perkin's Long Green was released in 1983-84 as Harbhajan Bhindi from College of Agriculture, Solan to commemorate the memory of Dr. Harbhajan Singh.

☆ At temperatures above 42°C, flower buds in most of the cultivars of okra may desiccate and drop causing yield losses.

☆ Pusa Sawani is adapted to larger pH range and has some tolerance to salinity.

☆ About 5,000 plants per hectare is recommended for branching types, and 66,000 plants per hectare for non-branching types.

☆ *Bhindi* does not require training or pruning. Varieties developed so far are upright growing and hence staking is also not required.

☆ Varieties like Arka Abhay and Pusa A-4 give quick branching after pruning.

☆ YVMV (Yellow Vein Mosaic Virus) is the most serious disease of okra. Its vector is whitefly (*Bemisia tabaci*).

☆ Jassid is the most serious pest of okra.

☆ Okra seed treatment by GA (400 ppm), IAA (20 ppm) or NAA (20 ppm) enhances germination.

☆ Ethephon (100-150 ppm) reduces vegetative growth and weakens apical dominance.

☆ Cycocel (1,000-1,500 ppm) reduces plant height.

☆ Post-harvest treatment with cycocel (100 ppm) enhances shelf-life of fruits.

☆ For processing industry and fresh fruit export, 6-8 cm long fruits are sorted out.

## Cucurbitaceous vegetables

### Cucumber

☆ The immature fruits of *Cucumis sativus* L. cucumber, *(gherkin)* are used as salad and for pickling.

☆ Cucumber is the second most widely cultivated cucurbit after watermelon.

☆ In temperate countries, cucumber is extensively grown in glasshouse.

☆ *C. hardwickii* might be a progenitor of the cultivated cucumber.

☆ The open pollinated cultivars of cucumber are monoecious in sex form.

☆ The first $F_1$ hybrid, Pusa Sanyog, developed in India in 1971, has a gynoecious female parent.

☆ Generally slicing cucumber have white spine colour and pickling cucumber have black spine colour.

☆ Variety, Himangi is resistant to bronzing.

☆ Pusa Sanyog is an $F_1$ hybrid of Japanese Gynoecious Line x Green Long of Naples.

☆ Poinsett is resistant to downy mildew, powdery mildew, anthracnose and angular leaf spot.

☆ Himangi is developed by selection from the cross Poinsett x Kalyanpur Ageti.

☆ Cucumber prefer slightly lower temperature than watermelon and muskmelon.

☆ In some parts of Maharashtra and South India, the real cucumber is called *kakri* which should not be confused with *kakri* of North India.

☆ In cucumber, application of ethrel (150-200 ppm) increases the number of female flower, fruit-set and in turn increases the fruit yield.

☆ Growth regulators like GA (1,500-2,000 ppm) and chemicals like silver nitrate (200-300 ppm) induce the male flowers on gynoecious cucumber.

☆ In cucumber, chemicals/plant regulators may be applied at 2-true-leaf stage.

☆ In cucumber, the yield could be doubled by using tropical gynoecious F₁ hybrids.

☆ Chilling injury as a physiological storage disorder of cucumber is noticed when fruits are exposed to temperature below -10⁰C for prolonged period.

**Pumpkin**

☆ Pumpkin have high carotene content and good keeping quality.

☆ The flowers of pumpkin are more nutritive than fruits.

☆ Pumpkin is specially known for its low cost production and long keeping quality.

☆ As pumpkin tolerates hotter conditions better than the other cultivated species of *Cucurbita,* it is most widely grown throughout the tropics of both hemisphere today.

☆ The genomic structure of *C. moschata* is AABB indicating it is an amphidiploid.

☆ Botanically, fig leaf gourd or Malabar gourd is *C. ficifolia* Boucha.

☆ Pumpkin (*C. moschata*) appears to be more closely related to xerophytic forms (*C. digita, C. palmata, C. cylindiata, C. foetodissima*) and is considered as the axis through which the species under xerophytic forms are related to each other.

☆ Mesophytic form (*C. lundelliana*) has wide spectrum of compatibility with the cultivated species particularly *C. moschata*.

☆ Flowers of pumpkin are unisexual, solitary or fasciculate and are lemon-yellow to deep orange in colour.

☆ The most popular cultivar in *C. moschata* is the Butternut squash with fruits relatively small and cylindrical with pronounced bulb surrounding the seed cavity and have bright orange flesh.

☆ Pumpkin is highly cross pollinated due to monoecious nature and entomophily.

☆ No inbreeding depression has been observed in pumpkin due to the fact that the original form of cucurbits is hermaphrodite.

☆ Arka Chandan is very rich in carotene content. (3331 iu/100 g of flesh).

☆ Pusa Vikas is highly suitable for cultivation in spring-summer season in North India.

☆ Short-day, low-night temperature and high relative humidity is best for pumpkin production.

☆ In pumpkin, high nitrogen under high temperature condition promotes maleness in flowering resulting in low fruit set and low yield.

☆ A growth regulator, ethrel (250 ppm) can be applied to increase the female flower production which help to increase the yield.

## Watermelon

☆ *Citrullus vulgaris* is the ancestor of cultivated watermelons.

☆ The citron or preserving melon (var. *citroides* of Bailey) can be considered as intermediate stage between primitive and bitter and cultivated and sweet forms of watermelon.

☆ The cultivar Arka Manik carries high resistance to powdery mildew, downy mildew and red pumpkin beetle. ·

☆ A seedless variety of watermelon, namely Pusa Bedana is an $F_1$ hybrid between Tetra 2 x Pusa Rasal.

☆ Cool nights and warm days are ideal for accumulation of sugar in the watermelon.

☆ Watermelon is sown in relay system just before digging of potatoes in late January or early February in north India.

☆ In watermelon, to increase the fruiting and the fruit yield, exogenous application of chemicals such as TIBA (25-250 ppm), boron (3-4 ppm), molybdenum (3-4 ppm) and calcium (20-25 ppm) is recommended. These chemicals/plant growth regulators must be applied at 2-true-leaf stage.

☆ The average sweetness in watermelon is recorded around 9-10% TSS.

☆ In Indian cultivars of watermelon, white heart seen at the central portion of rind shows poor quality. Similarly 'hollow heart' and 'fibrous flesh' signify over maturity.

## Muskmelon

☆ Mature fruits of muskmelon (*Cucumis melo* L.) are round in shape and 8-16 cm in diameter.

☆ *C. anguria, C. ficifolius* and *C. metuliferus* are resistant to nematode.

☆ Anthesis in muskmelon takes place at 5.30-6.30 a.m. at temperatures of 22-29°C.

☆ Muskmelon, snap melon (*phoot*) and long melon (*kakri*) easily cross with each other because they belong to same species. But none of them crosses with round melon (*Precitullus fistulosus* (Stocks) Pang.)

☆ Fruit set in monoecious lines of *C. melo* is 29-42%.

☆ Keeping quality of Arka Rajhans is excellent.

☆ Melons require tropical climate and fairly high temperatures of 35°-40°C during fruit development.

☆ Cool nights and warm days are ideal for accumulation of sugars in the fruits.

☆ Most of the cucurbits germinate well when the day temperature is above 25°C.

☆ Muskmelon is slightly more tolerant to soil acidity.

☆ Soaking of seeds in Ethephon at 480 mg/litre for 24 hours improves germination in muskmelon at low temperature.

☆ In muskmelon, the highest yield and good fruit quality are obtained with pruning (as the first hermaphrodite flower is borne on secondary branch arising from the eighth node, the secondary branches are pinched off up to the seventh node) plus staking.

☆ In muskmelon, application of ethrel (250 ppm) increases the fruiting and yield.

☆ Exogenous application of silver thiosulphate (300-400 ppm) induces the male flower in gynoecious muskmelon.

☆ The chemicals/plant growth regulators should be applied twice at 2-true-leaf stage and secondly at 4-true-leaf stage in muskmelon.

☆ Muskmelon is a climactric fruit.

☆ The fruits, when mature, slip out easily from the vine with a little pressure or jerk or if not, get separated the next day. This is called full slip stage. Most of the muskmelon cultivars behave in this way.

## Bottle gourd

☆ Bottle gourd has great export potential. The fruits exported should be light green to dark green in colour having a length of 30-40 cm and should be straight.

☆ Pusa Meghdoot is $F_1$ hybrid between Pusa Summer Prolific Long and Sel 2.

☆ Pusa Manjari is a round fruited $F_1$ hybrid between Pusa Summer Prolific Round and Sel 11.

☆ · Fruits of Pusa Naveen are perfectly cylindrical and straight without any crook-neck or curve.

☆ Pusa Summer Prolific Round is prolific bearer and heavy yielder.

☆ Fruits of Arka Bahar are devoid of crook-neck and have flesh with pleasant aroma.

☆ Samrat has good keeping quality and is good for box packing.

★ Pusa Hybrid 3 is very early in maturity.

★ Bottle gourd is a typical tropical plant which requires a hot and humid climate for the best growth.

★ Short days and humid climate promote femaleness in bottle gourd.

★ Pre-sowing treatment of seeds with 600 ppm of succinic acid for 12 hrs improves germination and seedling growth In bottle gourd.

★ In Maharashtra, bottle gourd is trained on bower systems. This increases yield over untrained vines.

★ The Samrat cultivar of bottle gourd has the highest yield potential on bower system.

★ The method of pruning and training developed at Mahatma Phule Krishi Vidyapeeth, Rahuri (Maharashtra) is commercially used by the bottle gourd growers in Maharashtra.

★ Fruit set can be increased by spraying the plant twice at the 2 and 4 true-leaf stages with ethrel (100-150 ppm), maleic hydrazide (400 ppm), triodobenzoic acid (50 ppm), boron (3-4 ppm) and calcium (20 ppm).

★ Maleic hydrazide at 400 ppm promotes the female flower production and increases fruit set and in turn the yield.

## Ash gourd

★ Ash gourd or wax gourd or white gourd or Chinese preserving melon is also known as *Petha kaddu*.

★ Ash gourd is mainly grown in North India, especially in Uttar Pradesh, where it is used in preparation of *petha* sweet. Agra *petha* is famous all over India.

✰ The surface of the fruit of ash gourd is covered with a bluish-white, waxy ash. The fruit has a long shelf-life.

✰ Ash gourd is an extensive trailing or climbing annual herb.

✰ Mudliar is a variety of ash gourd.

✰ Short-day, low-night temperature and humid climate are good for production of female flowers in ash gourd.

✰ Ash gourd fruit at maturity have a white waxy surface. There is a more wax bloom on ripening of the fruit and the peduncle withers.

## Sponge gourd

✰ *Luffa cylindrica* M.J. Roem. (Synonym : *L. aegyptica* Mill.) (sponge gourd, towel gourd, smooth loofah, vegetable sponge, dish cloth gourd) is an annual monoecious climber.

✰ The fruit is 20-50 cm long and almost cylindrical in shape.

✰ The flowers of sponge gourd open in early morning hours (4.0-8.0 am).

✰ The sponge gourd fruits contain higher protein and carotene than ridge gourd.

✰ Sponge gourd contains a gelatinous compound called *luffein*.

✰ Fruits of sponge gourd are smooth and are called *ghia tori*.

✰ Pusa Chikni, Phule Prajakta and Pusa Supriya are the varieties of sponge gourd.

☆ In sponge gourd, exogenous application of ethrel (250 ppm) has been found to be beneficial in female flower production.

## Ridge gourd

☆ *Luffa acutengula* (L.) Roxb. (angled loofah, ridge gourd, ribbed gourd) is an annual monoecious climber.

☆ The fruits of ridge gourd are 15-14 cm long and club-shaped. The surface is distinctly ribbed.

☆ Satputia is a cultivar of Bihar, which is hermaphrodite in sex form and produces smaller fruits in clusters.

☆ Ridge gourd contains a gelatinous compound called *luffein.*

☆ Fruits of ridge gourd are ribbed and are called *kali tori.*

☆ Pusa Nasdar, Konkan Harita, Punjab Sadabahar are the varieties of ridge gourd.

☆ The sex ratio can be regulated by exogenous application of growth regulators. NAA (200 ppm) increases the female flower production in ridge gourd and in turn increases the yield significantly.

## Pointed gourd

☆ Pointed gourd is commonly called *parwal* and it is a perennial cucurbit.

☆ Bengal-Assam area is the primary centre of origin of pointed gourd.

☆ *Trichosanthes dioica* Roxb. (pointed gourd) is a dioecious climber with perennial root stock.

☆ The fruit of pointed gourd is fusiform and 8-12 cm long.

☆ In order to plant one hectare of pointed gourd, about 2,000-2,500 cuttings are required.

☆ Pointed gourd is vegetatively propagated through vine cuttings and root suckers.

☆ In order to ensure maximum fruit set and yield only 10-12% male plants are maintained in the garden to ensure the source of polliniser. The remaining plants must be female ones.

☆ Training the pointed gourd crops gives high yields.

☆ During winter, pointed gourd becomes dormant and sprouts again in summer.

☆ Seed propagation is avoided in pointed gourd because of poor seed germination and since it is dioecious in nature, it produces male and female plants in equal proportion if they are planted from the seeds.

## Ivy gourd

☆ Ivy gourd is also known as coccinia or little gourd or scarlet gourd.

☆ Ivy gourd is most commonly used as a vegetable in southern and central India.

☆ The fruits of Ivy gourd are rich source of carbohydrates, proteins, vitamins A and C.

☆ Ivy gourd is dioecious in nature.

☆ Ivy gourd is a climbing perennial and has tuberous roots.

☆ Ivy gourd is propagated through stem cutting.

☆ To get good yield of Ivy gourd, 5-10% of the vines of male type should be planted in the field.

☆ Ivy gourd fruits are produced almost throughout the year.

☆ An average of 200-300 fruits weighing 3-4 kg can be harvested per vine per year.

## Snake gourd

☆ *Trichosanthes cucumerina* L. (Synonym: *T. anguina* L.) (Snake gourd) is cultivated widely in south and south-east Asia.

☆ Konkan Sweta is a variety of snake gourd.

☆ Snake gourd needs a warm and humid climate for best growth. High humidity is favourable for growth and fruit development.

☆ Training is essential for snake gourd crop. The bower system of training is best for this crop.

☆ Fruits of snake gourd are hand picked when they are still tender and about 1/4-1/3 of their full size.

## Bitter gourd

☆ Bitter gourd is also known as balsampear or bitter cucumber.

☆ Bitter gourd fruits are rich in iron.

☆ The cucurbitacin (bitter glucocide) may help in preventing spoilage of cooked vegetable of bitter gourd.

☆ *Momordica charantia* is incompatible with *M. dioica* and *M. balsamin*.

☆ Anthesis and anther dehiscence in bitter gourd occurs early in the morning.

★ White coloured varieties of bitter gourd are less bitter in taste and preferred in South India.

★ The characteristic bitter taste in the fruit is due to the bitter principle 'momodicidin'.

★ Phule Green is developed by cross between Green Long and Delhi Local.

★ Konkan Tara has good keeping quality and its shelf-life is 7-8 days under ambient temperature.

★ When temperature is above 36°C there is poor production of female flowers resulting in poor yield. The production of female flowers is increased by low temperature treatment (20°C) under short day.

★ Seed treatment with B9 at 3-4 ppm for two hours gives the highest number of female flowers per plant.

★ MH at 50-150 ppm and CCC at 50-100 ppm increases female:male ratio.

★ Ethrel at 25 ppm increases female flowers in bitter gourd.

## Round gourd

★ Round gourd is also known as Indian squash or round melon or tinda or squash melon.

★ The fruit has cooling effect and contains vitamin A.

★ Arka Tinda is an early summer season cultivar.

★ The optimum temperature for seed germination of round gourd is 27°C.

★ Fruits at very tender stage are harvested.

★ MH applied at 50 ppm at 2 and 4-leaf stage stimulates the elongation of main axis, induces the formation of female flower nodes, and increases the number of female flowers and fruit yield.

## Winter squash

☆ Winter squash (*vilayati kaddu, halwa kaddu*) is a monoecious, annual climber.

☆ Arka Suryamukhi is an improved variety. It is resistant to fruit fly.

☆ The plant of winter squash may tolerate frost.

☆ Winter squash is commonly grown in the hills.

☆ Pink Banana is a variety of winter squash.

## Summer squash

☆ Summer squash is also called vegetable marrow or field pumpkin.

☆ Pusa Alankar is an $F_1$ hybrid between EC 207050 and Sel.1.

☆ Punjab Chhappan Kaddu has a predominant female tendency, field resistance to downy mildew and red pumpkin beetle.

☆ Patty PAN, an introduction from USA, is a short-duration variety.

☆ The fruits of summer squash are harvested when they are at one-third maturity stage.

☆ *C. moschata* was used as a bridge to transfer disease resistance gene from *C. martinezii* to *C. pepo*.

☆ The three-way cross *C. pepo* x (*C. moschata* x *C. martinezii*) is being used to transfer resistance to powdery mildew and cucumber mosaic virus from *C. martinezii* to *C. pepo*.

## Long melon

☆ Long melon (*kakri* or *tar*) is more commonly grown in North India.

☆ Long melon can tolerate cool climate better than muskmelon and snap melon.

☆ In Lucknow and nearby areas, two types of long melon are very popular. They are *Laila ki Ungalian* (Fingers of Laila) and *Majnu ki Pasalian* (Ribs of Majnu). The former type is small and thin fruited, whereas, the latter has thin and long fruits.

☆ Arka Sheetal is an improved cultivar of long melon.

☆ Long melon fruits are harvested when they are very tender and at about half-grown stage.

☆ Green type of long melon fruits are comparatively good for transportation.

## Snap melon

☆ Botanically, snap melon is *Cucumis melo* var. *mamordica*.

☆ Snap melon (*Phoot* or *Kachra*) is not as popular as long melon but it is still grown during summer and rainy season in North India.

☆ Usually the fruits burst when the snap melon is ripe, therefore, it is popularly called *phoot*.

☆ Immature snap melon fruits having light green to dark green colour are harvested for vegetable and salad purpose.

## Chow-chow

☆ Chow-chow or cristophine chow-chow or chayote or askas or choco (*Sechium edule* Swartz.) is a herbaceous perennial climbing monoecious vine with large tuberous roots.

☆ Chow-chow is grown principally for its pear-shaped fruits, which are cooked in many ways.

☆ The fruits of chow-chow are fleshy and pyriform with longitudinal furrows.

☆ Chayote is generally raised by planting entire fruits. It can also be propagated by tuberous roots.

☆ Chow-chow show vaviparous germination.

☆ Chayote is the only cultivated member of cucurbitaceae family which is single-seeded.

## Kakrol

☆ Kakrol or Spine gourd (*Momordica dioica*) is a large, dioecious perennial climber.

☆ Kakrol roots are tuberous.

☆ Kakrol plant remains dormant in winter.

☆ The shape of kakrol fruit is ovate, 10 to 12 cm long, pointed, densely covered with conical spines.

☆ The young leaves of kakrol are also consumed as a leafy vegetable.

☆ Kakrol is rich in protein and vitamin C.

☆ Kakrol is generally propagated by its tuberous roots.

☆ It can also be propagated by seeds.

☆ In order to raise one hectare area of kakrol, about 5 lakh tuberous roots and about 3 to 4 kg seed are sufficient.

☆ About 10% of male plants should be planted for good fruit set.

☆ Kakrol seeds contain oil which is used as an illuminant.

## Kartoli

☆ The botanical name of kartoli or Sweet gourd is *Momordıca cochinchinensis.*

☆ Kartoli is a dioecious, perennial plant. Its vines are annual.

☆ The crop planted once will give yield for at least three to four years.

☆ The fruits are small, 2.5 to 5 cm long, ovoid ellipsoidal, densely echinate with soft spines.

☆ Kartoli goes into dormancy after giving fruits in the months of October to November.

☆ Kartoli is propagated by its tuberous roots.

# Tuber crops

## Potato

☆ Among the major potato growing countries of the world, China ranks first in area, followed by the Russian Federation, Ukarine and Poland. India ranks fifth in area in the world.

☆ The highest area and production of potato is in Uttar Pradesh followed by West Bengal and Bihar.

☆ The highest productivity of potato is in West Bengal followed by Gujarat.

☆ About 50% of potato produced in the world is utilized as human food.

☆ Potato is grown in India in almost all the states except Kerala.

☆ About 82% of the area under potato crop lies in the plains where the crop is grown during short days of winters from October-March.

★ About 10% of the area under potato lies in the hills where the crop is grown during long days of summer from April-September.

★ Potato is a dicot plant.

★ The widely cultivated potato belongs to tetraploid species *Solanum tuberosum*.

★ Potato is a self pollinated crop but is vegetatively propagated.

★ The cultivated tetraploid varieties of potato are highly heterozygous and most of them are pollen sterile.

★ Potato naturally flowers under cool climate and long-day conditions of more than 15 hrs light.

★ Potato flowers are hermaphrodite (bisexual).

★ Kufri Sinduri is a cross between Kufri Red and Kufri Kundan.

★ Kufri Badshah is a cross between Kufri Jyoti and Kufri Alankar.

★ Kufri Bahar is a cross between Kufri Red and Gineke.

★ Kufri Swarna is highly resistant to late blight and cyst nematode.

★ Kufri Jawahar is a cross between Kufri Neelamani and Kufri Jyoti.

★ Kufri Sutlej is a cross between Kufri Bahar and Kufri Alankar.

★ Kufri Pukhraj is resistant to early blight.

★ Kufri Chipsona 1 and Kufri Chipsona 2 are suitable for processing and producing light colour chips and finger fries.

☆ Maximum area under potato is in alluvial soil followed by hills, black and red soils.

☆ Potato is basically a crop of temperate region.

☆ Generally, the potato crop is raised in India when the maximum day temperature is below $35^0$C and night temperature not above $20^0$C.

☆ Most of the potato genotypes do not tuberize when the night temperature is more than $23^0$C.

☆ Potato crop is taken in autumn-winter/spring seasons in the plains and during summer/autumn in the hills.

☆ Potato is traditionally propagated through tubers.

☆ Potato tubers have a dormancy of nearly 8-10 weeks.

☆ The eyes on the surface of tubers contain axillary buds. When the dormancy is over, the axillary buds start germinating and produce sprouts. Such sprouted tubers, when planted in the soil, put up fast and vigorous growth.

☆ The multiple sprouting stage is the best stage for tuber planting because it generates emergence of several strong stems.

☆ Seed Plot Technique, essentially consists of raising the healthy seed crop during low aphid period available in northern plains during October-January and cutting the haulms and harvesting the crop before they cross damaging level.

☆ Currently, the Central Potato Research Institute (CPRI), Shimla has the sole responsibility of producing breeder's seed in the country.

☆ CPRI was established in 1949 with headquarters at Patna which was later shifted to Shimla in 1956.

★ The All India Coordinated Potato Improvement Project (AICPIP) started in 1971 with its headquaters at CPRI, Shimla (H.P.)

★ The International Potato Centre (CIP) is an autonomous scientific institution at Lima in Peru, established in 1971.

★ Seed plot technique was developed by Dr. Pushkarnath

★ Dehaulming in potato is done 10-12 days before harvesting.

★ Dormancy of potato can be broken by treating the potatoes with 1% thiourea + 1 ppm gibberelic acid.

★ For seed production, tuber having 30-40 g weight are the most economical and give the highest yield.

★ For main crop, cut tubers can be planted, each piece having two to three eyes and weighing at least 25 g.

★ Cyst nematode is the most widely occurring nematode of potato.

## Sweet potato

★ *Ipomoea batatas* L; the cultivated sweet potato, belongs to the family Convolvulaceae.

★ In the world scenario, more than 80% of the area under sweet potato cultivation and production lies in Asia.

★ China ranks first in area and production of sweet potato.

★ *I. batatas* is a hexaploid having 90 chromosomes.

★ *I. trifida* is thought to be its nearest wild relative and its most probable ancestor.

☆ Anthesis in sweet potato takes place very early in the morning between 4-5 a.m.

☆ Although sweet potato is often treated as an annual crop, the plant is a perennial vine.

☆ The flowers of sweet potato are solitary or cymose and vary in colour from white to purple.

☆ Sweet potatoes are moderately drought-tolerant.

☆ Sweet potato is usually vegetatively propagated through vine cuttings.

☆ Root-knot nematode (*Meloidogyne incognita*) and reniform nematode (*Rotylenchulus reniformis*) are the most widely occurring nematodes of sweet potato throughout the world.

☆ Pusa Harit is a heavy yielder and has very late bolting habit as it has low chilling requirement.

☆ Pusa Bharati has higher vitamin C and ß-carotene content.

## Yams

☆ Dioscorea species commonly known as yams, belong to the family Dioscoreaceae under monocotyledons.

☆ Two Asiatic yams, viz., *Dioscorea alata* (water yam or greater yam) and *Dioscorea esculenta* (lesser or Chinese yam) are the major food yams of India.

☆ *Dioscorea rotundata* (white yam) which is extensively cultivated in the African continent, is a recent introduction to India.

☆ Diosgenin present in many wild yam species has found good use in drugs for oral contraceptives.

☆ Yam plants have tuber or rhizome from which the vines and roots emerge annually during the growing season.

☆ In *D. alata*, bulbil is a minor storage organ while in *D. bulbifera*, the bulbil is the major storage organ.

☆ In *D. alata* no seed dormancy exists.

☆ Sree Shilpa is the first hybrid variety identified and released in *D. alata*.

☆ Yams are propagated vegetatively through tuber.

☆ Yam is also propagated through vine cuttings but tuber production by this method is slow.

☆ A seed tuber weight of 200-250 g is ideal for optimum production in *D. alata*. For *D. esculenta*, a seed tuber yam weight of 100-125 g is optimum.

☆ Yam tubers generally remain dormant for about 2-2¼ months in storage.

☆ Dormancy in yams can be broken artificially by quick dipping in 4-8% solution of ethylene chlorohydrin followed by dry storage.

☆ In yam, sprouting is delayed by soaking the tubers for 10 hrs in 0.1% solution of maleic hydraxide.

## Globe artichoke

☆ The Globe artichoke (*Cynara scolymus*) is a herbaceous perennial in which immature flower heads or buds are used as vegetable.

☆ Green Globe and Purple Globe are popular cultivars of artichoke

☆ Globe artichoke are propagated by seeds or by suckers or off-shoots.

☆ The small heads of globe artichoke are eaten raw or cooked, while large heads are eaten only after cooking. The thick receptacle known as 'heart' is used for canning.

## Cassava

☆ Cassava (*Manihot esculenta* Crantz.), variously called tapioca, manioc, mandioca and yaca, is a major starchy root crop of the tropics.

☆ Nigeria is the major country growing cassava. Kerala, where the crop was first introduced in India, accounts for 50% of area under cassava.

☆ The highest production and productivity of cassava is noted in Tamil Nadu.

☆ Cassava, belonging to the family Euphorbiaceae, is a native of Brazil.

☆ Cassava is highly heterozygous.

☆ Cassava is perennial shrub producing enlarged tuberous roots and variously branched stems.

☆ Cassava is a monoecious plant producing both male and female flowers on the same inflorescence and is cross pollinated.

☆ Cassava grown from the seeds develop a tap root system typical of dicots.

☆ Cassava grown from the stem cuttings develop many of the adventitious roots, of which a few differentiate into tubers.

☆ The main constituent of the cassava tuber is starch.

☆ The bitter principle of cassava is cyanogenic glucoside (HCN).

☆ The yellow colour of the cassava flesh is due to the

presence of carotene, the precursor of vitamin A.

* Nidhi is a variety of cassava

* Cassava is propagated by cuttings.

## Jerusalem artichoke

* Jerusalem artichoke (*Helianthus tuberosus*) is cultivated for its edible tubers.

* Jerusalem artichoke is a commercial source of levulose used as a sweetening agent by diabetic patients.

* Jerusalem artichoke yield is normally higher under long-day conditions.

* Jerusalem artichoke is vegetatively propagated by using free small sets or tubers with 2-3 eye buds, weighing approximately 50 g.

* The tubers are ready for harvest when the leaves begin to wither and die.

* The carbohydrate insulin is stored in the tubers. This is of interest because of its value in the diet of persons suffering from diabetes.

## Chinese water chestnut

* Chinese water chestnut (*Eleocharis dulcis* Burm. F) is a perennial leafless sedge commonly found in marshes and swamps throughout the world.

* The edible tubers of Chinese water chestnut are rich in starch.

* In India, it is sold in Kolkata under the name Singapuri Keysur.

* It is propagated through the small corms.

## Coleus

☆  Coleus (*Solenostemon rotundifolius* Poir.) commonly known as Chinese potato, produces edible tubers which can be used as a vegetable.

☆  The genus *Coleus* belongs to family Labitae.

☆  Coleus is believed to have originated in Africa.

☆  Sree Dhara is a variety of coleus released in Kerala.

☆  Propagation in coleus is by vine cuttings.

## Giant taro

☆  The fleshy aerial stems of the giant taro is the primary product.

☆  The giant taro originated in Sri Lanka.

☆  Three species, *Alocasia indica*, *A. macrorhiza* and *A. cucullata* are cultivated in Assam and West Bengal as food crops.

☆  Propagation in giant taro is by suckers, shoot tips and cormels.

## Queensland arrowroot or Purple arrowroot

☆  Queensland arrowroot (*Canna indica* L.) is perennial, herbaceous monocotyledon.

☆  The young rhizomes are eaten as vegetable and cooked tubers are delicious.

☆  *Canna edulis* appears to have originated in the Andean region of South America or Peruvian coast.

☆  Queensland arrowroot is normally propagated from the underground fleshy rhizomes.

## West Indian arrowroot

☆ West Indian arrowroot (*Maranta arundinacea* L.) is cultivated for its edible rhizome.

☆ The inflorescence in West indian arrowroot are terminal bearing with few white flowers.

☆ West Indian arrowroot originated in the northern South America and the lesser Antilles.

☆ Approximately 3,000-3,500 kg of bits are required to plant one hectare.

## Tannia

☆ Tannia (*Xanthosoma sagittifolius*) is a herbaceous edible species grown in India.

☆ Corms and cormels are usual planting material. Healthy cormels of bigger size and 20-25 cm long are commonly used.

☆ The setts from the top portion of the main corm with a thickness of 5-10 cm containing the apical bud are also used for propagation.

☆ In some parts of Maharashtra and Gujarat where only leaves are harvested, the main corms and cormels are left in the field. They act as new planting material.

## Elephant's foot

☆ Elephant's foot (*Amorphophallus companulatus*) or '*suran*' or '*jimmikand*', is gaining popularity because of its yield potential and culinary properties.

☆ Gajendra and Santragachi are the non-acrid, high-yielding varieties of elephant's foot.

☆ High-yielding, non-acrid varieties should be propagated vegetatively by using either cut tubers or small tubers.

☆ For commercial cultivation, whole or cut tubers 500-1,000 g are used for planting.

☆ Whole tubers should be preferred over cut tubers to minimize rotting.

☆ Elephant's foot tubers have long dormancy period which can be broken by treating them with thiourea (0.1%). $GA_3$ and ethrel are also effective in breaking the dormancy.

☆ Water stagnation during crop growth results in tubers that are very hard to cook. Hence, water stagnation should not be allowed at any stage of crop growth.

## Colocasia

☆ Colocasia (*Colocasia esculenta*) or taro is the most important and one of the oldest crops.

☆ Two types of colocasia, eddoe (*C. esculenta* var. *antiquorum*) and dasheen (*C. esculenta* var. *esculenta*) are commonly cultivated throughout India.

☆ The eddoe type is commonly called *arvi* and dasheen as *benda*. Eddoe type is most prevalent as vegetable.

☆ Africa ranks first in area and production of colocasia, followed by Asia.

☆ Colocasia tubers are rich in starch.

☆ Colocasia leaves and petioles are also used as vegetables.

☆ Colocasia tubers contain more protein, minerals, phosphorus and iron compared to other tuber crops.

☆ Satamukhi, Sree Rashmi and Sree Pallavi are improved varieties of colocasia.

☆ Colocasia is propagated vegetatively, mostly by small cormels weighing 20-25 g.

☆ One tonne planting material is enough for planting a hectare crop of colocasia.

☆ Water stagnation in the field results in tubers that become very hard to cook. This occurs both in dasheen as well as eddoe varieties. To maintain the cooking quality of tubers, proper drainage should be maintained in the field.

## Yam bean

☆ Yam bean (*Pachyrrhizus erosus*), popularly known as potato bean and *mishrikand*.

☆ Its starchy conical or turnip-shaped fleshy tubers are eaten. High sugar content in tuber imparts sweet taste when eaten raw.

☆ Mexican type cultivars have larger tubers (500-700g) in size, whereas local type cultivars have smaller tubers (200-300 g).

☆ The Mexican type cultivars are less sweet compared to local ones. So, they are less preferred in the Indian markets.

☆ The local type cultivars having smaller tubers are more sweet. The flesh is white with less fibre. There is no cracking.

☆ Rajendra Mishrikand 1, an improved variety, is very popular in north Bihar.

☆ Yam bean is propagated usually through seeds.

☆ The mature pods containing seeds are source of a toxic substance rotenone and are sometimes harmful for grazing cattle in the field.

☆ Cracking of tubers is the main problem due to unavailability of proper soil moisture. Delay in harvesting also causes cracking in tubers.

☆ Availability of proper soil moisture during 45-90 days check cracking. Application of potassium reduces cracking of tubers.

# Pot herbs or Greens Leafy vegetables

## Spinach beet or *Palak*

☆ Spinach or *Palak*, beet is one of the most common leafy vegetables of tropical and sub-tropical regions.

☆ Spinach beet is primarily used as pot herb.

☆ Spinach beet is a rich source of Vitamin A and C and also contains appreciable amount of protein, calcium and iron.

☆ Palak leaves contain low oxalic acid.

☆ Leaves of spinach beet might have been first used in Bengal and hence it is known as *Beta vulgaris* var. *bengalensis*.

☆ Leaves of cultivar Punjab Green are succulent and free from sourness.

☆ The cultivars Pusa Jyoti was developed by polyploidization of the culture as a result of induced mutagenesis using All Green as a source material.

☆ Pusa Harit was developed by hybridization between Sugar beet and Local Palak at IARI Regional Station, Katrain, H.P.

☆ Jobner Green was developed by selection from a spontaneous mutation detected from a local collection, Sl. No. 5.

☆ Pusa Bharati is the latest variety at the IARI, New Delhi.

## Spinach

★ Traditionally, spinach or *vilayati palak* is classified as pot herb. It is the green ones that are normally consumed as cooked vegetable.

★ Spinach is not commonly grown in India except in hilly areas.

★ Spinach is highly nutritious among the group of leafy salad vegetables.

★ Spinach is related to Swiss chard, sugar beet, table beet, pig weed and salt beet.

★ Virginia Savoy is a prickly seeded cultivar of spinach.

★ Spinach is a long-day crop.

★ Spinach tolerates frost better than most other vegetables.

★ Early Smooth Leaf is a smooth seeded cultivar of spinach.

★ Prickly seeded cultivars of spinach are best suited for the autumn-winter crop in the hills, whereas, smooth seed cultivars are suitable for spring-summer crop in hills and autumn sowing in plains.

## Fenugreek

★ Fenugreek is the third largest seed spice in India after coriander and cumin.

★ Fenugreek seeds substantially contain diosgenin which is used as starting material in the synthesis of sex hormones.

★ India is the major producer and exporter of fenugreek seed spice.

☆ Rajasthan claims the monopoly in its production, accounting for about 80% of fenugreek produced in the country.

☆ There are two species of the genus *Trigonella* which are of economic importance, viz., *Trigonella foenum-graecum*, the common *methi* and *T. corniculata*, the *Kasuri methi*.

☆ *T. corniculata* (*kasturi methi*) is a slow growing type and remains in rosette condition during most of the vegetative growth period.

☆ *T. foenum-graecum* (common *methi*) has white or light-violet coloured flowers.

☆ Each plant of common *methi* bears 10-15 twigs and 100-200 podes. Each pod contains 10-20 small, hard, yellowish-brown seeds which are about 3 mm long.

☆ Fenugreek is an annual herb. Leaves are light-green and are pinnately trifoliate.

☆ The flowers of fenugreek are papilionaceous.

☆ Fruits of fenugreek are legumes.

☆ The flowers of fenugreek are white or yellow in colour.

☆ Anthesis in fenugreek takes place between 9 a.m. and 6 p.m. with a peak at 11:30 a.m. and 12:30 p.m.

☆ Fenugreek is both a tropical and temperate crop.

☆ Fenugreek is capable of tolerating frost and freezing weather.

☆ Rajendra Kranti is a cultivar of fenugreek.

☆ A spray of ascorbic acid at 250-450 ppm improves the plant growth, enhances flowering, seed number and size of fenugreek.

☆ In fenugreek, 10-100 ppm of GA spray enhances the internodal length, height and number of leaves. It also induces early flowering.

☆ A spray of 0.05% sodium 2,3-Isobutyrate just before flowering improves the yield of fenugreek.

## Amaranth

☆ Amaranth is primarily used as pot herb.

☆ Amaranth is the most common leafy vegetable grown during summer and rainy season in India.

☆ Among leafy types, *Amaranthus tricolour* L. is the main cultivated species in India.

☆ Amaranth is a cross pollinated (anemophilous) crop.

☆ Pusa Kiran and Pusa Kirti are the varieties of amaranth.

☆ Amaranth can be grown in soil pH as high as 10.

☆ Amaranth is a warm season crop adapted to the conditions of hot, humid tropics, but it is also suitable for temperate climate during summer.

☆ The first cutting is done 3-4 weeks after sowing.

## Salad crops

### Lettuce

☆ Lettuce is a major salad crop in North America.

☆ *Lactuca serriola* seems to be the probable progenitor of cultivated lettuce.

☆ Punjab Lettuce No. 1 is a non-heading type cultivar of lettuce.

☆ Great Lakes and Imperial 859 are crisp head type cultivars of lettuce.

☆ Slobolt and Chinese Yellow are leaf type cultivars of lettuce.

☆ White Boston is butter head type cultivar of lettuce.

☆ Dark Green is cos type cultivar of lettuce.

☆ Great Lakes is resistant to tip burn.

☆ Lettuce seed may go into dormancy when subjected to high temperature in the dark and its exposure to chilling at $4^0$C-$6^0$C for 3-5 days results in breaking dormancy.

☆ High temperature promotes seed stalk and causes a bitter taste of leaves and induces tip-burn injury in lettuce.

☆ Leaf type lettuce cannot be stored under ordinary conditions because it loses the moisture very soon.

☆ It is better not to harvest the lettuce immediately after a rain or while the dew is on it, because the leaves being crisp and brittle break easily while handling.

## Parsley

☆ Parsley (*Petroselinum crispum*) is used as salad as well as for flavouring and garnishing crop.

☆ Parsley is a rich source of iron and Vitamin C.

☆ There are two groups of varieties, one with plain leaves and other with curled leaves commonly known as moss-curled.

☆ The variety Hamburg has tough stem, deeply cut plain leaves and moss-curled is dwarf with finely cut serrated and deeply curled leaves.

## Celery

☆ Celery (*Apium graveolens*) is cultivated for its fleshy leaf stalk.

☆ The exposure to high temperature causes bitterness in celery leaves.

☆ Celery crop bolts when temperature falls below $15^0C$.

☆ In celery, there are 2 types of varieties – self-blanched or yellow and green leaved. Green leaved varieties are preferred in the Indian market.

☆ The varieties, Florida Golden and Golden are self-blanched, whereas Wright Grove Giant, Fort Hook Emperor and Standard Beared are green-leaved.

☆ Celery is mainly propagated by seed.

☆ In celery, proper blanching is usually done to make the crop crisp, reduce acrid flavour, increase good flavour and tenderness.

## Perennial Vegetables

### Drumstick

☆ Drumstick or morning or horse radish tree or radish tree or west Indian ben (*Moringa oleifera* Lam) is one of the most popular tree vegetables in Indian households.

☆ Drumstick belongs to family Moringaceae.

☆ Drumstick is cultivated for its leaves, flower buds and tender fruits, all of which are used as nutritious vegetable.

☆ Drumstick seeds contain an oil called been or behen oil.

☆ *M. oleifera* is a deciduous tree of 8-10 m height.

☆ The flowers of drumstick are white or creamy white.

☆ Anthesis has been reported to commence as early as 4.30 a.m. and continues till 6.30 a.m. with a peak at 5.30 a.m.

☆ Drumstick is a tropical plant and grows well in the plains.

☆ Jaffna is a variety of drumstick.

☆ Perennial types of drumstick are propagated by limb cutting.

☆ Annual types of drumstick are propagated by seeds.

## Horse Radish

☆ The horse radish is a perennial crop growns for its pigment roots.

☆ The pungency in horse radish is due to the presence of an allyl isothiocyanate and butyl-thiocyanate, occurring in combination with the glucoside sinigrin.

☆ Horse radish is propagated by root cuttings. It is also propagated by crowns.

## Breadfruit

☆ The breadfruit (*Artocarpus altilis* Postperg) is native of Malaysia.

☆ In India, breadfruit is chiefly cultivated in southern regions, chiefly on west coast and western ghats of Nilgiris comprising of lower Palani hills, Wynad, Courtullain and Annamalais.

☆ The fruits of the seedless bread-fruit contain a high amount of carbohydrates (27.98%). It is rich in calcium.

☆ It is said to have been served as a staple food as bread for Britishers and hence the name 'Bread-fruit'.

☆ The bread-fruit plant is monoecious, but dicolinous, bearing separate male and female inflorescence on specialised laterals.

☆ Lateritic red loam soil in the west coast and western ghats is ideal for bread-fruit.

☆ Yellow Heart is a variety of breadfruit.

☆ Seeded types of bread-fruit are propagated through seeds.

☆ The propagation of seedless variety is possible through root cutting and air layering of root sucker.

☆ Seedlessness in bread-fruit is due to simulative parthenocarpy.

☆ One tree of bread-fruit bears 50-100 fruits weighing 25-50 kg annually.

## Chekurmanis

☆ Chekurmanis (*Sauropus androgynus* Meer) is a perennial small shruby leafy vegetable.

☆ Due to high nutritive value chekurmanis is commonly called 'Multi- Vitamin Greens'.

☆ The inflorescence of chekurmanis is auxiliary with small reddish flowers.

☆ Chekurmanis is highly cross-pollinated and entomophilous because of protogynous and monoecious nature of flower.

☆ The leaves and tender shoots of chekurmanis are mainly used as leafy vegetables or as salad.

☆ Chekurmanis is propagated by seeds and stem cuttings.

## Ceylon spinach or Water leaf

☆ Ceylon spinach or Water leaf (*Talinum triangulare* (Jacq) Wild.) is a soft mucilaginous leafy vegetable grown in the tropics.

☆ Ceylon spinach is a shade loving leafy vegetable.

☆ The leaves and tender shoots are used as vegetable in water Leaf.

☆ Ceylon spinach is native to Brazil.

☆ Water leaf is a herbaceous perennial.

☆ Flower of Ceylon spinach is pink in colour.

☆ Ceylon spinach is a short-day plant.

☆ Water leaf is propagated through seeds or herbaceous stem cutting.

## Basella

☆ Basella is commonly called Indian spinach or Malabar night shade.

☆ *Basella alba* (dark green and round-to-oval leaves) and *B. rubra* (red coloured leaves and stems) are different botanical forms of the species.

☆ Basella is believed to have originated in South Asia.

☆ Basella flowers are sessile, bisexual, rose coloured or yellowish-white.

☆ The succulent leaves with petiole and tender stems are eaten as vegetable.

☆ Basella can be propagated by seeds, as well as stem cuttings.

## Curry leaf

☆ Curry leaf is an important perennial tree vegetable-cum-spice crop of India.

☆ Ksenigin and Murayin are crystalline glucoside present in curry leaves and flowers respectively.

☆ Curry leaf believed to have originated in the Terai tract of Uttar Pradesh.

☆ Suwasini is a variety of curry leaf.

☆ Curry leaf is propagated through seeds.

☆ Polyembryony has been reported in curry leaf.

☆ Fruit of curry leaf contains two seeds per fruit and such seeds are separated by a thin papery seed coat.

## Asparagus

☆ Asparagus is a perennial dioecious herb.

☆ The tender shoots called 'spears' are used as vegetables.

☆ The tender shoots of asparagus contain white crystalline substance, asparagine, which is used in medicine as diuretic in cardiac dropsy and chronic gout.

☆ Asparagus (*Asparagus officinalis* L.) belongs to family Liliaceace.

☆ Female flowers of asparagus plants produce berries with one to three seeds per fruit.

☆ The male flowers of asparagus are yellowish-green in colour.

☆ Asparagus is propagated through seed or crown.

☆ Blanching is practiced to blanch the young spears and get 'white asparagus' for canning.

## Sea kale

☆ Sea kale is a hardy perennial.

☆ Sea kale belongs to the family cruciferae.

☆ Sea kale is grown for its young leaves and shoots.

## Rhubarb

☆ Rhubarb is perennial with large fleshy underground rhizome having a fibrous root system.

☆ The economic part of rhubarb is the large thick leaf stalk or petiole.

☆ Rhubarb is protandrous. Lower flowers are fertilized by pollen from upper flowers.

☆ Victoria and Linnaeus are the oldest varieties of rhubarb.

☆ McDonald, Ruby Valentine, Sunrise, Strawberry and Cherry Red are new varieties of rhubarb.

☆ Rhubarb is a cold resistant plant.

☆ Rhubarb is propagated by the division of crowns.

☆ Rhubarb is harvested from the second year of planting.

☆ Production of rhubarb stalks during the winter is called 'Forcing of rhubarb' in regions where the climate is suitable for production of vigorous crowns.

☆ Rhubarb is resistant to dry conditions.

☆ Rhubarb is tolerant to soil acidity.

# SEVENTEEN

## Objective Types

### A. Multiple Choice

1. _____ is a cultivar of bottle gourd.
   - (a) Pusa Meghdoot
   - (b) CO-1
   - (c) Australian Green
   - (d) Straight Eight

2. _____ is a hybrid between sugar beet and local palak.
   - (a) Jobner Green
   - (b) All Green
   - (c) Pusa Jyoti
   - (d) Pusa Harit

3. _____ seed per hill should be planted in snake gourd.
   - (a) 1 or 2
   - (b) 2 or 3
   - (c) 3 or 4
   - (d) 4 or 5

4. _____ tubers or setts will be needed to raise the yams on one hectare area.
   - (a) 14,300
   - (b) 15,300
   - (c) 13,300
   - (d) 16,000

5. _____ variety of *Brassica pekinensis* is a useful leaf vegetable for plains and hills of North India.
   (a) Chinese
   (b) American
   (c) Canadian
   (d) Japanese

6. _____ variety of lima bean need support.
   (a) Pole type
   (b) Semi-pole type
   (c) Bush type
   (d) All

7. _____ is a single seeded cucurbitaceous vegetable.
   (a) Chayote
   (b) Kakrol
   (c) Kartoli
   (d) Ivy gourd

8. A cucurbitaceous crop considered to be perennial is_____
   (a) *Parval*
   (b) Chow-chow
   (c) Snake gourd
   (d) None of the above

9. A *guar* meal contains about _____ protein.
   (a) 11.3%
   (b) 22.2%
   (c) 33.3%
   (d) 44.4%

10. A pea variety which gives 50% produce during first picking is_____.
    (a) Arkel
    (b) Meteor
    (c) Mattar Ageta-6
    (d) None of the above

11. A stem vegetable is _____.
    (a) Carrot
    (b) Knol-khol
    (c) Sweet potato
    (d) Radish

12. A triploid watermelon variety, Pusa Bedana, was released in_____.
    (a) 1970
    (b) 1969
    (c) 1972
    (d) 1989

13. About _____kg seed is required to sow one hectare area of round gourd.
    (a) 3-5
    (b) 5-8
    (c) 8-12
    (d) 13-15

14. About 3 kg of seed or _____ root tubers of pointed gourd are enough to plant an area of one hectare.
    (a) 20,000
    (b) 30,000
    (c) 40,000
    (d) 50,000

15. According to All India Medical Science report, the per capita vegetable requirement in India is_____.
    (a) 500 g
    (b) 400 g
    (c) 200 g
    (d) 300 g

16. All cole crops belong to the family_____.
    (a) Cruciferae
    (b) Umbelliferae
    (c) Brassicaceae
    (d) None of the above

17. All cucurbits are propagated by seeds except_____.
    (a) Pointed gourd
    (b) Little gourd
    (c) Kakrol
    (d) All of the above

18. Although French bean is a self-pollinated crop, the percentage of cross-pollination may vary from_____.
    (a) 2 to 8
    (b) 10 to 15
    (c) 15 to 20
    (d) None of the above

19. Amaranth belongs to the family_____.
    (a) Chenopodiaceae
    (b) Amaranthaceae
    (c) Portulacaceae
    (d) None of the above

20. Amaranth is _____ crop.
    - (a) Self-pollinated
    - (b) Cross-pollinated
    - (c) Often-cross-pollinated
    - (d) None

21. Amaranth is _____ plant.
    - (a) $C_3$
    - (b) $C_4$
    - (c) CAM
    - (d) None

22. Amaranth leaves are rich in_____
    - (a) Vitamin A
    - (b) Vitamin B
    - (c) Vitamin C
    - (d) All of the above

23. Among all the members of cucurbitaceous crops _____ is richest in iron content.
    - (a) Bitter gourd
    - (b) Watermelon
    - (c) Muskmelon
    - (d) Wax gourd

24. Among the root crops, the application of potassic fertilizers is essential in_____.
    - (a) Radish
    - (b) Turnip
    - (c) Carrot
    - (d) None of the above

25. Among the various forcing structures, temperature can be controlled more easily in_____.
    (a) Greenhouses
    (b) Hot beds
    (c) Cold frames
    (d) None of the above

26. An early maturing variety of potato which does not degenerate rapidly is_____.
    (a) Up-to-Date
    (b) Kufri Sindhuri
    (c) Kufri Chandramukhi
    (d) None of the above

27. An edible podded variety of garden pea is_____.
    (a) Mattar Ageta-6
    (b) Mithi Phali
    (c) Bonneville
    (d) None of the above

28. An $F_1$ hybrid of cucumber recommended by I.A.R.I. is _____.
    (a) Pusa Sanyog
    (b) Poinsette
    (c) Straight Eight
    (d) None of the above

29. An isolation distance of _____ between varieties of watermelon is considered enough for pure seed production.
    (a) 100-200 m
    (b) 200-300 m
    (c) 300-375 m
    (d) 400-500 m

30. An isolation distance of _____between two varieties of amaranth will maintain high genetic purity of seeds.

   (a) 100 m

   (b) 200 m

   (c) 300 m

   (d) 400 m

31. Arka Abhay is a variety of _____.

   (a) Brinjal

   (b) Chilli

   (c) Tomato

   (d) Bhindi

32. Arka Jeet is a popular cultivar of_____.

   (a) Muskmelon

   (b) Watermelon

   (c) Bottle gourd

   (d) Bitter gourd

33. Arka Jyoti is an improved variety of_____.

   (a) Muskmelon

   (b) Watermelon

   (c) Bottle gourd

   (d) Snap melon

34. Arka Manik, a variety of watermelon, is said to be resistant to_____.

   (a) Downy mildew

   (b) Powdery mildew

   (c) Wilt

   (d) All of the above

35. Arka Suryamukhi is an improved variety of_____.
    (a) Summer squash
    (b) Winter squash
    (c) Pumpkin
    (d) Musk melon

36. Arkel variety of garden pea is most suitable for_____.
    (a) Early season
    (b) Mid-season
    (c) Late season
    (d) None of the above

37. Asparagus is a_____.
    (a) Monoecious
    (b) Dioecious
    (c) Staminate
    (d) Hermaphrodite

38. Asparagus is propagated by_____.
    (a) Seeds
    (b) Crowns
    (c) Seedling
    (d) All of the above

39. Asparagus is_____.
    (a) Annual
    (b) Biennial
    (c) Perennial
    (d) All

40. Asparagus starts giving sizeable crop after a period of_____.

    (a) One year

    (b) Two years

    (c) Three years

    (d) None of the above

41. Aster yellow disease of celery is caused by_____

    (a) Fungus

    (b) Bacterium

    (c) Virus

    (d) None of the above

42. Average monthly temperature required for better growth and fruit set in tomato is_____.

    (a) 21°C to 23°C

    (b) 30°C to 35°C

    (c) 35°C to 40°C

    (d) 5°C to 10°C

43. AVRDC is situated in_____.

    (a) Tokyo

    (b) Taiwan

    (c) Tasmania

    (d) Turkmenistan

44. *Badi chaulai* belongs to the species_____.

    (a) Blitum

    (b) Tricolor

    (c) Caudatus

    (d) None of the above

45. Bean variety Pusa Parvati is evolved through_____.
    (a) X-rays
    (b) Gamma rays
    (c) EMS
    (d) MMS

46. Beet leaf is_____.
    (a) Self-pollinated
    (b) Cross-pollinated
    (c) Often-cross-pollinated
    (d) None

47. Beetroot fruit (capsule) contains_____.
    (a) 2 to 6 seeds
    (b) 8 to 10 seeds
    (c) 12 to 14 seeds
    (d) 16 to 20 seeds

48. Before transplanting, seedlings of asparagus raised from seed are allowed to grow in the nursery bed for a period of_____.
    (a) 4 months
    (b) 8 months
    (c) 12 months
    (d) None of the above

49. Best pod-set in French bean is obtained with plants kept for four hours after pollination at_____.
    (a) 5-15°C
    (b) 15-25°C
    (c) 25-35°C
    (d) None of the above

50. Best temperature for storing tomato fruits is_____.

    (a) 12°C - 15°C

    (b) 4.5°C

    (c) 5°C - 10°C

    (d) 15°C - 20°C

51. Bhindi variety tolerant to salinity is_____.

    (a) Kalyanpur Green

    (b) Type-3

    (c) Pusa Sawani

    (d) Pusa Dwarf

52. Bitter taste in brinjal fruit is due to_____.

    (a) Anti-vitamin E factor

    (b) CN glycocides

    (c) Solasodine

    (d) Trypsin inhibitors

53. Black heart is a physiological disorder of_____.

    (a) Tomato

    (b) Chilli

    (c) Cabbage

    (d) Potato

54. Black heart of potato is developed when tubers are stored at a temperature of_____.

    (a) -5°C

    (b) 0°C

    (c) 5°C

    (d) None of the above

55. Black leg disease of cabbage is caused by_____.
    (a) Bacterium
    (b) Virus
    (c) Fungus
    (d) None of the above

56. Black rot in cauliflower is caused by_____.
    (a) Fungal disease
    (b) Bacterial disease
    (c) Viral disease
    (d) None of the above

57. Black rot of cabbage is transmitted by_____.
    (a) Mechanical means
    (b) Seed
    (c) Vectors
    (d) None of the above

58. Blossom-end rot in chilli is due to_____.
    (a) Fungal disease
    (b) Viral disease
    (c) Non-parasitic cause
    (d) None of the above

59. Blossom-end rot in tomato is caused by the deficiency of_____.
    (a) Calcium
    (b) Magnesium
    (c) Boron
    (d) None of the above

60. Bottle gourd has been growing wild in_____.
    (a) Coast of Malabar
    (b) China
    (c) South America
    (d) None of the above

61. Bottle gourd originated in_____.
    (a) Africa
    (b) America
    (c) India
    (d) None of the above

62. Brinjal belongs to the species_____.
    (a) Khasianum
    (b) Sisymbrifolium
    (c) Melongena
    (d) None of the above

63. Brinjal is a native of_____.
    (a) Africa
    (b) South America
    (c) India and Africa
    (d) Korea

64. Brinjal seedlings are transplanted at a spacing of
    _____.
    (a) 60 x 30 - 45 cm
    (b) 75 - 90 x 60 - 70 cm
    (c) 50 - 60 x 50 - 60 cm
    (d) None of the above

65.. Brinjal variety MDU 1 was evolved by_____.

    (a) X-rays

    (b) Gamma rays

    (c) EMS

    (d) MMS

66. Brinjal variety Manjri Gota is recommended for the state of_____.

    (a) Punjab

    (b) Maharashtra

    (c) Bihar

    (d) U.P.

67. Broad bean is also known as_____.

    (a) French bean

    (b) Faba bean

    (c) Cluster bean

    (d) Hyacinth bean

68. Broad bean is susceptible to_____.

    (a) Saline soil

    (b) Acidic soil

    (c) Alkali soil

    (d) Saline-Alkali soil

69. Brown-heart in turnip is caused by deficiency of_____.

    (a) B

    (b) Mo

    (c) Ca

    (d) Mn

70. Brown-heart is a physiological disorder of_____.

    (a) Radish

    (b) Carrot

    (c) Turnip

    (d) Beet

71. Bud, blossom and fruit drop in chilli, in **general, is** due to _____.

    (a) Deficiency of nitrogen

    (b) Humid climate

    (c) Unfavourable temperature and water supply

    (d) None of the above

72. Bulking rate is very high in the potato variety_____.

    (a) Kufri Chandramukhi

    (b) Kufri Alankar

    (c) Kufri Naveen

    (d) None of the above

73. Bursting of cabbage heads may take place due to _____.

    (a) Sudden heavy irrigation after a dry spell

    (b) Continuous moisture supply

    (c) High temperature

    (d) None of the above

74. Bushy varieties of vegetable-marrow should be grown at a spacing of_____.

    (a) 45 x 45 cm

    (b) 60 x 45 cm

    (c) 125 x 45 cm

    (d) None of the above

75. Cabbage belongs to the botanical variety_____.
    (a) Botrytis
    (b) Capitata
    (c) Gongylodes
    (d) None of the above

76. Cabbage heads store well at_____.
    (a) 15°C at 60-70% R.H.
    (b) 5°C at 70-80% R.H.
    (c) 0°C at 90-95% R.H.
    (d) None of the above

77. Cabbage is a heavy feeder of_____.
    (a) N, K
    (b) N, P
    (c) N, P, K
    (d) P, K

78. Cabbage is cross pollinated owing to_____.
    (a) Floral morphology
    (b) Protandry
    (c) Protogyny
    (d) Self-incompatibility

79. Calabreeze is the variety of_____.
    (a) Green sprouting
    (b) Heading type broccoli
    (c) Cauliflower
    (d) None of the above

80. California Wonder is an important variety of_____.
    (a) Hot pepper
    (b) Sweet pepper
    (c) Bird pepper
    (d) None of the above

81. Carrot belongs to the family_____.
    (a) Cruciferae
    (b) Malvaceae
    (c) Umbelliferae
    (d) None of the above

82. Carrot is a heavy feeder particularly of _____.
    (a) N
    (b) P
    (c) K
    (d) NK

83. Carrot is pollinated by_____.
    (a) Insects
    (b) Wind
    (c) Water
    (d) All of the above

84. Carrot is rich in_____.
    (a) Carotene
    (b) Protein
    (c) Fat
    (d) Carbohydrate

85. Carrot is_____.

    (a)  Self-pollinated

    (b)  Cross-pollinated

    (c)  Often-cross-pollinated

    (d)  All of the above

86. Carrot roots become longer _____.

    (a)  Above 10°C

    (b)  Below 10°C

    (c)  Above 25°C

    (d)  Below 25°C

87. Carrot roots become shorter _____.

    (a)  Above 25°C

    (b)  Below 25°C

    (c)  Above 10°C

    (d)  Below 10°C

88. Carrot roots become shorter under_____.

    (a)  Higher temperature

    (b)  Lower temperature

    (c)  High humidity

    (d)  Low humidity

89. Carrot seed will not germinate below _____ °C.

    (a)  7.9

    (b)  6.9

    (c)  5.9

    (d)  3.9

90. Carrot yellow disease is caused by_____.

    (a) Fungus

    (b) Bacterium

    (c) Virus

    (d) None of the above

91. Cauliflower curds can be stored for 30 days at_____.

    (a) 0°C with 85-90 per cent R.H.

    (b) 10°C with 70-80 per cent R.H.

    (c) 15°C with 60-70 per cent R.H.

    (d) None of the above

92. Cauliflower originated in_____.

    (a) Mediterranean coast

    (b) South and Central America

    (c) Iran (Persia)

    (d) None of the above

93. Cauliflower requires an optimum monthly average temperature of_____.

    (a) 5-10°C

    (b) 10-15°C

    (c) 15-20°C

    (d) None of the above

94. Celery belongs to the family_____

    (a) Composite

    (b) Umbelliferae

    (c) Cruciferae

    (d) None of the above

95. Celery seed takes nearly _____ to germinate.
    (a) 10 days
    (b) 20 days
    (c) 30 days
    (d) 40 days

96. Challenger is a variety of_____.
    (a) Indian bean
    (b) Lima bean
    (c) Cluster bean
    (d) Hyacinth bean

97. Chayote is generally raised by planting_____.
    (a) Seed
    (b) Tuber
    (c) Vine cutting
    (d) Entire fruit

98. Chemical used for controlling root-knot nematodes in brinjal is_____.
    (a) Aldrin
    (b) Nemagon
    (c) Chloropyrophos
    (d) None of the above

99. Chenopodiaceae is the family of_____.
    (a) Potato
    (b) Sweet potato
    (c) Okra
    (d) Spinach

100. *Chhoti chaulai* is suitable for growing in_____

    (a)  Spring

    (b)  Rainy season

    (c)  Both the spring and rainy season

    (d)  None of the above

101. Chilli Guchhedar is highly suitable for_____.

    (a)  Drying and canning

    (b)  Salad purpose

    (c)  Drying only

    (d)  None of the above

102. Chillies can be grown from sea level to an altitude of _____.

    (a)  1000 m

    (b)  2000 m

    (c)  3000 m

    (d)  4000 m

103. Chillies Hybrid-1 (CH-1) was evolved by_____.

    (a)  CSAUA & T (Kanpur)

    (b)  PAU (Ludhiana)

    (c)  CCSHAU (Hissar)

    (d)  IARI (New Delhi)

104. Cluster bean belongs to the genus_____.

    (a)  Phaseolus

    (b)  Dolichos

    (c)  Cyamopsis

    (d)  None of the above

105. Cluster bean contains mucilaginous substance called
_____.
    (a)  Allantoin
    (b)  Allantonic acid
    (c)  Mannogalacton
    (d)  Glycoprotein

106. Cluster bean is sown at a spacing of_____.
    (a)  75 x 60 cm
    (b)  60 x 45 cm
    (c)  45 x 20 cm
    (d)  None of the above

107. Cluster bean is_____.
    (a)  Self-pollinated
    (b)  Cross-pollinated
    (c)  Often-cross-pollinated
    (d)  None

108. Cluster bean produce an amide_____.
    (a)  Aspargine
    (b)  Allantoin
    (c)  Glycoprotein
    (d)  Allantonic acid

109. Colour of Pusa Sweti variety of turnip is_____.
    (a)  Red
    (b)  Golden
    (c)  White
    (d)  None of the above

110. Conical head type of cabbage cultivars have _____maturity.
  (a) Early
  (b) Mid
  (c) Late
  (d) None

111. Contender is a variety of_____
  (a) French bean
  (b) Broad bean
  (c) Lima bean
  (d) None of the above

112. Cowpea is highly susceptible to_____ .
  (a) Zn
  (b) B
  (c) Mo
  (d) Mn

113. Cowpea is probably a native of_____.
  (a) India
  (b) China
  (c) Central Africa
  (d) None of the above

114. Cowpea variety highly resistant to mosaic virus is _____.
  (a) Pusa Dofasli
  (b) Cowpea 263
  (c) Pusa Komal
  (d) None of the above

115. Cracking of tomato fruit is brought about by_____.
    (a)  Restricted watering
    (b)  A period of drought followed by sudden watering
    (c)  Over-watering
    (d)  Insufficient watering

116. Cross-pollination in *palak* is brought about by_____.
    (a)  Insects
    (b)  Water
    (c)  Wind
    (d)  None of the above

117. Cucumber mosaic virus is spread by_____.
    (a)  Seeds
    (b)  Aphids
    (c)  Whitefly
    (d)  None of the above

118. Cucumber plant grows well at soil reaction (pH) between _____.
    (a)  4.5 and.5.5
    (b)  6.5 and 8.5
    (c)  5.5 and 6.7
    (d)  None of the above

119. *Cucurbita moschata* is cross-fertile with_____.
    (a)  *Cucurbita pepo*
    (b)  *Cucurbita ficifolia*
    (c)  *Cucurbita maxima*
    (d)  None of the above

120. Cultivated chilli originated in_____.
    (a) Peru
    (b) Iran
    (c) China
    (d) None of the above

121. Cultivated tomato belongs to the species_____
    (a) Hirsutum
    (b) Peruvianum
    (c) Pimpinellifolium
    (d) Esculentum

122. Curds of Dania variety of cauliflower are available during the period of_____.
    (a) Mid-January to April
    (b) January to February
    (c) Mid-December to mid-January
    (d) None of the above

123. Curing of harvested sweet potato tubers takes place if the tubers are kept for 10 days at_____
    (a) 20°C and 90% R. H.
    (b) 40°C and 70% R. H.
    (c) 80°C and 30% R. H.
    (d) None of the above

124. Daily mean temperature favourable for successful production of brinjal is_____.
    (a) 5°C to 10°C
    (b) 10°C to 15°C
    (c) 13°C to 21°C
    (d) None of the above

125. Deficiency of molybdenum in cauliflower leads to
_____.
    (a) Brown rot
    (b) Whiptail
    (c) Riceyness
    (d) None of the above

126. Delayed harvesting in radish may cause_____.
    (a) Forking
    (b) Pithiness
    (c) Deformed roots
    (d) All of the above

127. Depending upon the soil type and weather, pea crop requires total irrigations numbering_____.
    (a) 3-4
    (b) 8-10
    (c) 15-16
    (d) None of the above

128. Depending upon the type of seed, some seed-borne diseases are effectively controlled by soaking the seed for 20-30 minutes in hot water with a temperature of _____.
    (a) 30°C
    (b) 50°C
    (c) 70°C
    (d) None of the above

129. Directorate of Vegetable Research is situated in
_____.
    (a) Bhopal
    (b) New Delhi
    (c) Varanasi
    (d) Ludhiana

130. During cropping period of brinjal, the average number of irrigations required is_____.

    (a)  5

    (b)  8

    (c)  16

    (d)  None of the above

131. Early blight of tomato is caused by_____.

    (a)  Bacterium

    (b)  Fungus

    (c)  Virus

    (d)  None of the above

132. Early cauliflower is ready for harvesting in the period of _____.

    (a)  Mid-September to mid-November

    (b)  Mid-November to mid-January

    (c)  Mid-January to April

    (d)  None of the above

133. Earthing of potato crop is done_____.

    (a)  20 days after sowing

    (b)  40 days after sowing

    (c)  50 days after sowing

    (d)  None of the above

134. Edible part of cucumber (pepo) is _____.

    (a)  Pericarp

    (b)  Fleshy receptacle

    (c)  Mesocarp, endocarp, placentae

    (d)  Placentae

135. Edible part of melon (pepo) is _____.
    (a) Endosperm
    (b) Mesocarp
    (c) Receptacle
    (d) Seed coat

136. Edible part of tomato is _____.
    (a) Pericarp and placentae
    (b) Succulent thalamus
    (c) Mesocarp
    (d) Endocarp

137. Edible portion of artichoke is _____.
    (a) Root
    (b) Fruit
    (c) Flower bud
    (d) None of the above

138. Elephant foot corms (buds) weighing 100-200 g are used for planting of the _____ year.
    (a) First
    (b) Second
    (c) Third
    (d) Fourth

139. Elephant's foot corms at the end of _____ year weigh 6.8-9.0 kg.
    (a) Second
    (b) Third
    (c) First
    (d) Fourth

140. Elephant's foot is rich in vitamins_____.

   (a)  A and B

   (b)  B and C

   (c)  C and D

   (d)  Only B

141. Exceptionally large potato tubers may have more chances of_____.

   (a)  Black-heart

   (b)  Brown-heart

   (c)  Hollow-heart

   (d)  All of the above

142. $F_1$ hybrid variety of summer squash is_____.

   (a)  Early Yellow Prolific

   (b)  Punjab Chhappan Kaddu-1

   (c)  Pusa Alankar

   (d)  Patty PAN

143. Favism is related to_____.

   (a)  Indian bean

   (b)  Broad bean

   (c)  Cluster bean

   (d)  Cowpea

144. First cutting of *kasuri methi* after sowing is done in about_____

   (a)  25 - 30 days

   (b)  35 - 40 days

   (c)  45 - 50 days

   (d)  None of the above

145. First intergenic cross between radish and cabbage was made by_____.
    (a) Jennings
    (b) Karperenko
    (c) Hull
    (d) Richy

146. First picking in bitter gourd can be done after_____.
    (a) 50-60 days
    (b) 70-80 days
    (c) 80-90 days
    (d) None of the above

147. First picking in okra can be done after_____
    (a) 30 - 35 days
    (b) 45 - 50 days
    (c) 60 - 65 days
    (d) None of the above

148. Flat head or drum head and savoy types of cabbage cultivars have _____maturity.
    (a) Early
    (b) Mid
    (c) Late
    (d) None

149. Fleshy edible part of knol-khol is an enlargement of _____.
    (a) Root
    (b) Stem
    (c) Leaf
    (d) None of the above

150. Flower initiation and higher yields may be expected with the treatment of onion bulbs by_____.

    (a) NAA

    (b) IAA

    (c) IBA

    (d) $GA_3$

151. Flower of brinjal may be_____.

    (a) Long-styled

    (b) Medium-styled

    (c) Pseudo-short styled

    (d) All of the above

152. Flowers of beet leaf are_____.

    (a) Staminate

    (b) Pistillate

    (c) Hermaphrodite

    (d) All

153. For a balanced diet, the dieticians have recommended daily consumption of leafy vegetables to the tune of_____.

    (a) 50 g

    (b) 116 g

    (c) 250 g

    (d) None of the above

154. For canning, tomato fruits are picked at _____.

    (a) Immature green stage

    (b) Mature green stage

    (c) Half-ripe or pink stage

    (d) Red ripe stage

155. For controlling an attack of nematodes, grow a resistant variety like_____.
    (a) S-12
    (b) Punjab Kesri
    (c) Punjab Chhuhara
    (d) PNR-7

156. For controlling phomopsis blight in brinjal, temperature for hot water treatment for 30 minutes is_____.
    (a) 30°C
    (b) 40°C
    (c) 50°C
    (d) None of the above

157. For controlling sprouting of onions in storage, it is advantageous to apply pre-harvest foliar spray of maleic hydrazide (MH-40) at the concentration of_____.
    (a) 500 ppm
    (b) 1500 ppm
    (c) 2500 ppm
    (d) None of the above

158. For controlling stem-fly attack, apply_____.
    (a) Thimet 10G
    (b) Sevin granules
    (c) Thiodan
    (d) None of the above

159. For controlling weeds effectively in cauliflower field, apply stomp in moist soil_____.
    (a) Four days before transplanting
    (b) One day before transplanting
    (c) One day after transplanting
    (d) None of the above

160. For distant marketing, melon fruits are picked at
    _____.
    (a) Full-slip stage
    (b) Half-slip stage
    (c) Green mature stage
    (d) None of the above

161. For distant transportation, tomato fruits are picked
    at _____.
    (a) Immature green stage
    (b) Mature green stage
    (c) Turning stage
    (d) Red ripe stage

162. For farms located at a distant place from the market,
    one should follow the crop rotation of_____.
    (a) Potato-Onion-Green manure
    (b) Cauliflower-Tomato-Okra
    (c) *Palak*-Knol-khol-Onion (green)-Chilli
    (d) None of the above

163. For getting an early crop of brinjal, the sowing may
    be preferably done in_____.
    (a) Silt loam
    (b) Sandy loam
    (c) Clayey soil
    (d) None of the above

164. For getting an early crop of vegetable marrow, sowing
    of seeds under protection is done in_____.
    (a) October-November
    (b) Mid-December to mid-January
    (c) Mid-January to mid-February
    (d) None of the above

165. For getting good fruit yield in *parval*, planting ratio of female to male plants should be_____.
    - (a)   15-20 to 1
    - (b)   30-40 to 1
    - (c)   1 to 1
    - (d)   None of the above

166. For getting high seed yield, bulbs of onion are planted at a spacing of_____.
    - (a)   60 x 45 cm
    - (b)   45 x 45 cm
    - (c)   45 x 30 cm
    - (d)   None of the above

167. For getting higher bulb yield, grow onions at a spacing of _____.
    - (a)   30 x 45 cm
    - (b)   30 x 7.5 cm
    - (c)   15 x 7.5 cm
    - (d)   None of the above

168. For getting higher number of female flowers in bottle gourd, spray the plants at 2-4 leaf stage with boron at a concentration of_____.
    - (a)   20 ppm
    - (b)   10 ppm
    - (c)   3 ppm
    - (d)   None of the above

169. For growing carrot, radish and turnip for table purpose, sowing is done at a spacing of_____.
    - (a)   45 x 30 cm
    - (b)   30 x 15 cm
    - (c)   45 x 7.5 cm
    - (d)   None of the above

170. For induction of greater number of female flowers in watermelon, spray seedlings at 2-4 leaf stage with 2, 4, 5- tri-iodobenzoic acid at_____.

    (a) 10-15 ppm

    (b) 25-50 ppm

    (c) 50-100 ppm

    (d) None of the above

171. For keeping the sugar content low, potato tubers should be stored at a temperature of_____.

    (a) 5°C

    (b) 10°C

    (c) 15°C

    (d) None of the above

172. For long storage, potato tubers should be kept at _____.

    (a) 5-10°C

    (b) 10-15°C

    (c) 15-20°C

    (d) None of the above

173. For longer storage of cucumber fruits the temperature should be kept at_____.

    (a) 5°C

    (b) 10°C

    (c) 25°C

    (d) None of the above

174. For planting of one hectare area of pointed gourd about _____ cuttings are required.
    (a) 1000-1500
    (b) 1500-2000
    (c) 2000-2500
    (d) 60000-65000

175. For planting one hectare area of pointed gourd about _____ kg seed is required.
    (a) 20-25
    (b) 25-30
    (c) 30-35
    (d) 35-40

176. For preparation of beverage called *Kanji*, the most suitable carrots are_____.
    (a) Red carrots
    (b) Orange carrots
    (c) Black carrots
    (d) None of the above

177. For processing tomato the fruits are picked at_____.
    (a) Immature stage
    (b) Pink stage
    (c) Hard ripe stage
    (d) Over-ripe stage

178. For raising early crop of cabbage, sowing of seeds in nursery is done in_____.
    (a) August-September
    (b) May-June
    (c) September-October
    (d) None of the above

179. For raising late varieties of cauliflower, seedlings are transplanted at a spacing of_____.

    (a)   60 x 45 cm

    (b)   45 x 45 cm

    (c)   45 x 30 cm

    (d)   None of the above

180. For raising seedlings of cucurbits, it is desirable to use polythene bags of the size_____.

    (a)   4 x 6 cm

    (b)   10 x 15 cm

    (c)   25 x 35 cm

    (d)   None of the above

181. For seed production of cucumber, maintain _____ isolation between two cultivars.

    (a)   800 m

    (b)   1000 m

    (c)   1200 m

    (d)   1600 m

182. For seed production of snowball varieties of cauliflower in the Kulu Valley, optimum time of sowing of the crop is _____.

    (a)   Last week of August

    (b)   Mid-May to end of June

    (c)   Mid-September to end of October

    (d)   None of the above

183. For vegetable seeds in cold storage at 40 to 50°F, the relative humidity should not be higher than_____.

    (a)   45 per cent

    (b)   60 per cent

    (c)   70 per cent

    (d)   None of the above

184. French bean is a native of_____.
    (a) Asia
    (b) Mediterranean region
    (c) South and Central America
    (d) None of the above

185. French bean variety resistant to anthracnose disease is _____.
    (a) Contender
    (b) Kentucky Wonder
    (c) Tweed Wonder
    (d) None of the above

186. Fresh unshelled peas can be stored well for two weeks at _____.
    (a) 5°C and 85% R.H.
    (b) 0°C and 90-95% R.H.
    (c) 15°C and 60-70% R.H.
    (d) None of the above

187. Fruit of okra is rendered unmarketable due to the attack of_____.
    (a) Aphids
    (b) Mites
    (c) Painted bug
    (d) None of the above

188. Fruit setting in brinjal is usually in the flower having _____.
    (a) Long and medium style
    (b) Short and medium style
    (c) Medium style
    (d) Short style

189. Fruits of vegetable-marrow are ready for first picking after _____.

    (a)  40-50 days

    (b)  60-80 days

    (c)  80-100 days

    (d)  None of the above

190. Fruits of wild forms of bottle gourd are bitter in taste due to_____.

    (a)  Solanin

    (b)  Tomatine

    (c)  Cucurbitacin

    (d)  None of the above

191. Fungicide used for controlling powdery mildew disease of pea is_____.

    (a)  Captan

    (b)  Indofil M-45

    (c)  Karathane

    (d)  None of the above

192. Garden beet belongs to the family_____.

    (a)  Cruciferae

    (b)  Chenopodiaceae

    (c)  Umbelliferae

    (d)  None of the above

193. Garden beet belongs to the genus_____.

    (a)  Brassica

    (b)  Beta

    (c)  Raphanus

    (d)  None of the above

194. Garden beet goes to seeding before attaining marketable size at temperature_____.

    (a)  20°C

    (b)  15°C

    (c)  Below 10°C

    (d)  None of the above

195. Garden pea belongs to the sub species_____.

    (a)  Arvense

    (b)  Hortense

    (c)  Elatius

    (d)  None of the above

196. Garden pea was probably distinguished from field pea in 1536, but their common use began after _____.

    (a)  1586

    (b)  1600

    (c)  1675

    (d)  1700

197. Garlic belongs to the species_____.

    (a)  Cepa

    (b)  Porrum

    (c)  Sativum

    (d)  None of the above

198. Garlic bulb is a_____.

    (a)  Multiple bulb

    (b)  Compound bulb

    (c)  Clove

    (d)  All of the above

199. Garlic bulbs contain _____ cloves.

    (a)  10-20

    (b)  20-25

    (c)  25-30

    (d)  7-10

200. Garlic freezes at the average temperature of _____ °C.

    (a)  - 5

    (b)  - 4

    (c)  - 3

    (d)  -2

201. Garlic having largest bulb size_____.

    (a)  Creole

    (b)  Madrasi

    (c)  Tabiti

    (d)  Jamunanagar Local

202. Globe artichoke is propagated by means of_____.

    (a)  Suckers

    (b)  Root cuttings

    (c)  Seeds

    (d)  None of the above

203. Golden nematode is a serious pest of_____.

    (a)  Tomato

    (b)  Potato

    (c)  Brinjal

    (d)  Cabbage

204. Great Lakes is a variety of_____
   - (a) Crisp head lettuce
   - (b) Butter head lettuce
   - (c) Leaf type lettuce
   - (d) None of the above

205. Greening of potato results in_____.
   - (a) Increase in nutritional quality
   - (b) Decrease in nutritional quality
   - (c) Increase in disease resistance
   - (d) Decrease in disease resistance

206. Growth habit of Henderson Bush variety of lima bean is _____.
   - (a) Pole type
   - (b) Semi-pole type
   - (c) Bush type
   - (d) None of the above

207. Haploid plants have been produced by anther culture in tomato by_____.
   - (a) Greshof and Doy (1972)
   - (b) Tyagi and Prasad (1965)
   - (c) Mehta and Upadhyaya (1960)
   - (d) H.B. Singh (1963)

208. Hardening of cabbage seedling may be done by withholding irrigations for _____days prior to transplanting.
   - (a) 4-6
   - (b) 6-8
   - (c) 8-10
   - (d) 10-12

209. Harvesting of *chaulai* is done_____after sowing.
    (a) 2 weeks
    (b) 3-4 weeks
    (c) 5-6 weeks
    (d) None of the above

210. Haulms in potato seed crop are cut during_____.
    (a) End of December or first week of January
    (b) Second fortnight of January
    (c) First fortnight of February
    (d) None of the above

211. Healthy and stocky seedlings of cauliflower are transplanted at _____ cm distance.
    (a) 60 x 30-45
    (b) 60 x 60
    (c) 45 x 45
    (d) 40 x 15

212. Hermaphrodite cultivar of ridge gourd _____.
    (a) Pusa Nasdar
    (b) Satputia
    (c) Co.1
    (d) Jhinga Torai

213. Heterosis in bottle gourd was reported in a number of crosses by_____.
    (a) I.D. Tyagi (1973)
    (b) V.S. Seshadri (1988)
    (c) G.S. Gill (1980)
    (d) U.K. Verma (1985)

214. High quality of pea seed should have a minimum germination percentage of _____.
    (a) 75
    (b) 85
    (c) 95
    (d) 65

215. Highest pod yield in French bean is obtained at a soil pH of_____.
    (a) 5.3-6.0
    (b) 6.0-7.0
    (c) 7.5-8.5
    (d) None of the above

216. Horse radish is propagated by_____.
    (a) Seeds
    (b) Root cuttings
    (c) Stem cuttings
    (d) None of the above

217. How can one protect the stem-end of radish root from becoming green?
    (a) By thinning
    (b) By staking
    (c) By earthing
    (d) All of the above

218. Hyacinth bean or *Sem* originated in_____.
    (a) America
    (b) India
    (c) China
    (d) None of the above

219. Hybrid variety of brinjal most suitable for *bhartha* making is _____.
    (a) BH-2
    (b) BH-1
    (c) Pusa Anmol
    (d) None of the above

220. Hybrid variety of chilli recommended for cultivation is_____.
    (a) Punjab Lal
    (b) CH-1
    (c) Punjab Surkh
    (d) None of the above

221. Immediately after harvest, the tubers have a rest or a dormant period of_____.
    (a) 1 to 2 months
    (b) 2 to 3 months
    (c) 3 to 4 months
    (d) None of the above

222. In carrot, seed-to-seed method is preferred for _____ production.
    (a) Nucleus seed
    (b) Foundation seed
    (c) Certified seed
    (d) All of the above

223. In cauliflower, browning occurs due to deficiency of_____.
    (a) Nitrogen
    (b) Molybdenum
    (c) Boron
    (d) Sulphur

224. In Central India, sweet potato vine cuttings are planted during_____.

    (a) May-June

    (b) June-July

    (c) October-November

    (d) None of the above

225. In certain cucurbits pinching of male buds before anthesis in the female parent ensures hybrid seed production was suggested by_____.

    (a) Choudhary and Singh (1971)

    (b) Tyagi and Singh (1965)

    (c) Seshadri and More (1988)

    (d) V. Swarup (1964)

226. In cucurbits, which is dioecious in nature?

    (a) Bottle gourd

    (b) Pointed gourd

    (c) Muskmelon

    (d) Watermelon

227. In curry leaf, the plant parts mostly eaten are_____

    (a) Leaves

    (b) Tender stems

    (c) Flowers and fruits

    (d) None of the above

228. In general, potato needs _____ mm of irrigation water for optimum production.

    (a) 500

    (b) 600

    (c) 700

    (d) 800

229. In green vegetables, the limiting amino acids is_____.
    (a) Methionine
    (b) Arginine
    (c) Lysine
    (d) Tryptophan

230. In India, seed of cabbage can be produced in_____.
    (a) Hills
    (b) Deccan plateau
    (c) Plains
    (d) None of the above

231. In India, seeds of leek are produced in_____.
    (a) Plains
    (b) Higher altitudes in the hills
    (c) Coastal areas
    (d) None of the above

232. In northern India, muskmelon is generally sown during the months of_____.
    (a) November-December
    (b) December-January
    (c) February-March
    (d) None of the above

233. In northern India, sweet potato vine cuttings are planted during_____
    (a) March-April
    (b) June-July
    (c) August-September
    (d) None of the above

234. In onion, bolting takes place due to_____
    (a) High temperature
    (b) Low temperature
    (c) High nitrogen
    (d) Deficiency of nitrogen

235. In order to ensure proper fruit-set, it is essential to plant cuttings from male plants in between the female plants to the tune of_____.
    (a) 10 per cent
    (b) 5 per cent
    (c) 2 per cent
    (d) None of the above

236. In order to plant one hectare of muskmelon about _____kg seed is required.
    (a) 1-2
    (b) 2-3
    (c) 3-4
    (d) 4-5

237. In order to predict the days taken for obtaining a given quality or maturity level of pea after sowing, it is essential to have a knowledge of_____.
    (a) Degree days
    (b) Maturity standards
    (c) Variety
    (d) None of the above

238. In order to sow one hectare area of Indian bean about _____ kg seed is enough.
    (a) 10-20
    (b) 20-30
    (c) 30-40
    (d) 5-10

239. In order to sow one hectare crop of fenugreek____
____ kg seed is required.

    (a)  5-10
    (b)  10-15
    (c)  15-20
    (d)  20-25

240. In our diet the staple vegetable is_____.

    (a)  Potato
    (b)  Brinjal
    (c)  Tomato
    (d)  Chillies

241. In potato, Hollow heart is due to_____.

    (a)  Excessive watering and fertilization
    (b)  Excessive watering and high temperature
    (c)  Excessive watering and low temperature
    (d)  Excessive watering and haulm cutting

242. In squash melon, 100 g of seed contain about ____
____ seeds.

    (a)  100
    (b)  200
    (c)  250
    (d)  350

243. In the case of potato, the seed rate per hectare is
_____.

    (a)  1 quintal
    (b)  5 quintals
    (c)  10 quintals
    (d)  15 quintals

244. In the control of nematode disease of tomato, it is advisable to grow resistant variety *viz.,*_____.
     (a) Punjab Kesri
     (b) Pusa Ruby
     (c) Punjab NR-7
     (d) None of the above

245. In the plains of northern India, most of the varieties of pea are suitable for sowing from_____.
     (a) Mid-August to September
     (b) October-November
     (c) After mid-November
     (d) None of the above

246. In the U.S.A., muskmelon varieties resistant to powdery mildew have been developed by using a resistant donor from_____.
     (a) Iran
     (b) Afghanistan
     (c) India
     (d) None of the above

247. In the U.S.A., netted varieties of muskmelon are known as_____.
     (a) Honey Dews
     (b) Casabas
     (c) Cantaloupes
     (d) None of the above

248. In tomato, for fruit setting in adverse condition spray_____.
     (a) 2, 4-D
     (b) PCPA
     (c) Ethrel
     (d) Cycocel

249. In turnip, earthing may be done after _____ days of sowing.
    (a) 5-10
    (b) 10-15
    (c) 15-20
    (d) 20-25

250. In watermelon, the most important judging indication for maturity is_____.
    (a) Days after pollination
    (b) Dull sound when fruit is thumped
    (c) Metallic sound when fruit is thumped
    (d) None of the above

251. In yams, poisonous alkaloids, volatile acids and calcium oxalate are present in the _____.
    (a) Leaves
    (b) Roots
    (c) Tubers
    (d) Flowers

252. India is the _____ largest producer of vegetable next ot China.
    (a) First
    (b) Second
    (c) Third
    (d) Fourth

253. Indian bean is also known as_____.
    (a) French bean
    (b) Cluster bean
    (c) Hyacinth bean
    (d) Broad bean

254. Indian bean is_____.
   (a)   Self-pollinated
   (b)   Cross-pollinated
   (c)   Often-cross-pollinated
   (d)   None

255. Indian cauliflowers are primarily _____.
   (a)   Self-incompatible
   (b)   Self-compatible
   (c)   Self-sterile
   (d)   Cross-incompatible

256. Isolation distance for certified seed production in radish is_____.
   (a)   1000 m
   (b)   1600 m
   (c)   800 m
   (d)   200 m

257. It is essential to plant about _____ per cent male plants of pointed gourd for obtaining higher yields.
   (a)   5
   (b)   10
   (c)   15
   (d)   20

258. Ivy gourd is propagated by_____.
   (a)   Seed
   (b)   Root cutting
   (c)   Tuberous root
   (d)   Stem cutting

259. Japanese White is a variety of_____.
    (a) Carrot
    (b) Radish
    (c) Turnip
    (d) None of the above

260. Jerusalem artichoke is cultivated for its_____.
    (a) Succulent leaves
    (b) Flower buds
    (c) Tubers
    (d) None of the above

261. Kakrol is generally propagated by_____.
    (a) Bulbets
    (b) Tuberous root
    (c) Vine cutting
    (d) Tubers

262. Kakrol is _____.
    (a) Monoecious
    (b) Dioecious
    (c) Andromonoecious
    (d) Gynodioecious

263. *Kasuri methi* belongs to the species_____.
    (a) Foenum-graecum
    (b) Corniculata
    (c) Oleracea
    (d) None of the above

264. Knol-khol is propagated by_____.
    (a) Cutting
    (b) Bulb
    (c) Corm
    (d) Seed

265. Kohlrabi (Knol-khol) belongs to the botanical variety
_____.
    (a) Italica
    (b) Gongylodes
    (c) Gemmifera
    (d) None of the above

266. Kufri Badshah variety of potato is resistant
to_____.
    (a) Viral disease
    (b) Late blight
    (c) Wart
    (d) None of the above

267. Kufri Sinduri is a cross between_____.
    (a) Kufri Red and Adina
    (b) Kufri Red and Phulwa
    (c) Utimus and Adina
    (d) Kufri Red and Kufri Kundan

268. Leaf curl virus in nature is spread by_____.
    (a) Aphids
    (b) Leaf hopper
    (c) Whitefly
    (d) None of the above

269. Leaf or cutting type of lettuce belongs to the
variety_____.
    (a) Capitate
    (b) Crispa
    (c) Longifolia
    (d) None of the above

270. Leafy vegetable which is very popular in **Kashmir** is_____
    - (a) Spinach
    - (b) *Karam sag*
    - (c) *Sarson ka sag*
    - (d) None of the above

271. Leek is grown for its_____.
    - (a) Bulbs
    - (b) Cloves
    - (c) Branched stem and leaves
    - (d) None of the above

272. Leek is mainly propagated by_____.
    - (a) Seed
    - (b) Bulb
    - (c) Root cutting
    - (d) Leaf cutting

273. Lettuce does well in a relatively cool growing season with a monthly mean temperature of_____.
    - (a) 8 to 10°C
    - (b) 12 to 15°C
    - (c) 15 to 18°C
    - (d) None of the above

274. Lettuce is a_____.
    - (a) Self-pollinated crop
    - (b) Often-cross-pollinated crop
    - (c) Highly cross-pollinated crop
    - (d) None of the above

275. Lettuce performs best when grown in a soil with pH
of_____.

   (a)   4.5 to 5.5

   (b)   5.8 to 6.6

   (c)   7.0 to 8.5

   (d)   None of the above

276. Lettuce seed does not germinate properly when the
soil temperature is above_____.

   (a)   20°C

   (b)   25°C

   (c)   30°C

   (d)   None of the above

277. Lima bean belongs to the species_____.

   (a)   Vulgaris

   (b)   Lunatus

   (c)   Sinensis

   (d)   None of the above

278. Lima bean can be planted _____ in a year.

   (a)   Once

   (b)   Twice

   (c)   Thrice

   (d)   All of the above

279. Lima bean originated in or near_____.

   (a)   Egypt

   (b)   Guatemala

   (c)   India

   (d)   None of the above

280. Lima beans are ready for harvesting in_____
   - (a)   85-90 days
   - (b)   60-65 days
   - (c)   95-100 days
   - (d)   40-55 days

281. Little leaf disease of brinjal is caused by_____.
   - (a)   Mycoplasma
   - (b)   Bacteria
   - (c)   Fungi
   - (d)   None of the above

282. Little leaf disease of brinjal is spread by_____.
   - (a)   Aphids
   - (b)   Whitefly
   - (c)   Leaf-hopper
   - (d)   None of the above

283. Long and large-fruited varieties of bitter gourd are generally grown in_____
   - (a)   Summer season
   - (b)   Rainy season
   - (c)   Winter season
   - (d)   None of the above

284. Long melon is more commonly grown in_____.
   - (a)   North India
   - (b)   South India
   - (c)   East India
   - (d)   West India

285. Lower protein content in French bean is caused by the deficiency of_____.
    (a) Iron
    (b) Zinc
    (c) Calcium
    (d) None of the above

286. Lycopene development in tomato is adversely affected when temperature is above_____.
    (a) 30°C
    (b) 25°C
    (c) 20°C
    (d) 10°C

287. Main season crop of *petha* is harvested in the months of _____
    (a) June-July
    (b) September-October
    (c) December-January
    (d) None of the above

288. Main season pea varieties are sown at a spacing of _____.
    (a) 30 x 75 cm
    (b) 30 x 15 cm
    (c) 30 x 10 cm
    (d) None of the above

289. Melons can be stored for a week or more at_____.
    (a) 0°C to 1°C and 90% R.H.
    (b) 5°C and 80% R.H.
    (c) 10°C and 70% R.H.
    (d) None of the above

290. Minimum isolation distance between two tomato varieties for producing genetically pure breeder seed is _____.
    (a)  100 m
    (b)  200 m
    (c)  50 m
    (d)  400 m

291. More commonly cultivated species of *Luffa* in Europe and America is_____.
    (a)  Cylindrica
    (b)  Acutangula
    (c)  Egyptica
    (d)  None of the above

292. More tuberous root formation in sweet potato occurs if cuttings of vines are prepared from_____
    (a)  Upper portion
    (b)  Lower portion
    (c)  Middle portion
    (d)  None of the above

293. Most common sex expression in cucurbits is_____.
    (a)  Monoecism
    (b)  Dioecism
    (c)  Andromonoecism
    (d)  None of the above

294. Most of the cucurbits are monoecious and annual in habit except_____.
    (a)  Pointed gourd
    (b)  Chayote
    (c)  Kakrol
    (d)  All of the above

295. Mountain spinach is mainly grown on small scale during winters in_____.

    (a)   Maharashtra and Karnataka

    (b)   Gujarat and Madhya Pradesh

    (c)   Tamil Nadu and Kerala

    (d)   Himachal Pradesh and J & K

296. Much hoeing is not advisable in carrot as the smell of leaves attract_____.

    (a)   Wire worm

    (b)   Carrot fly

    (c)   Eel worm

    (d)   Aphids

297. Mucilage, a sticky substance in okra, is generally extracted from_____.

    (a)   Flowers

    (b)   Buds

    (c)   Stem and root

    (d)   Leaves

298. Muskmelon crosses with_____.

    (a)   Cucumber

    (b)   Watermelon

    (c)   Long melon

    (d)   None of the above

299. Muskmelon grows well in a soil with pH range between _____.

    (a)   4.5 and 5.5

    (b)   6.0 and 6.7

    (c)   7.5 and 8.5

    (d)   None of the above

300. Muskmelon originated in_____.
   (a) America
   (b) China
   (c) North-West India
   (d) None of the above

301. Muskmelon variety most suitable for distant transportation is_____.
   (a) Hara Madhu
   (b) Punjab Rasila
   (c) Punjab Sunehri
   (d) None of the above

302. Nucleus seed of beetroot can only be produced when crop is grown at _____km isolation under open condition.
   (a) 1
   (b) 2
   (c) 3
   (d) 4

303. Number of irrigations generally applied to chilli is _____.
   (a) 3-4
   (b) 6-8
   (c) 15-16
   (d) None of the above

304. Number of irrigations required for raising a good crop of onion ranges from_____.
   (a) 2-4
   (b) 5-10
   (c) 10-15
   (d) None of the above

305. Number of pickings done in chilli is_____.
    (a) 6-8
    (b) 10-12
    (c) 12-14
    (d) None of the above

306. Nursery sowing of lettuce is done in the months of_____.
    (a) August-September
    (b) September-October
    (c) October-November
    (d) None of the above

307. Oblong fruited hybrid variety of bottle gourd is_____.
    (a) Pusa Meghdoot
    (b) Pusa Summer Prolific Long
    (c) Pusa Manjari
    (d) None of the above

308. Okra crop is sown at a spacing of_____.
    (a) 75 x 45 cm
    (b) 60 x 45 cm
    (c) 45 x 75 cm
    (d) None of the above

309. Okra is_____ .
    (a) Self-pollinated
    (b) Cross-pollinated
    (c) Often-cross-pollinated
    (d) Both b and c

310. Okra seeds fail to germinate below _____ °C.

    (a)  5

    (b)  10

    (c)  15

    (d)  20

311. On weight basis _____ kg of cuttings of sweet potato are required for planting one hectare area.

    (a)  300

    (b)  400

    (c)  500

    (d)  600

312. One gramme carrot seeds contain about _____ seeds.

    (a)  550

    (b)  650

    (c)  750

    (d)  850

313. One gramme of celery seed holds about _____ seeds.

    (a)  1570

    (b)  2070

    (c)  2270

    (d)  2470

314. One gramme of onion seed, generally contains _____ seeds.

    (a)  180

    (b)  240

    (c)  300

    (d)  360

315. One gramme of radish seeds contain about ___
    _____. seeds.
    (a)  700-1000
    (b)  500-600
    (c)  600-675
    (d)  1100-1150

316. One gramme of turnip seed contains _____
    seeds.
    (a)  460-500
    (b)  360-400
    (c)  560-660
    (d)  260-360

317. One hundred gram of okra seed will contain about
    _____ seeds.
    (a)  600
    (b)  700
    (c)  800
    (d)  900

318. Onion cultivar suitable for growing in *kharif* season
    _____.
    (a)  Nasik Red
    (b)  N-53
    (c)  No.404
    (d)  All

319. Onion cultivar tolerant to drought is _____.
    (a)  Hissar-2
    (b)  Pusa Red
    (c)  Pusa Ratnar
    (d)  Baswant 780

320. Onion seeds are viable for _____ years.
    (a) 1-2
    (b) 2-3
    (c) 3-4
    (d) 4-5

321. Onion variety suitable for *kharif* crop is_____.
    (a) Pusa Red
    (b) Pusa Ratnar
    (c) Agrifound Dark Red
    (d) Agrifound Light Red

322. Optimum pH of the soil for maximum production of cauliflower is between_____.
    (a) 6.0 and 7.0
    (b) 4.5 and 5.5
    (c) 5.5 and 6.6
    (d) None of the above

323. Optimum pH range for better growth of okra is_____.
    (a) 4.5 - 5.5
    (b) 6.0 - 6.8
    (c) 7.5 - 8.5
    (d) None of the above

324. Optimum pH range for growth of sweet potato is_____
    (a) 4.5 - 5.5
    (b) 5.8 - 6.7
    (c) 7.5 - 8.5
    (d) None of the above

325. Optimum pH range of the soil for cabbage is between
    _____.
    (a) 6.0 and 7.0
    (b) 7.0 and 8.5
    (c) 5.5 and 6.5
    (d) None of the above

326. Optimum soil pH required for maximum growth of spinach is_____.
    (a) 5.0 - 6.0
    (b) 6.0 - 7.0
    (c) 7.0 - 8.0
    (d) None of the above

327. Optimum soil temperature for obtaining good yield of sweet potato is_____.
    (a) 15-20°C
    (b) 20-30°C
    (c) 30-40°C
    (d) None of the above

328. Optimum temperature and humidity for storage of sweet potato tubers are_____
    (a) 5°C and 95-100% R.H.
    (b) 10°C and 90-95% R.H.
    (c) 15°C and 85-90% R. H.
    (d) None of the above

329. Optimum temperature for okra seed's germination is _____ °C.
    (a) 20
    (b) 30
    (c) 15
    (d) 35

330. Optimum temperature for the growth of melon is about _____.
    (a)  15°C
    (b)  25°C
    (c)  30°C
    (d)  None of the above

331. Optimum time for planting onion bulbs for seed production is_____.
    (a)  September
    (b)  October
    (c)  December-January
    (d)  None of the above

332. Orach or chakwat (*Atriplex hortensis* L.) or mountain spinach is mainly grown on small scale during winters in _____.
    (a)  Maharashtra and Karnataka
    (b)  Gujarat and Madhya Pradesh
    (c)  Tamil Nadu and Kerala
    (d)  Himachal Pradesh and J & K

333. Orange-coloured varieties of carrot are rich source of _____.
    (a)  Carotene
    (b)  Lycopene
    (c)  Anthrocyanin
    (d)  None of the above

334. Original home of okra is_____.
    (a)  India
    (b)  Africa
    (c)  America
    (d)  None of the above

335. Out of the total cultivated area in India, vegetable crops occupy_____.
    (a)  0.5 per cent
    (b)  2.2 per cent
    (c)  5 per cent
    (d)  None of the above

336. *Palak* can be grow in slight _____ soil condition.
    (a)  Saline
    (b)  Alkaline
    (c)  Saline-alkaline
    (d)  Acidic

337. Pant Anupam, a variety of French bean, is_____.
    (a)  Suitable to grow in hills
    (b)  Tolerant to angular leaf spot and blight
    (c)  Mosaic
    (d)  All of the above

338. Pea edible podded cultivar is_____.
    (a)  Arkel
    (b)  Bonneville
    (c)  Perfection New Line
    (d)  Sylvia

339. Pear shaped variety of tomato is_____.
    (a)  Punjab Kesri
    (b)  PNR-7
    (c)  Punjab Chhuhara
    (d)  Punjab Tropic

340. Peas are grown during _____ season in plains of north India.

   (a) Kharif

   (b) Rabi

   (c) Zaid

   (d) All of the above

341. Per hectare seed rate of broad bean is_____ .

   (a) 20-30 kg

   (b) 30-40 kg

   (c) 50-60 kg

   (d) 70-100 kg

342. Perfection is a cultivar of_____.

   (a) Fenugreek

   (b) Asparagus

   (c) Celery

   (d) Basella

343. *Petha* is a native of_____.

   (a) Malaysia

   (b) India

   (c) America

   (d) None of the above

344. *Phoot* and *kakri* belong to the species_____.

   (a) Melo

   (b) Sativus

   (c) Anguria

   (d) None of the above

345. Photo-thermo-insensitive cultivar of cowpea is_____.
    (a) Pusa Dofasali
    (b) Pusa Rituraj
    (c) Philippines Early
    (d) FS-68

346. Plant part of celery consumed as salad is_____
    (a) Stem
    (b) Leaf blade
    (c) Petiole
    (d) None of the above

347. Plant parts of asparagus used for eating purpose are_____
    (a) Spears
    (b) Leaves
    (c) Tender stem
    (d) None of the above

348. Plant regulator which improves fruit set in tomato is _____.
    (a) Gibberellic acid
    (b) 2, 4- Dichlorophenoxy acetic acid
    (c) Maleic hydrazide
    (d) Ethephon

349. Planting times of three crops of potato grown in the Nilgiri hills are_____.
    (a) October, December, March
    (b) April, August and January
    (c) September, February and July
    (d) None of the above

350. Pods of Indian bean for cooking are usually harvested when they are _____.

    (a)  Fully ripe

    (b)  Under-ripe

    (c)  Over ripe

    (d)  Tender

351. Poi or Indian spinach belongs to the family_____.

    (a)  Chenopodiaceae

    (b)  Leguminoseae

    (c)  Basellaceae

    (d)  None of the above

352. Poinsette is the most popular variety of_____.

    (a)  Muskmelon

    (b)  Long melon

    (c)  Cucumber

    (d)  None of the above

353. Pointed gourd is a _____.

    (a)  Monoecious plant

    (b)  Hermaphrodite plant

    (c)  Dioecious plant

    (d)  Polygamous plant

354. Pointed gourd is propagated by_____.

    (a)  Tuberous root

    (b)  Seed

    (c)  Leaf cutting

    (d)  Vine cutting

355. Pollination and fertilization processes in tomato are impaired due to night temperature below_____.
    (a) 13°C
    (b) 5°C
    (c) 10°C
    (d) 20°C

356. Pollination in beetroot is_____.
    (a) By insect
    (b) By wind
    (c) By water
    (d) All of the above

357. Pollinator of radish_____.
    (a) Bees
    (b) Honey bees
    (c) Wind
    (d) Water

358. Possibility of producing seedless watermelon was demonstrated in_____.
    (a) U.S.A.
    (b) Japan
    (c) Holland
    (d) None of the above

359. Potato crop is planted at a spacing of_____.
    (a) 45 x 30 cm
    (b) 60 x 15-20 cm
    (c) 75 x 10 cm
    (d) None of the above

360. Potato is _____ plant for its tuber.

    (a) Short-day

    (b) Long-day

    (c) Day-neutral

    (d) None

361. Potato is a native of_____.

    (a) Europe

    (b) South America

    (c) Africa

    (d) None of the above

362. Potato is a rich source of vitamin_____.

    (a) A

    (b) B

    (c) C

    (d) D

363. Potato is mainly propagated by_____.

    (a) Tubers

    (b) Corm

    (c) Seeds

    (d) Setts

364. Potato tuber production is the maximum at _____°C.

    (a) 15

    (b) 20

    (c) 25

    (d) 30

365. Potato variety which is not resistant to wart disease is _____.
    (a) Kufri Sherpa
    (b) Kufri Jyoti
    (c) Kufri Muthu
    (d) Kufri Sinduri

366. Powdery mildew disease is a serious problem of which of the following crops_____.
    (a) Capsicum
    (b) Okra
    (c) Cucurbits
    (d) Pea

367. Pruning and training of tomato plants to a single stem helps in getting_____.
    (a) High total yield
    (b) High marketable yield
    (c) High early yield
    (d) Increased fruit size

368. Puffy fruit in tomato is caused by_____.
    (a) Viral attack
    (b) Physiological disorder
    (c) Bacterial attack
    (d) None of the above

369. Pumpkin as well as squashes do best at a soil pH of _____.
    (a) 4.0-5.5
    (b) 6.0-6.5
    (c) 7.5-8.5
    (d) None of the above

370. Pumpkin originated in_____.
    (a) India
    (b) Africa
    (c) America
    (d) None of the above

371. Pumpkin varieties mature in_____.
    (a) 70-80 days
    (b) 80-90 days
    (c) 110-120 days
    (d) None of the above

372. Purple Vienna is _____ variety of knol-khol.
    (a) Early
    (b) Mid
    (c) Late
    (d) All season

73. Pusa Chetki variety of radish is most suitable for sowing during_____.
    (a) March-August
    (b) October-November
    (c) After November
    (d) None of the above

4. Pusa Chikni is a variety of_____.
    (a) Sponge gourd
    (b) Ridge gourd
    (c) Bottle gourd
    (d) Bitter gourd

375. Pusa Jyoti is a variety of_____
    (a) *Palak*
    (b) *Methi*
    (c) Spinach
    (d) None of the above

376. Pusa Kesar is a variety of _____.
    (a) Carrot
    (b) Radish
    (c) Turnip
    (d) Chilli

377. Pusa Naubahar variety of cluster bean is suitable for _____.
    (a) Summer crop
    (b) Rainy season crop
    (c) Both summer and rainy season
    (d) None of the above

378. Pusa Rasraj is_____.
    (a) A pure line variety of muskmelon
    (b) A hybrid variety of muskmelon
    (c) A hybrid variety of watermelon
    (d) A seedless variety of watermelon

379. Quality of pea is ascertained by_____.
    (a) Refractometer
    (b) Tenderometer
    (c) Potentiometer
    (d) None of the above

380. Radish belongs to the genus_____.
    (a) Brassica
    (b) Raphanus
    (c) Daucus
    (d) None of the above

381. Radish is_____.
    (a) Cross-pollinated
    (b) Self-pollinated
    (c) Often-cross-pollinated
    (d) All of the above

382. Rainy season crop of cucumber is sown during the months of_____.
    (a) April-May
    (b) June-July
    (c) August-September
    (d) None of the above

383. Rainy season crop of okra is planted during_____
    _____
    (a) April-May
    (b) June-July
    (c) July-August
    (d) None of the above

384. *Raphano brassica* was produced by_____.
    (a) Blackeslee
    (b) Rimpu
    (c) Nagaheru
    (d) Karperenko

385. Rat-tail radish (*Raphanus sativus* var. *caudatus*) is exclusively grown for its_____.

    (a)  Long thin pods

    (b)  Tender roots

    (c)  Succulent leaves

    (d)  None of the above

386. Removal of terminal buds and lower leaves may result in_____.

    (a)  Early and bad quality sprouts

    (b)  Late and bad quality sprouts

    (c)  Late and good quality sprouts

    (d)  Early and good quality sprouts

387. Rhubarb is grown for its_____

    (a)  Thick leaf stalks

    (b)  Tubers

    (c)  Tender stem

    (d)  None of the above

388. Root system of sweet potato is_____

    (a)  Shallow

    (b)  Medium

    (c)  Deep

    (d)  None of the above

389. Root-knot nematode is a serious pest of_____.

    (a)  Tomato

    (b)  Potato

    (c)  Brinjal

    (d)  Cabbage

390. Root-to-seed method is preferred for _____ production.
     (a) Nucleus seed
     (b) Foundation seed
     (c) Certified seed
     (d) All of the above

391. Round head type of cabbage cultivars are _____ maturing.
     (a) Early
     (b) Mid
     (c) Late
     (d) None

392. Salt tolerant variety of onion _____.
     (a) Punjab Selection
     (b) No.404
     (c) Nasik Red
     (d) N-53

393. Sarda melon variety of muskmelon grown in Afghanistan is available in India in the months of _____.
     (a) May-June
     (b) August-September
     (c) October-November
     (d) None of the above

394. Satisfactory yield of bottle gourd is usually obtained at soil pH of _____.
     (a) 4.5 - 5.5
     (b) 6.0 - 7.0
     (c) 7.5 - 8.5
     (d) None of the above

395. *Sechium edule* is_____.
    (a) Annual
    (b) Biennial
    (c) Perennial
    (d) All

396. Seed of radish remain viable up to_____.
    (a) 1-2 years
    (b) 2-3 years
    (c) 3-4 years
    (d) 4-5 years

397. Seed production in beetroot is taken up in hills above
    _____.
    (a) 1200 m
    (b) 1400 m
    (c) 1600 m
    (d) 1800 m

398. Seed production of which crop is not done in plains?
    (a) Early cauliflower
    (b) Cabbage
    (c) Onion
    (d) Carrot

399. Seed rate of colocasia is _____Q/ha.
    (a) 8-12
    (b) 14-16
    (c) 16-18
    (d) 20-22

400. Seedlings of leek are transplanted in/on_____.
    (a) Raised beds
    (b) Flat beds
    (c) Trenches
    (d) None of the above

401. Seedlings or crowns of asparagus are planted at a spacing of about_____.
    (a) 1 m x 30 cm
    (b) 1.5 - 2.0 m x 45 - 60 cm
    (c) 3.0 - 4.0 m  x 60 - 75 cm
    (d) None of the above

402. Seed-propagated pointed gourd have male to female ratio of_____.
    (a) 10:1
    (b) 1:10
    (c) 50:1
    (d) 50:50

403. Seeds of tomato for spring-summer crop are sown in nursery in the months of_____.
    (a) October-November
    (b) June-July
    (c) January-February
    (d) December-January

404. Sex expression in *parval* plant is_____.
    (a) Monoecious
    (b) Dioecious
    (c) Andromonoecious
    (d) None of the above

405. Shape of head of cabbage variety Pride of India is _____.
    (a) Flat
    (b) Round
    (c) Pointed
    (d) None of the above

406. Single fruit of *palak* contains _____ seeds.
    (a) 2-3
    (b) 4-5
    (c) 6-7
    (d) 8-9

407. Snake gourd may not be successfully grown above an altitude of_____.
    (a) 500 m
    (b) 1000 m
    (c) 1500 m
    (d) 2000 m

408. Snowball-16 is a variety of_____.
    (a) Cabbage
    (b) Cauliflower
    (c) Potato
    (d) Tomato

409. Soil temperature below which seed germination of French bean does not take place is_____.
    (a) 10°C
    (b) 15°C
    (c) 20°C
    (d) None of the above

410. Sowing of *kulfa* is done in_____
    (a) Early spring
    (b) Rainy season
    (c) Winter season
    (d) None of the above

411. Sowing of seeds of early cauliflower is done from
_____.

    (a)  Mid-March to mid-April

    (b)  Mid-May to end of June

    (c)  July to August

    (d)  None of the above

412. Soybean is a native of_____.

    (a)  America

    (b)  Africa

    (c)  South-East Asia

    (d)  None of the above

413. Spacing at which dwarf varieties of cowpea should
be sown is_____.

    (a)  45 x 45 cm

    (b)  30 x 30 cm

    (c)  45 x 15 cm

    (d)  None of the above

414. Spacing for rainy season crop of *chaulai* is_____

    (a)  20 x 25 cm

    (b)  30 x 35 cm

    (c)  45 x 30

    (d)  None of the above

415. Spacing required for growing determinate varieties
of tomato is_____.

    (a)  75 x 60 cm

    (b)  75 x 30 cm

    (c)  90 x 60 cm

    (d)  120 x 60 cm

416. Spinach is rich in_____.
    (a) Vitamin A
    (b) Vitamin B
    (c) Vitamin C
    (d) Vitamin D

417. Spinach leaves provide high contents of_____.
    (a) Iron
    (b) Calcium
    (c) Both
    (d) None

418. Spinach originated in_____.
    (a) Europe
    (b) South-West Asia
    (c) America
    (d) None of the above

419. Spinach produces seed stalks in about _____ and seed can be harvested in about 150 days.
    (a) 45 days
    (b) 55 days
    (c) 65 days
    (d) 75 days

420. Spinach produces seed stalks in about_____.
    (a) 45 days
    (b) 55 days
    (c) 65 days
    (d) 75 days

421. Spring summer crop of bottle gourd is sown in
    (a) October-November
    (b) December-January
    (c) February-March
    (d) None of the above

422. Sprouting of onion can be checked in storage by application of growth inhibitor like M.H.-40 at the rate of _____.
    (a) 1500 ppm
    (b) 500 ppm
    (c) 1000 ppm
    (d) 2500 ppm

423. Sprouts of Brussels sprouts resemble_____.
    (a) Miniature cabbage
    (b) Heads of broccoli
    (c) Curds of cauliflower
    (d) None of the above

424. Squash melon is believed to have originated in_____.
    (a) China
    (b) Malaysia
    (c) India
    (d) None of the above

425. Squash melon seed should be sown when the temperature (air) warms upto _____ °C.
    (a) 15.6
    (b) 20.6
    (c) 24.6
    (d) 26.6

426. Sweet potato contains _____ per cent starch.
    (a) 10
    (b) 12
    (c) 14
    (d) 16

427. Sweet potato is propagated by_____.
    (a) Corm
    (b) Setts
    (c) Tubers
    (d) Vine cuttings

428. Sweet potato originated in_____.
    (a) India
    (b) Malaysia
    (c) Tropical America
    (d) None of the above

429. Swiss chard is closely related to_____.
    (a) *Palak*
    (b) Spinach
    (c) Mustard
    (d) None of the above

430. Systemic fungicide used for the control of downy mildew of muskmelon is_____.
    (a) Indofil M-45
    (b) Blitox
    (c) Ridomil MZ
    (d) None of the above

431. Temperate type radish variety is_____.
    (a) Pusa Chetki
    (b) Pusa Reshmi
    (c) Pusa Himani
    (d) Pusa Desi

432. The _____ is probably a native of South Asia and perhaps also of Africa, Australia and Polynesia.
    (a) Smooth sponge gourd
    (b) Bottle gourd
    (c) Pointed gourd
    (d) Bitter gourd

433. The _____ of celery are used as salad.
    (a) Fruits
    (b) Flowers
    (c) Shoot
    (d) Petioles

434. The altitude above sea level at which seeds of beetroot are produced is_____.
    (a) 2000 m
    (b) 1200 m
    (c) 400 m
    (d) None of the above

435. The altitude above sea-level upto which chilli can be grown is_____.
    (a) 200 m
    (b) 400 m
    (c) 6000 m
    (d) None of the above

436. The attack of pea leaf miner is more serious during the months of_____.
    (a) November
    (b) December
    (c) January-February
    (d) None of the above

437. The attack of pea stem fly is maximum in_____.
    (a) September
    (b) November
    (c) January
    (d) February

438. Pollen of which type of bean allergy to some people _____.
    (a) Broad bean
    (b) Cluster bean
    (c) Sword bean
    (d) None of the above

439. The bean used for extraction of gum is_____.
    (a) Cluster bean
    (b) Hyacinth bean
    (c) Yard-long bean
    (d) None of the above

440. The bean which can withstand sufficiently cold temperature is_____.
    (a) French bean
    (b) Broad bean
    (c) Hyacinth bean
    (d) None of the above

441. The best soil reaction (pH) for growing carrot is_____.
   (a)  6.5
   (b)  8.5
   (c)  Below 6.5
   (d)  None of the above

442. The best soil reaction (pH) for tomato cultivation is
   _____.
   (a)  Below 5.0
   (b)  8.0 and above
   (c)  6.0-7.0
   (d)  7.0-8.0

443. The best storage temperature for garlic is 0°C to 2.2°C
   with _____ per cent of relative humidity.
   (a)  60
   (b)  50
   (c)  40
   (d)  30

444. The best temperature for colour development in
   carrot is _____.
   (a)  10-15°C
   (b)  15-20°C
   (c)  20-25°C
   (d)  None of the above

445. The best temperature for the growth of cucumber
   is_____.
   (a)  10-15°C
   (b)  18-24°C
   (c)  25-35°C
   (d)  None of the above

446. The best time for transplanting cabbage seedling in the field is_____.

    (a) Morning

    (b) Afternoon

    (c) Evening

    (d) Night

447. The bitterness in colocasia corns is due to_____.

    (a) Calcium oxalate

    (b) Calcium carbonate

    (c) Calcium chloride

    (d) Potassium oxalate

448. The brinjal plant is _____ to soil acidity (pH 6.8 to 5.5).

    (a) Sensitive

    (b) Highly tolerant

    (c) Moderately tolerant

    (d) Both b and c

449. The centre of origin of tomato is_____.

    (a) Peru and Mexico

    (b) Mediterranean

    (c) China

    (d) India

450. The chillies are rich in vitamin_____.

    (a) A

    (b) C

    (c) A and C

    (d) None

451. The classification of vegetables, which even, though not in all, in majority of cases fulfils the basic requirement, is _____.
    (a) Based on parts used as food
    (b) Botanical classification
    (c) Based on methods of culture
    (d) None of the above

452. The cold storage temperature for melons Is 0°C and relative humidity _____ per cent.
    (a) 50-60
    (b) 60-70
    (c) 40-50
    (d) 80-90

453. The colour of Nantes variety of carrot is_____.
    (a) Red
    (b) Purple
    (c) Orange
    (d) None of the above

454. The colour of Pusa Sunehri variety of sweet potato is_____
    (a) White
    (b) Yellow
    (c) Light orange
    (d) None of the above

455. The common pumpkin grown throughout India is __ _____.
    (a) *C. maxima*
    (b) *C. moschata*
    (c) *C. pepo*
    (d) None of the above

456. The concentration of gibberellic acid at which number of female flowers in cucumber increases is_____.
    (a)  10-25 ppm
    (b)  25-100 ppm
    (c)  100-150 ppm
    (d)  None of the above

457. The country where maximum genetic diversity occurs in brinjal is_____.
    (a)  India
    (b)  Iran
    (c)  Peru and Bolivia
    (d)  China

458. The cowpea mosaic virus (CMV) is transmissible by _____.
    (a)  Sap
    (b)  *Aphis* spp
    (c)  *Myzus persicae*
    (d)  All of the above

459. The cowpea variety insensitive to day length is_____.
    (a)  Pusa Dofasli
    (b)  Punjab Barsati
    (c)  Pusa Phalguni
    (d)  None of the above

460. The crop with which turnip crosses easily is_____.
    (a)  Radish
    (b)  Mustard
    (c)  Cabbage
    (d)  None of the above

461. The distance of planting of knol-khol is _____ cm.
    (a) 45 x 15-20
    (b) 30 x 15-20
    (c) 45 x 45
    (d) 45 x 10-12

462. The economic yield from asparagus is obtained in the form of_____.
    (a) Fruits
    (b) Flowers
    (c) Leaves
    (d) Shoot

463. The edible portion of knol-khol is_____.
    (a) Root
    (b) Head or curd
    (c) Swollen stem
    (d) Bud

464. The flowers in brinjal are_____.
    (a) Hermaphrodite
    (b) Staminate
    (c) Pistillate
    (d) Solitary and hermaphrodite

465. The flowers of ridge gourd are_____.
    (a) White
    (b) Deep yellow
    (c) Pale yellow
    (d) Purple

466. The flowers of sponge gourd are_____.
    (a) White
    (b) Orange
    (c) Purple
    (d) Deep yellow

467. The flowers of sponge gourd open in_____.
    (a) Afternoon
    (b) Morning
    (c) Evening
    (d) Night

468. The fruits of summer squash are usually harvested when they are _____ grown.
    (a) 1/3rd
    (b) 1/4th
    (c) 2/3rd
    (d) 3/4th

469. The genus to which husk tomato belongs is_____.
    (a) Lycopersicon
    (b) Capsicum
    (c) Physalis
    (d) Solanum

470. The green chillies contain _____ which has medicinal value.
    (a) Capsaicin
    (b) Resin
    (c) Coumarin
    (d) Rutin

471. The highest production of chillies is in _____.

    (a) Maharashtra

    (b) Tamil Nadu

    (c) Andhra Pradesh

    (d) Andaman

472. The ideal soil for potato is one which is having a pH range of_____.

    (a) 4.5 to 5.0

    (b) 5.2 to 6.4

    (c) 6.5 to 7.5

    (d) None of the above

473. The insect which causes the development of purple blotch disease of onion is_____.

    (a) Aphid

    (b) Whitefly

    (c) Thrip

    (d) None of the above

474. The isolation distance for producing certified seed of cole crops is maintained at _____.

    (a) 0.5 km

    (b) 1.0 km

    (c) 1.5 km

    (d) 2.0 km

475. The keeping quality of pumpkin fruit is_____.

    (a) Bad

    (b) Good

    (c) Fair

    (d) Excellent

476. The largest area of chillies is in _____.
    (a) Maharashtra
    (b) Tamil Nadu
    (c) Andhra Pradesh
    (d) Andaman

477. The length of sweet potato cutting should normally be _____ with 4-5 sound (healthy) buds.
    (a) 40 cm
    (b) 30 cm
    (c) 20 cm
    (d) 10 cm

478. The loss due to sprouting in garlic is not more than_____.
    (a) 0.1%
    (b) 0.2%
    (c) 0.3%
    (d) 0.5%

479. The minimum temperature required for germination of seed of okra is_____
    (a) 10°C
    (b) 20°C
    (c) 30°C
    (d) None of the above

480. The most ancient type of garden is_____.
    (a) Kitchen garden
    (b) Market garden
    (c) Vegetable forcing
    (d) None of the above

481. The most common cucurbit grown under glasshouses is_____.
    (a) Bottle gourd
    (b) Watermelon
    (c) Cucumber
    (d) None of the above

482. The most common insect attacking radish is_____.
    (a) Red pumpkin beetle
    (b) Aphid
    (c) Thrip
    (d) None of the above

483. The most common leafy vegetable grown during summer in India is_____.
    (a) Amaranth
    (b) Portulaca
    (c) Indian spinach
    (d) None of the above

484. The most commonly grown garden beet variety is _____.
    (a) Detriot Dark Red
    (b) Pusa Kanchan
    (c) Pusa Chandrima
    (d) None of the above

485. The most favourable pH for pea is_____.
    (a) 6.0-7.5
    (b) below 5.5
    (c) 7.5-8.5
    (d) None of the above

486. The most important state growing chillies is_____.

    (a)   Karnataka

    (b)   Uttar Pradesh

    (c)   Andhra Pradesh

    (d)   None of the above

487. The most nutritive type of broccoli is_____.

    (a)   Purple coloured

    (b)   White coloured

    (c)   Green coloured

    (d)   None of the above

488. The most popular of the salad crops is_____

    (a)   Lettuce

    (b)   Celery

    (c)   Parsley

    (d)   None of the above

489. The most popular variety of watermelon is_____.

    (a)   Asahi Yamato

    (b)   Shipper

    (c)   Sugar Baby

    (d)   None of the above

490. The most serious disease of okra is_____.

    (a)   Yellow vein mosaic

    (b)   Powdery mildew

    (c)   Root rot

    (d)   None of above

491. The most serious virus disease of bean is_____.

(a) Common bean mosaic

(b) Yellow bean mosaic

(c) Curly top

(d) None of the above

492. The most serious virus disease of tomato in the rainy and autumn seasons is_____.

(a) TMV

(b) CMV

(c) Leaf curl

(d) Tomato spotted wilt virus

493. The most suitable time of transplanting sweet pepper is _____.

(a) March-April

(b) Mid-February

(c) April-May

(d) None of the above

494. The most troublesome insect-pest of lettuce is_____

(a) Leaf miner

(b) Aphid

(c) Painted bug

(d) None of the above

495. The native place of garden beet is_____.

(a) Asia

(b) America

(c) Europe

(d) None of the above

496. The number of seeds of bitter gourd per 100 g are about _____ .
    (a) 550
    (b) 450
    (c) 400
    (d) 650

497. The optimum temperature for brinjal germination is _____.
    (a) 5°C
    (b) 10°C
    (c) 15°C
    (d) 20°C

498. The optimum temperature for germination of pea seed is _____.
    (a) 5°C
    (b) 15°C
    (c) 22°C
    (d) None of the above

499. The optimum temperature for seed germination of round gourd is_____.
    (a) 24°C
    (b) 25°C
    (c) 26°C
    (d) 27°C

500. The physiological disorder caused by boron deficiency in tomato is_____.
    (a) Puffiness
    (b) Cracking
    (c) Blossom-end rot
    (d) Silvering

501. The pigment responsible for red colour in carrot is_____.
    (a) Carotene
    (b) Anthocyanin
    (c) Lycopene
    (d) Quercetin

502. The primary centre of origin of *parval* is_____.
    (a) China
    (b) Iran
    (c) India
    (d) None of the above

503. The primary centre of origin of watermelon is_____.
    (a) India
    (b) Iran
    (c) Africa
    (d) None of the above

504. The primary objective of greenhouse cultivation of vegetables is _____.
    (a) To get early yield
    (b) To get off season yield
    (c) To protect the crop from diseases
    (d) To protect the crop from excessive rains

505. The probable centre of origin of carrot is_____.
    (a) China
    (b) Punjab and Kashmir
    (c) Russia
    (d) None of the above

506. The probable main centre of pea is_____.
    (a) China
    (b) India
    (c) Ethiopia
    (d) None of the above

507. The pungency in chillies is due to_____ .
    (a) Cucurbitacin
    (b) Resins
    (c) Coumarin
    (d) Capsicin

508. The quantity of leafy vegetables recommended by dieticians to be consumed daily for a balanced diet is_____.
    (a) 70 g
    (b) 115 g
    (c) 300 g
    (d) None of the above

509. The red colour in fruits at the ripening stage in chillies is due to the pigment_____ .
    (a) Capsanthin
    (b) Quercetin
    (c) Anthrocynin
    (d) Catechol

510. The seed of wax gourd germinate within _____ days.
    (a) 6-8
    (b) 8-10
    (c) 10-12
    (d) 12-14

511. The seed rate of long melon is _____ kg/ha.
    (a) 1-2
    (b) 2-3
    (c) 3-4
    (d) 4-5

512. The seed rate per hectare of okra for rainy season is _____ kg.
    (a) 8-10
    (b) 10-12
    (c) 12-14
    (d) 14-16

513. The seed rate per hectare of okra for spring-summer and winter crop is _____ kg.
    (a) 5-10
    (b) 10-15
    (c) 15-20
    (d) 20-25

514. The seed required for one hectare sowing of carrot is_____
    (a) 1-2 kg
    (b) 10-15 kg
    (c) 4-5 kg
    (d) 15-20 kg

515. The seed requirement for one hectare planting of beet (*palak*) is_____.
    (a) 5-9 kg
    (b) 15-20 kg
    (c) 25-30 kg
    (d) 2-5 kg

516. The spring summer crop of French bean in the plains of northern India is sown in_____.
    (a) December-January
    (b) January-February
    (c) March-April
    (d) None of the above

517. The sweet potato crop should be earthed up when the crop is about_____.
    (a) 2½ months old
    (b) 3 months old
    (c) 1 month old
    (d) None of the above

518. The time of spraying Stomp weedicide in onion field is _____.
    (a) Four days before transplanting
    (b) One day after planting
    (c) Within seven days from transplanting
    (d) None of the above

519. The type of pollination in radish is_____.
    (a) Self-pollination
    (b) Cross-pollination
    (c) Often-cross-pollination
    (d) None of the above

520. The underground stem of *Nelumbium* is a popular vegetable in _____.
    (a) North India
    (b) South India
    (c) East India
    (d) West India

521. The usual spacing at which knol-khol seedlings are transplanted in the field is_____.
    (a) 60 x 45 cm
    (b) 45 x 30 cm
    (c) 30 x 20 cm
    (d) None of the above

522. The usual spacing followed for transplanting chilli seedlings in northern India is_____.
    (a) 60 x 30 cm
    (b) 75 x 30 cm
    (c) 45 x 30 cm
    (d) None of the above

523. The usual weight of seed piece of potato is _____.
    (a) 30-40 g
    (b) 40-50 g
    (c) 50-60 g
    (d) None of the above

524. The young plants of *guar* contain _____ which may cause toxicity to the animals.
    (a) Alkaloid
    (b) Bitter principle
    (c) Neurotoxin
    (d) HCN

525. The young potato plant grows the best at_____ °C.
    (a) 16
    (b) 20
    (c) 24
    (d) 28

526. The young tender half grown fruits of snake gourd are ready for harvest in about _____ weeks after planting for vegetable purpose.
    (a) 1-3
    (b) 3-5
    (c) 5-7
    (d) 7-9

527. Time of application of Stomp for controlling weeds in cabbage field is _____.
    (a) One day before transplanting
    (b) Four days before transplanting
    (c) A week before transplanting
    (d) None of the above

528. Time taken for harvesting in temperate type radish in contrast to tropical type is _____.
    (a) Less
    (b) More
    (c) Comparable
    (d) None of the above

529. Time taken for harvesting of pods after the first bloom in French bean is _____.
    (a) 5-7 days
    (b) 7-10 days
    (c) 14-20 days
    (d) None of the above

530. To raise seedlings for planting one hectare of cabbage about _____ of seeds are required.
    (a) 300-400 g
    (b) 400-500 g
    (c) 500-600 g
    (d) 100-200 g

531. To sow one hectare area, the seed rate of cluster bean required is_____.
     (a) 10-20 kg
     (b) 5-10 kg
     (c) 40-50 kg
     (d) 30-40 kg

532. To sow one hectare of knol-khol about _____ kg of seed is enough for direct sowing.
     (a) 1.5-2
     (b) 2-2.5
     (c) 2.5-3.5
     (d) 3.5-4

533. Tomato seed sufficient to raise crop of one hectare area is_____.
     (a) 500-800 g
     (b) 400-500 g
     (c) 200-300 g
     (d) 800-900 g

533. Tomato variety CO3 was evolved by_____.
     (a) X-rays
     (b) Gamma rays
     (c) EMS
     (d) MMS

534. Tomato variety Pusa Lal Meeruti was evolved by_____.
     (a) X-rays
     (b) Gamma rays
     (c) EMS
     (d) MMS

535. Tomato variety RKM 1 is evolved through_____.
    (a) X-rays
    (b) Gamma rays
    (c) EMS
    (d) MMS

536. *Trigonella corniculata (kasuri methi)* produces flowers of _____colour.
    (a) White
    (b) Red
    (c) Purple
    (d) Orange to yellow

537. *Trigonella foenum-gracecum* (common *methi*) produces flowers of _____ colour.
    (a) White
    (b) Red
    (c) Purple
    (d) Orange to yellow

538. Triploid watermelon variety is_____.
    (a) New Hampshire Midget
    (b) Pusa Bedana
    (c) Charleston Grey
    (d) None of the above

539. Tropical carrots are sown in the plains from_____.
    (a) Mid-August to beginning of September
    (b) November-December
    (c) July-August
    (d) None of the above

540. Tuber production in potato is maximum at a temperature of_____.

    (a)  10°C

    (b)  15°C

    (c)  20°C

    (d)  None of the above

541. Tuber production stops totally at a temperature of _____.

    (a)  25°C

    (b)  30°C

    (c)  35°C

    (d)  None of the above

542. Turnip belongs to the family_____.

    (a)  Umbelliferae

    (b)  Chenopodiaceae

    (c)  Cruciferae

    (d)  Convolvulaceae

543. Turnip is a _____.

    (a)  Root vegetable

    (b)  Stem vegetable

    (c)  Salad vegetable

    (d)  Leguminous vegetable

544. Type of flowers in ridge gourd_____.

    (a)  Andromonoecious

    (b)  Hermaphrodite

    (c)  Gynoecious

    (d)  All of the above

545. Type of vegetable gardening followed on the Dal Lake of the Kashmir Valley is_____.

    (a) Floating garden

    (b) Hydroponics

    (c) Kitchen garden

    (d) None of the above

546. Types of vegetable gardens according to the purpose for which they have been developed, are_____.

    (a) Two

    (b) Four

    (c) Six

    (d) None of the above

547. Under good management, once planted crop of pointed gourd can give economical returns for at least _____ years.

    (a) 2

    (b) 3

    (c) 4

    (d) 1

548. Variety of bottle gourd without 'crook neck'_____.

    (a) Pusa Summer Prolific Long

    (b) Pusa Summer Prolific Round

    (c) Pusa Meghdoot

    (d) Arka Bahar

549. Variety of okra resistant to yellow vein mosaic virus is_____

    (a) Pusa Makhmali

    (b) Pusa Sawani

    (c) Punjab No.8

    (d) None of the above

550. Variety of sponge gourd recommended by IARI is_____

   (a) Pusa Nasdar

   (b) Pusa Chikni

   (c) Yard Long

   (d) None of the above

551. Variety of winter squash recommended by I.I.H.R., Bangalore is_____.

   (a) Arka Chandan

   (b) Arka Suryamukhi

   (c) Large Red

   (d) None of the above

552. Variety tolerant to the attack of fruit borer is _____.

   (a) Punjab Neelam

   (b) Punjab Barsati

   (c) Jamuni Gola

   (d) None of the above

553. Varsha Upahar is a variety of _____.

   (a) Okra

   (b) Cabbage

   (c) Chilli

   (d) Cowpea

554. Vegetables are packed in cans and processed by heat in_____.

   (a) Retourt

   (b) Water bath

   (c) Sun

   (d) Any other means

555. Vegetables are rich source of_____.
  (a) Carbohydrates
  (b) Proteins
  (c) Vitamins
  (d) None of these

556. Vegetables are subjected to drying after_____.
  (a) Blanching
  (b) Without blanching
  (c) Sulphuring
  (d) Any other means

557. Very important vegetable included in solanaceous fruit group is_____.
  (a) Tomato
  (b) Potato
  (c) Okra
  (d) Sweet potato

558. Vine cuttings of sweet potato are planted at a spacing of_____.
  (a) 60 x 30 cm
  (b) 100 x 60 cm
  (c) 150 x 60 cm
  (d) None of the above

559. Viral diseases of potato are spread by_____.
  (a) White flies
  (b) Jassids
  (c) Aphids
  (d) None of the above

560. Virus diseases of chilli are spread by_____.
    (a) White flies
    (b) Thrips
    (c) Aphids
    (d) None of the above

561. Vitamin C content in French bean is enhanced by the application of minor element viz., _____.
    (a) Zinc
    (b) Iron
    (c) Copper
    (d) None of the above

562. Watermelon belongs to the species_____.
    (a) Sativus
    (b) Vulgaris
    (c) Lanatus
    (d) None of the above

563. Watermelon can be planted by_____.
    (a) Shallow pit method
    (b) Deep pit method
    (c) Ridge method
    (d) All

564. Watermelon is one of the few vegetables that grow well on acidic soil with pH of_____.
    (a) 4.0
    (b) 5.0
    (c) 6.0
    (d) None of the above

565. Watermelon seeds do not germinate satisfactorily below _____.
    (a) 15°C
    (b) 21°C
    (c) 35°C
    (d) None of the above

566. Wax gourd is a _____ crop.
    (a) Tender
    (b) Semi-hardy
    (c) Hardy
    (d) All of the above

567. When knol-khol is grown for seeds, it becomes _____.
    (a) Annual
    (b) Biennial
    (c) Perennial
    (d) All

568. When *parval* is propagated through seeds, the female and male plants are obtained in the ratio of_____.
    (a) 1 : 1
    (b) 2 : 1
    (c) 3 : 1
    (d) None of the above

569. When snake gourd fruit becomes ripe the colour changes to_____.
    (a) Dark green
    (b) Pale green
    (c) Light yellow
    (d) Orange yellow

570. Where do we have to contact to produce fresh seeds of potato?
    (a) New Delhi
    (b) Shimla
    (c) Lucknow
    (d) Darjeeling

571. Which is the correct sequence of the floral parts of pumpkin ?
    (a) Ovary, corolla and stigmatic lobes
    (b) Corolla, stigmatic lobes and ovary
    (c) Stigmatic lobes, ovary and corolla
    (d) Corolla, ovary and stigmatic lobes

572. Which of the following crop is self-pollinated?
    (a) Bottle gourd
    (b) Cauliflower
    (c) Radish
    (d) Tomato

573. Which of the following is a biennial vegetable?
    (a) Muskmelon
    (b) Tomato
    (c) Onion
    (d) Chilli

574. Which of the following is a stem vegetable?
    (a) Carrot
    (b) Sweet potato
    (c) Knol-khol
    (d) Radish

575. Which of the following is an early variety of cauliflower?

    (a) Pusa Ketaki

    (b) Synthetic Second

    (c) Snowball-16

    (d) None of the above

576. Which of the following is an often-cross-pollinated crop ?

    (a) Tomato

    (b) Brinjal

    (c) Pea

    (d) Onion

577. Which of the following is fruit and vegetable crop?

    (a) Sweet potato

    (b) Okra

    (c) Potato

    (d) Spinach

578. Which of the following is recommended isolation distance for certified seed production of cabbage ?

    (a) 200 m

    (b) 400 m

    (c) 800 m

    (d) 1000 m

579. Which of the following is self pollinated?

    (a) Bottle gourd

    (b) Brinjal

    (c) Radish

    (d) Cauliflower

580. Which of the following is the early variety of cabbage?

    (a)   Pride of India

    (b)   Early Drum Head

    (c)   Chieftain

    (d)   None of the above

581. Which of the following is the variety of watermelon?

    (a)   Pusa Joyti

    (b)   Arka Jyoti

    (c)   Arka Nishant

    (d)   None

582. Which of the following is used to check sprouting of onion under storage ?

    (a)   N.A.A.

    (b)   M.H.

    (c)   G.A.

    (d)   P.C.P.A.

583. Which of the following seed rate of TPS is recommended for planting one hectare area?

    (a)   150 g

    (b)   250 g

    (c)   350 g

    (d)   450 g

584. Which one is not correctly matched ?

    (a)   Ginger - Rhizome

    (b)   Garlic - Bulb

    (c)   Potato - Tuber

    (d)   Sweet potato - Stolon

585. Which one of the following cucurbitaceous vegetable ranks first in terms of nutritive value especially iron, phosphorus and ascorbic acid ?
    (a) Pointed gourd
    (b) Bitter gourd
    (c) Sponge gourd
    (d) Snake gourd

586. Which one of the following group of crops is most suitable and recommended for inter-cropping in the young orchards?
    (a) Short-duration fodders
    (b) Short-duration fruit crops
    (c) Polars and Eucalyptus
    (d) Short-duration legume vegetables

587. Which one of the following is a major disease of onion in India ?
    (a) Leaf blight
    (b) Smut
    (c) Purple blotch
    (d) White rot

588. Which one of the following is not a variety of garlic ?
    (a) Agrifound White (G-41)
    (b) Agrifound Parvati (G-313)
    (c) Yamuna Safed (G-1)
    (d) Agrifound Light Red

589. Which one of the following soil is best for vegetable cultivation?
    (a) Sandy
    (b) Sandy loam
    (c) Clay loam
    (d) Clay

590. Which one of the following types of vegetable garden normally has the highest crop intensity?
    (a) Truck garden
    (b) Kitchen garden
    (c) Vegetable garden for processing'
    (d) Market garden

591. Which one of the following vegetable gives maximum calories per 100 gram of edible matter?
    (a) Spinach
    (b) Watermelon
    (c) Sweet potato
    (d) Tomato

592. Which one of the following vegetable produces maximum seeds per fruit ?
    (a) Tomato
    (b) Brinjal
    (c) Chilli
    (d) Potato

593. Which one of the following vegetables is of Indian origin?
    (a) Chilli
    (b) Chow-chow
    (c) Bitter gourd
    (d) Turnip

594. Which one of the following vegetables is the richest source of protein?
    (a) Pea
    (b) Fenugreek
    (c) Pointed gourd
    (d) Cucumber

595. Which variety of *Brassica pekinensis* is a useful leaf vegetable for plains and hills of North India.

    (a)  Chinese

    (b)  American

    (c)  Canadian

    (d)  Japanese

596. Whip-tail in cauliflower is caused by_____.

    (a)  Excess of boron

    (b)  Deficiency of boron

    (c)  Excess of molybdenum

    (d)  Deficiency of molybdenum

597. White Vienna is _____ season variety of knol-khol.

    (a)  Early

    (b)  Mid

    (c)  Late

    (d)  All season

598. Who studied the heterosis in brinjal?

    (a)  B.S. Tomar

    (b)  H. Singh and T.S. Kalder

    (c)  N. Basavaraja

    (d)  All

599. Wind pollination plays an important role in seed production of_____.

    (a)  Radish

    (b)  Turnip

    (c)  Garden beet

    (d)  None of the above

600. Yellow coloured vegetables are rich source of
_____.

    (a)   Vitamin E

    (b)   Vitamin C

    (c)   Vitamin A

    (d)   Vitamin B

## B. Fill in the Blanks

1.  _____ temperature, especially _____, cause bolting in spinach.

2.  _____ and _____ are non-parasitic troubles of chilli that sometimes cause considerable loss.

3.  _____ is an $F_1$ hybrid of cucumber recommended by I.A.R.I.

4.  _____ is the most ancient type amongst the various types of gardens.

5.  _____ seeded varieties of garden pea are sweeter.

6.  _____ skinned tubers of sweet potato generally store better than _____ skinned ones.

7.  _____ disease of cabbage can be controlled by hot water treatment as in _____.

8.  _____ gardens are situated near some town or city market.

9.  _____ in turnip is due to deficiency of boron.

10. _____ is a giant-leaved cultivar of *palak*.

11. _____ is a wrinkle-seeded variety suitable for early season.

12. _____ is considered to be the place of origin of the watermelon.

13. _____ is more important than _____ in seed stalk development.

14. _____ is probably the place of origin of dolichos bean.

15. _____ is the German name for cabbage-turnip.

16. _____ is the most common leafy vegetable grown in India.

17. _____ is the most common storage disease of onion bulb.

18. _____ is the most commonly grown variety of green sprouting broccoli.

19. _____ is the most serious virus disease of tomato in India.

20. _____ is the non-bulb forming member of the onion family and is grown for its branched _____ .

21. _____ is the single typical member of cucurbitaceae family having single seed in fruit.

22. _____ types of flowers have been described in brinjal.

23. _____ variety is resistant to YVMV and is recommended for sowing in the rainy season.

24. _____ variety of tomato is pear shaped and withstands long transportion.

25. _____, a cucurbitaceous crop, is considered as perennial vegetable.

26. _____ and training of tomato plants is a common practice in certain parts of USA.

27. _____ cultivar of beet leaf is suitable for hills.

28. _____ (*Bari chaulai*) and _____ (*Chhoti chaulai*) are grown as leafy vegetable.

29. _____ fruits of wax gourd are harvested for vegetable whereas _____ fruits are taken for preparing sweet known as _____.

30. _____ humidity and _____ temperature are required for the storage of beet leaf.

31. _____ and _____ cause flower drop and poor fruit set in brinjal.

32. _____ is an improved variety of round gourd developed by IIHR, Bangalore.

33. _____ of tomatoes is caused by *Stemphylian solani* Weber and *Alternaria solani* Ellis.

34. _____ is a seedless variety of watermelon.

35. _____ and _____ are improved varieties of wax gourd.

36. _____ is the most important insect-pest of sweet potato.

37. _____, a winter squash variety, has been recommended by I.I.H.R., Bangalore.

38. _____ developed fruits of tomato are seedless.

39. _____ vegetables contain less of vitamin C than fresh ones.

40. _____ types of turnip are relatively sweeter and more palatable.

41. _____ is one of the most dangerous insects causing virus disease in potato.

42. _____ tubers of sweet potato store better than the white-skinned ones.

43. _____ can tolerate cool climate better than musk melon and snap melon.

44. _____ is the variety of okra.

45. A cauliflower crop often shows deficiency symptoms of _____ and _____ when grown either on an alkaline or highly acidic soil.

46. A daily supply of _____ kg of fresh vegetables can be assured from a kitchen garden of the size of 50 sq.m.

47. A.hermaphrodite variety known as _____ is available in ridge gourd.

48. A pre-harvest spray of potato with _____ controls sprouting.

49. A relatively _____ temperature as well as _____ photoperiod is essential for bulb formation in most of the commercial varieties of onion grown in Índia.

50. *Abelmoschus* genus is distinct from *Hibiscus* in having a deciduous type of calyx whereas it is _____ in the genus *Hibiscus*.

51. About _____ quintals per hectare of seed yield of Indian bean can easily be obtained.

52. According to dieticians, an individual should consume about _____ g of leafy and other vegetables and _____ g of root vegetables daily for a balanced diet.

53. All beans except _____ are susceptible to frost and are grown as a summer crop.

54. All cole crops have developed from wild cliff cabbage known as _____ from which the name cole is derived.

55. All the vegetables come under the sub-community _____.

56. All vitamins are found in small or large quantities in common _____ crops.

57. Although a ripe tomato is 94 per cent water, it is a good source of vitamin _____, and an excellent source of vitamin C.

58. Amaranthus has _____ photosynthetic cycle.

59. Among diseases, _____ is an important disease causing considerable loss to colocasia.

60. Among the most important states of India, only four viz, _____ , _____, _____ and _____ account for three-fourths of the total area under chilli.

61. Among the pole types of French bean _____ is the most commonly grown variety.

62. An edible podded variety of pea is_____.

63. Arka Chandan is the cultivar of _____.

64. Artichoke, commonly known as globe artichoke, is a perennial crop grown for its _____.

65. Ash gourd (*Benincasa hispida* Cong.) is considered to be of great _____ value.

66. Asparagus is cultivated for its tender shoots commonly known as _____.

67. Asparagus plants are _____ in sex expression.

68. Asparagus starts yielding a sizeable crop after about _____ years and gives an economic yield for about _____ years.

69. At a temperature of about _____, tuber production stops totally.

70. Based on their temperature requirements, all vegetables are roughly placed in two groups viz,. _____ and _____ vegetables.

71. Beet leaf is sown _____ times in a year.

72. Beetroot belongs to the family _____

73. Beetroot can be stored at _____ temperature and _____ relative humidity.

74. Best curing of tubers of sweet potato is done at a temperature of _____ and relative humidity of _____ maintained for ten days.

75. Best pod set in French bean is obtained with plants kept at _____ °C for four hours after pollination

76. Best temperature and humidity conditions for storing potatoes in cold storage are _____ °C and _____, respectively.

77. Bitter gourd has heating and _____ effect on the body.

78. Bitter gourd is generally considered to be an old-world species with its native home in the _____ and Asia.

79. Bitter gourd responds well to training on _____ during rainy season.

80. Bordeaux mixture (4 : 4: 50 or 4 : 2 : 50), _____ and Copramat check defoliation in tomatoes.

81. Boron deficiency causes _____ in sugar beet.

82. Botanical name of long melon is _____ var _____.

83. Botanical name of round melon is _____ var. _____.

84. Botanical variety _____ of *Raphanus sativus* is exclusively grown for its long thin pods.

85. Botanically bitter gourd is known as _____.

86. Bottle gourd cannot tolerate _____.

87. Bottle gourd might have originated in _____.

88. Brinjal crop yielding _____ of produce removes from soil 175 + 40 + 300 kg/ha of N+$P_2O_5$ + $K_2O$ and 30 + 10 kg/ha of MgO and sulphur, respectively.

89. Brinjal grows well in _____.

90. Brinjal is _____ in nature but is grown as _____.

91. Brinjal is a native of _____.

92. Broad bean is an important food crop in _____.

93. Broccoli and cauliflower are grown for their _____ as edible parts.

94. Broccoli is an Italian word from the Latin _____ meaning an arm or branch.

95. Brussels sprout gets its name from the city of Brussels in _____.

96. By following succession cropping and inter-cropping _____ of land may be made to supply adequate vegetables for an average family of five members.

97. Cabbage is a _____ pollinated crop.

98. Cabbage is heavy feeder of nitrogen and _____.

99. Cabbage, cauliflower, broccoli, *palak*, okra, onion and muskmelon are slightly tolerant to acidic soils with a pH range of _____.

100. Carrot belongs to the family _____.

101. Carrot is an excellent source of vitamin _____ and is rich in sugar.

102. Carrot is cultivated as _____ for its roots and _____ for seeds.

103. Carrot, bitter gourd, onion and tomato comprise good source of_____.

104. Cauliflower has a high requirement of boron and _____.

105. Cauliflower is sensitive to high _____.

106. Cauliflower with leaves attached can be stored for 30 days at _____ °C and _____ per cent R. H.

107. Celery is a biennial crop and its seeds are produced in _____.

108. Celery thrives well in a soil with pH of _____.

109. Chayote is _____ but its vines are _____.

110. Chillies are rich in vitamins, especially in vitamins _____ and _____.

111. Cluster bean is grown at a spacing of _____ cm.

112. Coccinia, a semi-perennial cucurbit, is propagated by _____ 25-30 cm long and 1.5-2.0 cm thick.

113. Commercial triploid seeds are available in some advanced countries like _____.

114. Cowpea is a _____ season crop.

115. Cowpea pods, when they are not picked at the right stage become _____.

116. Cracking of the skin of tomato fruits has been associated with deficiency of _____.

117. Cucumber can tolerate strongly _____.

118. Cucumber grows best at a temperature between _____ and _____.

119. Cucumber grows well at a row to row spacing of _____m and plant to plant spacing of _____ cm.

120. Cucumber is one of the oldest _____ vegetable crops having its origin probably in India.

121. Cucumber mosaic is readily transmitted by _____.

122. Cucurbits are _____.

123. Cultivated cowpea consists of three main groups namely (i) _____ (ii) _____ and (iii) _____.

124. Cuttings from _____ portion of sweet potato vines should be preferred for planting.

125. Deficiency of molybdenum is the main cause of _____.

126. Dieticians recommend inclusion of _____ grammes of vegetables in our daily diet.

127. Dormancy of potato tubers can be broken by treating the tubers with _____.

128. Double bean is a native of _____ and abudantly grown by American Indians.

129. Drying of the _____ at the stem end of watermelon fruit is taken as a sign of maturity.

130. During dry weather, round gourd should be irrigated at _____ days' interval.

131. Each cutting of sweet potato should have at least _____ nodes.

132. Early sowing in cauliflower is done from _____ to _____.

133. Early varieties of cauliflower are those which produce curds in the plains of northern India from _____ to _____.

134. Edible fruited species of the genus *Lycopersicon* are _____ and _____.

135. Egg-plant (brinjal) contains vitamins A and B and is an _____ vegetable.

136. Elephant's foot botanical name is _____.

137. Elephant's foot is common in _____ in valleys of Tapti, Purna and Ambika.

138. Every adult should consume at least _____ gm of vegetables per day.

139. Exclusion of light from stalks while plants are still growing makes them devoid of chlorophyll, and is referred to as _____.

140. Faba bean is tolerant to water stress due to higher _____ accumulation.

141. Faba bean reached India probably from _____ and was grown in north.

142. Fenugreek crop should be protected well from the incidence of _____ mildew.

143. Fenugreek is a legume crop which fixes _____ from the _____.

144. Fenugreek is a rich source of _____, _____, _____ and _____.

145. Fibrous framework of leaves, stems, bulbs, tubers and roots of vegetables yields _____ which satisfies appetite and prevents constipation.

146. First earthing-up in potato should be done when the plants are about _____ in height.

147. For better fruit-set in tomato, application of a mixture of _____ urea and 2, 4-D at _____ ppm as whole plant spray when the first few flower clusters appear has been found most effective and economical.

148. For economic production of hybrid seeds of onion, _____ lines are used as female parent.

149. For fixing the dates of sowing and harvest, the help of the knowledge of _____ or _____ is taken.

150. For protecting tomato seedlings in the field from frost, cover them with polythene bags of the size _____ and of _____ gauge thickness.

151. For seed production of radish in the plains, _____ type of cultivars should be selected.

152. For spreading varieties of brinjal, row to row distance should be _____ cm and that of plant to plant _____ cm.

153. Formation of new roots in plants of beans and cucurbits is _____ and there is a tendency for the roots to be _____ in these plants, which make them less effective in absorbing water.

154. French bean is probably a native of _____.

155. French bean, an ancient crop, is probably a native of _____.

156. Fruits of chilli hybrid variety _____ released by P.A.U., are very suitable for salad and drying.

157. Fruits of pointed gourd should be harvested when they are _____ and _____.

158. Fruits of round gourd are harvested at _____ stage.

159. Fruits of sponge gourd should be harvested when they are _____.

160. Ganga, Kaveri and Yamuna are _____ of cabbage.

161. Garlic belongs to the family _____.

162. Garlic is _____ season crop.

163. Garlic is propagated by _____ which are detached individually from the bulb.

164. Garlic is supposed to have originated from Central Asia and _____ regions.

165. Generally the yield of fresh green chillies is _____ times higher than that of dry chillies.

166. Generally, _____ is not harmed by downy mildew but the disease causes much damage to _____.

167. Germination percentage of peas is reduced if the moisture percentage of seeds increases _____ during storage period.

168. Gibberellic acid at higher concentrations induces _____ but at lower concentrations of 10-25 ppm increases _____.

169. Globe artichoke is propagated by means of _____ from the old root-stocks.

170. Golden Acre is a _____ or _____ type.

171. Green spears of asparagus have _____ nutritive value than blanched spears.

172. Growing of vegetables out of their normal season is known as _____.

173. Harvested brinjal fruits can be kept for seven to ten days in good condition at _____ °F and _____ per cent relative humidity.

174. Heading type of broccoli is more like _____.

175. Heavy irrigation to cabbage after development of heads causes _____ within 24 hours.

176. Henderson Bush is an important variety of _____.

177. High _____ is favourable for growth and development of plants and fruit formation of snake gourd.

178. Higher seed yield in radish is expected from _____ method.

179. Hill potatoes are nearly free from _____ _____ pathogen.

180. I.A.R.I., has released an $F_1$ hybrid _____ of summer squash.

181. If tomato fruits are exposed to intense sunlight, it may cause a disorder known as _____.

182. Immature tomato blossoms drop rapidly during _____ due to increased transpiration.

183. In _____, after first shower, when shoots come out from root tubers, it is the best planting time for sweet gourd in Assam.

184. In _____ method of seed production in cabbage, selected true-to-type fully matured plants are uprooted during November-December and are reset in new beds.

185. In Bihar, Himachal Pradesh and Nillgiri hills of India, potato is grown as a _____ crop.

186. In early varieties of tomatoes concentric cracking, _____ is more common owing to their light, open foliage.

187. In floating gardening, a floating base is first made from the roots of _____ grass which grow wild in some parts of the Dal Lake.

188. In floating gardening, all intercultural operations and occasional sprinkling of water are done from _____.

189. In hills, the best time for sweet potato planting is during _____.

190. In India, potato has been cultivated since its introduction in the early part of _____ century.

191. In most of the cucurbits, an ideal variety is one which has _____ female to male flower ratio.

192. In Nasik division of Maharashtra, Ootacamund region of Tamil Nadu and parts of Kerala, cabbage is grown during _____ also.

193. In Nilgiri hills and Tamil Nadu, potato is rotated with _____ in alternate years.

194. In northern India, vine cuttings of sweet potato are planted during _____.

195. In *palak* _____ pollination is effective.

196. In parts of South America, sweet potato is also called _____ or _____.

197. In radish, higher genetic purity can be maintained when seed production is done by _____ _____ method.

198. In the plains of India, for the *rabi* crop, seeds are sown from _____ to _____.

199. In the plains of India, French beans are sown in _____ for the autumn crop and in _____ for the spring-summer crop.

200. Rainy season vegetables are generally sown in _____.

201. Indian bean is grown throughout tropical regions of Asia, _____ and America.

202. It is believed that small seeded pea originated from _____, and the large seeded ones from Mediterranean.

203. It is estimated that potato in Nilgiri Hills of Tamil Nadu was introduced in _____.

204. Ivy gourd is propagated by _____.

205. Jerusalem artichoke is cultivated for its _____.

206. June-July sown cauliflower should be transplanted in new beds before their final transplantation into the _____.

207. Kakrol seed contains oil which is used as an _____.

208. Kartoli is propagated by _____.

209. Knol-khol (*Brassica oleracea* var _____) is little known in India.

210. Late varieties of cauliflower give curds from_____ _____ to _____.

211. Leaves of amaranthus are _____ in nature.

212. Leaves of Swiss chard are _____ than those of beet leaf.

213. Leek belongs to the species _____.

214. Leek is included in the group of _____ crops.

215. Lettuce belongs to family _____.

216. Lettuce does well in a relatively cool growing season with a monthly average temperature of _____.

217. Lettuce mosaic is transmitted through _____.

218. Lettuce seed does not germinate properly when the soil temperature is above _____.

219. Long melon is usually sown in _____ and overwintered under artificial protection from frost.

220. Long melon thrives best in _____ and _____ climate.

221. Long melon variety Selection-3 is free from _____ principles.

222. Main crop of potato in hills of Himalayan region is planted in _____.

223. Manjari Gota is a variety of brinjal recommended for

    _____.

224. *Methi* is _____ season crop.

225. Minimum monthly average temperature requirement of tomato and sweet pepper is _____.

226. Most commonly occurring disease in nursery is ____

    _____.

227. Most economic utilization of space can be obtained by making use of fence on the periphery for training_____ during the summer and rainy season and _____ in winter.

228. Most ideal variety of radish grown in summer and rainy season only is _____.

229. Most of the cucurbits are _____ in sex expression and a few are _____.

230. Most of the French bean varieties are _____.

231. Most of the members of cucurbitaceae family contain _____.

232. Mostly orange colour in tomato is due to the presence of _____ which is a precursor of _____.

233. Muskmelon is a native of _____ India.

234. Muskmelon is planted from _____ in plains and _____ in hills.

235. Muskmelon variety _____ is grown in Afghanistan and is available in India in October-November.

236. No vegetable belongs to the division _____.

237. Normally, snake gourd is trained on _____, _____ and _____.

238. Nursery of lettuce is sown in _____.

239. Odour in garlic is due to _____.

240. One of the most important onion growing states in India is _____.

241. Onion and other bulb crops belong to the family ___ _____.

242. Onion bulbs and green onion are good source of vitamin _____.

243. Onion does not thrive well in areas receiving more than _____ of annual average rainfall.

244. Onion is propagated by _____ and _____ as well.

245. Onion seedlings are transplanted at a spacing of _____ cm.

246. Onion seedlings for *kharif* crop are transplanted in _____ and produce good bulbs in _____ under North Indian conditions.

247. Onion, okra, asparagus and summer squash supply _____.

248. Only _____ and _____ styled brinjal flowers marked by a swollen ovary at the base, bear fruits.

249. Optimum pH range for growth of okra is _____ _____.

250. Optimum pH range for onion cultivation is between _____ and _____.

251. Optimum temperature for growth of muskmelon is about _____.

252. Outbreak of late blight disease in potato is commonly noticed in _____, _____ and _____ weather.

253. Pepo is a type of berry in which the outer wall, which is the _____, becomes hard as in cucurbitaceae family.

254. Per capita estimated consumption of potato in _____ is about 13 kg per annum.

255. Perennial and annual types of Indian beans are known as Typicus and _____, respectively.

256. Perkin's Long Green is recommended for _____ _____ areas.

257. Pink colour of skin of the radish root is due to _____ _____.

258. Planting root materials are usually collected from old planting of pointed gourd during _____.

259. Pointed gourd cannot be grown successfully in _____ climate.

260. Pointed gourd is a _____ cucurbitaceous crop.

261. Pointed gourd is a _____, therefore, male and female plants are _____.

262. Pointed gourd is a native of India and _____ is considered to be the primary centre of its origin.

263. Pointed gourd is easily digestible, diuretic, laxative, envigorates the _____, and is useful in disorders of the circulatory system.

264. Pointed gourd is mainly propagate by _____ and _____ cuttings.

265. Pointed gourd is one of the rare cucurbits which is propagated by _____.

266. Poor _____ application may result in bolting in onion.

267. Potato belongs to the family _____.

268. Potato crops remove large amount of potassium followed by _____.

269. Potato is a native of South America and was cultivated by _____.

270. Potato is mainly propagated by_____ _____ and _____ techniques.

271. Potato requires _____ conditions for its tuber development.

272. Potato stored at -1 to 0°C suffers from internal breakdown known as _____.

273. Potato was introduced in Europe by the early _____ exporters during sixteenth century.

274. Potato, when grown in alkaline soils, is attacked by _____ disease.

275. Pre-harvest foliage sprays of _____ may be helpful in controlling the sprouting of onion in storage.

276. Premature seeding and or failure of the leaves to form a _____ are common defects of early cabbage.

277. Pungency in radish is due to _____.

278. Punjab Selection, Pusa Ratnar and Arka Kalyan are the varieties of _____ onion.

279. Pusa Bedana is a seedless variety of watermelon obtained as hybrid between _____ and _____.

280. Pusa Chetki variety of radish is suitable for sowing from _____ to _____ in the plains.

281. Pusa Dofasali variety of cowpea is suitable for growing in both _____ and _____ seasons.

282. Pusa Early Bunching and Kasuri Selection are the cultivars of _____.

283. Pusa Himani variety of radish is suitable for sowing from _____ to _____ in the plains.

284. Pusa Lal variety of sweet potato is a selection from a Japanese variety _____.

285. Pusa Rituraj can be grown both in _____ and _____.

286. Pusa Sadabahar and Pusa Naubahar are the varieties of cluster bean mainly grown for _____ purposes.

287. Pusa Sanyog is a _____ of cucumber.

288. Radish is found growing wild in _____ region.

289. Rhubarb grows best in _____ climatic conditions.

290. Rhubarb is grown for its _____.

291. Rhubarb originated from the colder parts of Asia, probably _____.

292. Ripe watermelon fruit when thumped with finger(s) gives out a _____ sound as against metallic and ringing sound by an immatured one.

293. Ripening of muskmelon fruits begins first from the _____ accompanied by a change in skin colour.

294. River-bed cultivation of ridge gourd requires _____ _____ irrigations.

295. River-bed cultivation of wax gourd usually requires _____ watering than upland cultivation.

296. Round gourd prefer to grow in has _____ _____ soil rich in _____.

297. Row to row and plant to plant spacing of _____ cm is kept for the early varieties of cauliflower.

298. S-48, Arka Pragati and Hissar-2 are the varieties of white _____.

299. Scientific name of potato tuber moth is _____.

300. Seed is generally raised from the _____ crop.

301. Seed production of _____ type of turnip cultivars should be done in plains and for _____ types it should be done in the hills.

302. Seedlings of lettuce are transplanted at a spacing of _____.

303. Seeds of Brussels sprout are only produced in _____ at an altitude of _____ m and above

304. Seeds of leek are produced in India at _____ in the hills.

305. Seeds of okra will not germinate below _____.

306. Sex expression in snake gourd is _____ and in pointed gourd is _____.

307. Short days are beneficial for _____ _____ in potato.

308. Single fruit of *palak* contains _____ seeds.

309. Snake gourd is found growing wild in India and the Indian _____ is thought to be its source of origin.

310. Some of the varieties of _____ bean are used for extraction of gum.

311. Spacing recommended for dwarf varieties of tomato is _____ cm.

312. Specialised gardens away from the market but having good means of transport are known as _____.

313. Spinach is a _____ season crop.

314. Spinach is cultivated as an _____ for its leaves and as a _____ for obtaining seed.

315. Sponge gourd is probably indigenous to _____.

316. Spraying of tomato plants at _____ stage using 25-50 ppm $GA_3$ improves the quality of fruits.

317. Sprouts of Brussels sprout resemble _____ and are borne on the axils of leaves.

318. Squash melon fruits become ready for harvesting after _____ days of sowing.

319. Squash melon is believed to have originated in ____ _____.

320. Summer crop of okra is sown from _____ to _____.

321. Summer squash is also known as _____.

322. Summer squash like pumpkin is of _____ origin and was first of the squashes to be introduced in India from Europe.

323. Sweet potato contains 16 and 4 per cent of starch and sugar, that is, _____ per cent alcohol producing material.

324. Sweet potato cuttings are planted at a spacing of _____ cm in rows _____ cm apart.

325. Sweet potato is believed to have originated in _____.

326. Sweet potato is one of the most _____ resistant vegetable.

327. Sweet potato is very tolerant to _____ soil.

328. Tetraploid watermelon is produced by treating the seedlings with _____.

329. The _____ of pumpkin are more nutritive than _____.

330. The age of transplanting asparagus is _____ old seedlings or crown.

331. The botanical name of common *sitaphal* grown throughout India is _____.

332. The botanical name of watermelon is _____.

333. The cole crops which are grown for the consumption of leaves or stem are _____ and _____.

334. The colour of the outer skin of onion bulb is due to _____.

335. The commercial production of cabbage is done by _____ method.

336. The critical stages for irrigation in onion are _____ and _____.

337. The crops which are highly tolerant to acidic soils (pH 6.8 to 5.0) are _____ and _____.

338. The cultivated carrot probably originated in the hills of Punjab and Kashmir, with a secondary centre of distribution in _____ and North Africa.

339. The defects noticed in cauliflower are premature heading, _____ of plants and production of undersized head.

340. The flowers of ridge gourd open in the _____ while those of the sponge gourd do so in the _____.

341. The fruit of round gourd _____ effect and contain vitamin A.

342. The fruits of Ivy gourd have _____ keeping quality.

343. The green leaves of fenugreek are used as _____ and also as _____.

344. The green pods of cluster bean are rich source of _____, C, and _____.

345. The important viral diseases which attack chilli are _____, _____ and _____.

346. The leaves or seed stalks of onion fall down from the point of attack by _____.

347. The mature corms of elephant's foot store well under _____.

348. The most favourable range of pH for pea cultivation is between _____ and _____.

349. The most important viral disease of okra is _____ _____.

350. The most serious disease of cluster bean is _____.

351. The most serious insect-pest of cucumber is_____.

352. The most serious virus disease of bean is _____.

353. The name cauliflower has originated from the Latin words _____ and _____ which means cabbage and flower, respectively.

354. The netted varieties of muskmelon are commonly known as _____ in the U.S.A.

355. The onion originated from the region comprising N-W India, Afghanistan, Tajikistan, Uzbekistan (former USSR) and _____.

356. The origin of potato is _____.

357. The original home of okra is _____.

358. The original home of radish is probably _____ _____ and_____.

359. The original race of pea came from _____.

360. The place of origin of celery extends from Sweden to Algeria, _____ and in Asia to Caucasus, Beluchistan, Egypt and the mountains of India.

361. The potato variety Kufri Kuber has been developed at Potato Breeding Station (now CPRI) Shimal in _____.

362. The pungency in onion is due to the presence of volatile oil _____.

363. The pungency is garlic is due to _____.

364. The rat-tail radish is grown for _____ which are used as _____.

365. The red colour in tomato is due to pigment, _____ _____.

366. The seed of watermelon does not germinate satisfactorily below _____.

367. The seed rate of celery is about _____ g per hectare.

368. The seed rate of garlic varies from 340 to 570 kg/ha of _____.

369. The seed rate of knol-khol to raise seedlings for one hectare is _____ grammes.

370. The seed rate of lettuce for nursery and direct sowing is 500-750 and _____ g, respectively.

371. The seed rate of radish is _____ kg/ha.

372. The seeds of snake gourd and bitter gourd take _____ days to germinate.

373. The tomatoes crops supplied with rich nutrition fail to produce _____.

374. The tubers of potato are borne underground at the _____.

375. The type of vegetable garden known as _____ is seen on the Dal Lake of the Kashmir Valley.

376. The usual size of seed piece in potato is _____ gm with _____ mm diameter.

377. The usual spacing for chillies in northern India is _____ cm.

378. The usual spacing for khol-rabi is kept about _____ _____ from row to row and _____ from plant to plant.

379. The vegetables commonly grown under glasshouses or glass frames are _____ and _____.

380. The word truck in truck gardening has no relationship with a truck but is derived from a French word _____ meaning _____.

381. The yield of radish is _____ Q/ha.

382. The young plants of potato grow best at a temperature of _____ and later growth is favoured at a temperature of _____.

383. There are _____ types or groups of lettuce.

384. There are at least _____ types of beans.

385. There are two types of tomatoes. (i) _____ (ii) _____type.

386. Thorough curing of onion bulbs for _____ weeks is required before being placed in storage

387. To ensure proper fruit-set, about _____ per cent cuttings from male plants are planted at random in between the female plants of pointed gourd.

388. To sow one hectare area of knol-khol about _____ _____ kg of seed is required for raising seedlings.

389. Tomato fruits for canning are picked when they are _____.

390. Tomato seed sufficient to raise one hectare area is _____ gramme.

391. Tomato was perhaps introduced into India by the _____ though there is no definite record of when and how it came to India.

392. Tomatoes do best in a soil that has a soil reaction with pH from _____ to _____.

393. Transplanting of leek is done in _____ about 30 to 45 cm deep.

394. Transplanting of onion seedlings earlier than _____ gives more number of bolters.

395. Tuber production in potato is maximum at a temperature of _____ and decreases with the rise of temperature.

396. Turnip greens are good source of minerals such as _____ and iron and vitamin A.

397. Under ordinary conditions, round gourd fruits can be kept for _____ days only.

398. Usually, carrot is grown at a distance of _____ cm between two rows and _____ cm between plant to plant.

399. Usually, radish is grown at a distance of _____ cm between plants and _____ cm between rows.

400. Vegetable crops occupy only about _____ per cent of the total cultivated area of the country.

401. Vegetable growing is generally more lucrative than any other type of _____.

402. Vegetables viz., fenugreek, spinach and mountain spinach are rich in _____.

403. Vitamin A is abundantly found in _____ vegetables.

404. Watermelon belongs to the species_____.

405. Watermelon is grown in USA since _____.

406. Watermelon is one of the few vegetables that grow well on a soil having a pH even upto _____.

407. Wax gourd is popularly known as _____.

408. Welsh onion (*Allium fistulosum*) is probably of _____origin.

409. West is Indian bean is known as _____.

410. Western Asia and area around _____ are onion's secondary centres of development.

411. When colour of ground spot (white spot) where the watermelon fruit rests on the ground assumes _____, it is indicative of its maturity.

412. When muskmelon fruit on ripening part completely and easily from the stem leaving a circular depression, it is said to be at _____.

413. Where possible, a _____ garden is preferred to a square one.

414. White Iscle, Scarlet Globe and Pusa Himani are _____ of varieties of radish.

415. White Vienna, Purple Vienna and King of North are the cultivars of _____.

416. Wild forms of pointed gourd are found throughout _____.

417. Wild radishes are found in _____.

418. Wine is also made from _____ besides grapes.

## C. True or False

1. 1-5 ppm 2, 4-D spray can increase the yield of tomato by ten times.

2. 2219A male sterile line of cucumber has been exploited for hybrid production.

3. A large number of improved cultivars of snap melon are available for commercial cultivation in India.

4. A pre-harvest spray of maleic hydrazide controls sprouting of potato tubers in storage.

5. A relatively low temperature as well as short photoperiod is essential for bulb formation in most of the commercial varieties.

6. About 15-20 days old seedings of Brussels sprouts are most suitable for transplanting.

7. About 40,000-50,000 cuttings of sweet potato are required to plant one hectare.

8. Acridity of tubers in some varieties of colocasia is due to calcium oxide.

9. Alkaline soils create favourable conditions for scab disease.

10. All beans are susceptible to frost and are grown as a summer crop.

11. All bulb crops belong to the family Liliaceae.

12. All cole crops have a common ancestor.

13. All Green, Jobner Green and Pusa Jyoti are the improved cultivars of fenugreek.

14. All varieties can do better under forcing structure.

15. Almost all cucurbits are sensitive to waterlogging and freezing temperature.

16. Amaranthus does not do well on heavy, poorly-drained soil or on sandy soils which are poor in water holding capacity and poor in nutrients.

17. Amaranthus has $C_3$ photosynthetic cycle which indicates its high productivity.

18. Amaranthus is often-cross-pollinated crop.

19. Amaranthus seeds are very small, therefore, some sand is mixed to get uniform distribution of seed.

20. *Amaranthus tricolour* is *bari chaulai*.

21. Application of bacterium culture does not improve the yield of peas.

22. Application of maleic hydrazide (MH) at 25-50 ppm once or twice at the 2-leaf and again at the 4-leaf stage induces a greater number of female flowers.

23. Arka Harit and Pusa Do Mausami are the cultivars of cowpea.

24. Arka Jyoti is the hybrid of cucumber evolved by IARI.

25. Arka Komal is a variety of French bean developed by IIHR, Bangalore.

26. Arka Nishant, Punjab Safed and Pusa Reshmi are Asiatic or tropical type of radish varieties.

27. Arka Sheel and Arka Navneet are the promising hybrids of brinjal.

28. Artichoke is a popular vegetable in India.

29. Asparagus is propagated by suckers.

30. Asparagus plants are dioecious in character.

31. Asparagus starts yielding a sizeable crop after about three years.

32. At present, the cultivation of broad bean in India is not so popular on largescale.

33. Beetroot is a cool-to-warm season crop.

34. Beetroot is propagated by cormels.

35. Being a self-pollinated crop, sponge gourd does not need to maintain any isolation distance for its seed production from other cultivar.

36. Being perennial in nature, chayote requires judicious application of manure and fertilizers.

37. Best quality spears of asparagus are obtained from the second to fifth year.

38. Black scurf affected plant sometimes produce aerial tubers in the axils of leaves.

39. Blossom-end rot and sunscald are non-parasitic troubles of chilli that sometimes cause considerable loss.

40. Botanical classification is largely used in almost all textbooks.

41. Botanical name of cabbage is *Brassica oleracea* var. *capitata* and family cruciferae.

42. Botanically, there is no difference in sponge gourd and ridge gourd.

43. Botanically, tomato is known as *Lycopersicon melongena*.

44. Bottle gourd can withstand cold climate better than muskmelon and watermelon.

45. Bottle gourd is a monoecious crop.

46. Broad bean is warm season vegetable.

47. Broad bean require cool season but for ripening of pods higher temperatures are beneficial.

48. Brussels sprouts can be grown on very high acidic soils.

49. Brussels sprouts require hot season.

50. Cabbage is a cool season crop.

51. Cabbage is a cross-pollinated crop and does not cross with other members of the cole group.

52. Cabbage is a deep-rooted crop and is a poor feeder.

53. Carrot belongs to family Umbelliferae.

54. Carrot can tolerate slight acidic and alkali soil reactions.

55. Carrot grown on heavy soils tend to become rough and coarse because the roots fail to penetrate the hard soil.

56. Caterpillars are the most troublesome insect-pest on lettuce.

57. Cauliflower is grown for its white tender head or curd.

58. Cauliflower is sensitive to high acidity.

59. Cauliflower varieties are not responsive to temperature and photoperiod.

60. Celery is a water loving plant.

61. Celery is an annual plant

62. Celery plant is moderately sensitive to salinity.

63. Celery seedlings are transplanted in trenches as they facilitate earthing up and blanching.

64. Chayote responds well to lower planting.

65. Chillies can withstand frost up to some extent.

66. Chow-chow is considered a perennial cucurbit.

67. Cluster bean is highly tolerant to drought.

68. Colocasia corms store well for longer period of time.

69. Colocasia is a bulb crop.

70. Colocasia is grown only once in a year.

71. Colocasia matures in 80-90 days.

72. Colocasia require winter season.

73. Compared to cauliflower, Brussels sprouts require long growing period.

74. Cowpea is a native of India.

75. Cowpea is probably the native of Central Africa.

76. Cucumber is quick growing dioecious annual cucurbit.

77. Cucumber is used mostly for salad.

78. Cucumber mosaic virus is readily transmitted by aphids.

79. Cucumber requires cool season and low humidity for the formation of pistillate flowers.

80. Curing of sweet potato tubers is done best at 80°C and if 30 per cent R. H. is maintained for 10 days.

81. *Dioscorea batatas* is more commonly grown in the tropical and sub-tropical regions of India.

82. Direct contact of fertilizers (particularly nitrogen and potash) with seeds affect germination adversely.

83. During winter season, vines of pointed gourd die but grow again in spring which follows flowering and fruiting immediately.

84. Dwarf cultivar require support (staking).

85. Early blight is the most serious fungal disease of potato.

86. Early maturing potato varieties are favoured by relatively short days while late ones are better adapted to long days.

87. Early varieties of cauliflower, if sown late, produce button head and late varieties, if sown early, go on giving leafy growth and produce curd very late.

88. Earthing up of an individual plant of snake gourd does not give any support or prevent plants from direct contact with water.

89. Elephant's foot is propagated by small pieces of corms.

90. Elephant's foot is grown in flat beds and on broad ridges.

91. Elephant's foot is not as rich in minerals and vitamins A and B as potato.

92. Elephant's foot is propagated by corm.

93. Elephant's foot species are found in tropical As. and America, and about 14 species occur in India.

94. Excessive application of nitrogen in potato crop results in luxuriant and succulent growth such foliage is generally more liable to be affected by diseases.

95. Extent of cross-pollination is the same in both beans and peas.

96. Flowers of *palak* are hermaphrodite.

97. For distant transportation, pink fruits of tomato are picked.

98. For economic hybrid seed production in cucumber and muskmelon, gynoecious lines are used as female parents.

99. For getting higher yield of sponge gourd, balanced supply of nutrients and soil moisture are needed.

100. For improving the quality of leeks, blanching is essentially done.

101. For propagating sweet potato from vine cuttings, cuttings from the basal portion of the vines should be preferred.

102. For raising one hectare crop of sponge gourd about 20 to 25 kg seed is required.

103. For seed production of knol-khol, low temperature and long-day conditions are needed which are found in the hills.

104. For seed production, there is no need to have any isolation distance between two cultivars of okra.

105. French bean does well at places 3000-4000 m above mean sea level.

106. French bean is a good source of protein, calcium, iron and vitamins.

107. French bean is richer than the hyacinth bean in its nutritive value.

108. French bean is tolerant to acid soils, but pH should not be lower than 3.5.

109. Fruit setting in brinjal can be increased by use of growth enhancing substances and nutrition.

110. Fruits of sponge gourd should not be harvested when they are tender.

111. Fruits of winter squash are available for harvesting within 200 days after planting.

112. Fully matured, spongy and fibrous fruits of sponge gourd are most suitable for vegetable purposes.

113. Galls formed on the roots of beans are often confused with bacterial nodules.

114. Garden asparagus is closely related to ornamental asparagus.

115. Garden pea is grouped as *Pisum sativum* sub sp. *arvense*.

116. Garlic bulbs are cured for four to five weeks in shade before storing.

117. Garlic is a spice or condiment used for flavouring.

118. Garlic needs a richer soil than onion.

119. Great Lakes is a variety of leafy lettuce.

120. Green leaf stalks of celery are more nutritive than blanched ones.

121. Green spears of asparagus are more nutritive than blanched (white) spears.

122. Green tender fruits of winter squash are never harvested.

123. Green type of broccoli, which is more nutritive, is the most popular.

124. Hebbal Avare-3, Wal Konkan-1 and Arka Jay are the vegetable varieties of faba bean.

125. High moisture content and low nitrogen delay maturity.

126. High quality carrots are those which do not have a relatively large outer core.

127. High temperature above 12-15°C promotes seed stems and causes a bitter taste in the leaves.

128. High temperature reduces yield and quality of curds.

129. Higher seed yield of radish is expected from root-to-seed method.

130. Hoeing is required in cauliflower during early stage.

131. Hot weather and inadequate supply of moisture result in deterioration of root quality.

132. Hybrid varieties have advantages of increased adaptability to adverse environment, more resistance to diseases, better quality, earliness and increased yield.

133. If picking of okra fruits is delayed, the quality will be improved.

134. In beetroot, earthing up is usually done to cover the swollen roots.

135. In cold season vegetable plants, hardening is done to withstand possible burning due to sunshine, hot winds and dry soil at their planting season.

136. In cucumber, a small portion of the stem-end is cut cross wise and rubbed together to remove the black substance that comes out.

137. In green leafy vegetables, tryptophan is the limiting amino acid.

138. In India, cabbage seeds are raised in the Kashmir Valley.

139. In kohlrabi, fleshy edible portion is an enlargement of root.

140. In northern India, sweet potato vine cuttings are planted during April-May.

141. In *palak*, insect pollination is effective.

142. In pea, the germination is epigeal.

143. In plants like *Brassica*, temperature controls flowering.

144. In radish, for root-to-seed method, sowing is done early when roots have matured.

145. In radish, harvest when pods are fully ripe.

146. In radish, higher genetic purity can be maintained when seed production is done by seed-to-seed method.

147. In seed-to-seed method of seed production of cabbage, the plants are allowed to over-winter in their original position.

148. In sweet gourd, male and female flowers occur on the same plant.

149. In the Nilgiri hills, three potato crops are raised in succession.

150. In vegetable crops which are difficult to transplant, there is a tendency for the roots to be suberized or cutinized.

151. In winter squash, ripe fruits have no storage life compared with green fruits.

152. Indeterminate type of tomato plants exhibit 'self topping' growth habit.

153. India exports chillies and their products to Abu Dhabi, Australia, Canada, Japan, U.K. and USA.

154. India is the second largest producer of vegetables next to China.

155. Indian bean Hebbal Avare-1 has been evolved by the segregation of Local Avare x Red Typicus.

156. Indian bean is also known as French bean.

157. Indian bean is sown in temperate climate only.

158. It is better to avoid application of partially decayed organic matter in the current growing season of the carrot crop because many roots will tend to become malformed or forked.

159. It is easy to store by hanging the bunch of garlic bulbs with tops in well-ventilated sheds.

160. It is possible to recognize the sex of asparagus plants in the nursery.

161. It takes three to four years for the corms to be ready for harvesting.

162. Ivy gourd is a dioecious vine crop.

163. Ivy gourd is sensitive to waterlogging conditions.

164. Jerusalem artichoke is of great value in the diet of diabetic persons because its tubers store insulin carbohydrate.

165. Kakrol is a bush shrub.

166. Kakrol is a monoecious.

167. Kakrol is not dioecious.

168. Kartoli fruits are small, roundish and covered with soft spines.

169. Kartoli is a widely cultivated crop in India.

170. Kartoli is dioecious in nature.

171. Kartoli requires warm and humid climate.

172. Knol-khol is a hardy winter vegetable.

173. Knol-khol is not as popular as cabbage and cauliflower.

174. Kufri Chandramukhi has better keeping quality than
    • Kufri Red under ordinary conditions.

175. Leafy type lettuce is also sown directly in the field.

176. Leaves are eaten as salad in celery.

177. Leaves of amaranthus are non-perishable in nature.

178. Leek is a biennial crop and its seeds are produced in India at higher altitudes in the hills.

179. Leek is a bulb forming member of the onion family.

180. Leek is a hardy biennial but grown as an annual.

181. Leek is grown for its cloves.

182. Lettuce is a salad-cum-leafy vegetable of temperate climate.

183. Lettuce is probably a native of Europe and Asia, and has been in cultivation for at least 2,000 years.

184. Lettuce plant bolts quickly due to insufficient availability of soil moisture.

185. Lima bean is also known as double bean.

186. Lima bean is cross-compatible with French bean

187. Lima bean is cultivated on largescale in India.

188. Lima bean neither grows well in spring nor in rainy season.

189. Lima bean requires fertile soil as compared to French bean.

190. Little leaf disease of brinjal is caused by virus, spread by aphids.

191. Long fruited varieties of bitter gourd are generally grown in the summer season.

192. Long melon can be stored under ordinary conditions for long period without any deterioration in its quality.

193. Long melon can stand cool climate better than muskmelon.

194. Longevity of brinjal seed is four years.

195. Lower fertility, higher temperature and longer light period induce maleness in cucumber.

196. Mature leaves of colocasia are most suitable for vegetables.

197. Maturity of onion is indicated by complete drying of leaves.

198. Maturity of sweet potato tubers is proper if cut surface of tuber exposed to air, dries up soon.

199. Melchers *et al.* (1978) produced somatic hybrid 'Pomato' by fusion between potato and tomato.

200. Mild temperate climate is more suitable for French bean.

201. Molybdenum deficiency symptoms in cauliflower occur in highly alkaline soils.

202. Most of the cucurbits are monoecious and a few are dioecious.

203. Most of the French bean varieties are long-day types.

204. Mucilage, a sticky substance in okra, is generally extracted from flowers and buds.

205. Mucilaginous material in okra is due to glycoproteins.

206. Muskmelon fruit should not be picked at full slip stage for home consumption.

207. Muskmelon is drought resistant and susceptible to frost.

208. Muskmelon tolerates a slightly cooler weather than cucumber.

209. No variety of French bean has yet been reported to be resistant to yellow bean mosaic.

210. Non-ridged okra fruits are easy to thrash.

211. Occasionally, flowering is also seen in Elephant's foot.

212. Okra belongs to the genus *Hibiscus*.

213. Okra is sown in April-May in the hills.

214. One or more large central corms are termed as *kachalu* while large number of lateral cormels are termed as *arvi*.

215. Onion bulbs that are to be stored, are usually thoroughly cured for 3 or 4 weeks.

216. Onion can be cultivated as a rainfed crop between April and August even at 1525-2134 m elevations.

217. Onion crop transplanted earlier than December and January gives higher yield but the number of bolters may be more.

218. Onion has got medicinal values. Its use is the best remedy against sunstroke.

219. Onion is insensitive to high acidity.

220. Onion is one of the most important vegetable for foreign exchange earning.

221. Onion seed remain viable for more than one year.

222. Onion varieties which store well produce seed within the same year.

223. Only lettuce and celery are grown on a commercial scale in India.

224. Only pseudo-short-styled and true short-styled flowers in brinjal bear fruits.

225. Optimum temperature and R.H. for storage of sweet potato tubers are 5°C and 70 per cent, respectively.

226. Pea has high percentage of digestible protein and good content of vitamins and minerals.

227. Peas are very sensitive to drought.

228. Pointed gourd is an important summer crop in north Bihar and eastern U.P.

229. Pointed gourd is propagated by seed for taking a commercial crop.

230. Pointed gourd propagates by seeds alone.

231. Pointed gourd ratoon crop flowers 10-15 days earlier than the planted crop.

232. Potato has wide adaptability for climate and soil requirements, therefore, it is grown throughout the country.

233. Potato in India was introduced in the seventeenth century.

234. Potato is a good source of carbohydrates, calories and proteins.

235. Potato is a quick growing crop.

236. Potato is native of Ireland.

237. Potato tubers are borne underground at the root ends.

238. Potato tubers stored in cold storage should be kept for 12-24 hours at 15°C before exposure to atmosphere.

239. Potato varieties viz; Phulwa, DRR, Satha and Hellora, are not chance survivals of early introduction.

240. Premature production of seed stalks in onion is known as bolting.

241. Pulling of carrot roots becomes easy when bed is watered lightly about 24 hours or so prior to pulling.

242. Pungency in chilli is due to an alkaloid solasidine.

243. Pungency in onion is due to a volatile oil allyl propyl disulphide.

244. Punjab Chhuara has revolutionized tomato cultivation in Punjab state.

245. Punjab Chhuhara is a determinate variety of tomato.

246. Punjab-8 variety of okra has been evolved by irradiation of seeds.

247. Pusa Alankar is the hybrid of summer squash.

248. Pusa Chikni is not an improved variety of sponge gourd.

249. Pusa Early Prolific is a variety of Indian bean developed by IIHR, Bangalore.

250. Pusa Hybrid is the popular hybrid of bottle gourd.

251. Pusa Jyoti is a variety of watermelon.

252. Pusa Kesar, a cultivar of carrot, is a selection from cross between Local Red and Nantes Half Long.

253. Pusa Kranti and Azad Kranti are the hybrid varieties of brinjal.

254. Pusa Makhmali is a YVMV resistant variety of okra.

255. Pusa Naubahar is a variety of cluster bean suitable for sowing in both summer season and rainy season.

256. Pusa Purple Long is not a variety of brinjal.

257. Pusa Sawani variety of okra is tolerant to YVM disease.

258. Pusa Suyog is the popular hybrid of capsicum.

259. Radish can be grown throughout the year.

260. Radish is a quick growing crop.

261. Radish is suitable for inter-cropping.

262. Rhubarb is grown for its large thick leaf stalks and is propagated by the division of crown.

263. Ridge gourd belongs to cruciferae family.

264. Ridge gourd can never be grown on flat land.

265. Ridge gourd is affected by waterlogging condition.

266. Ridge gourd prefers to grow in heavy black cotton soils.

267. Ridge gourd require cool climate.

268. Ripe seeds of okra are sometimes used as a substitute for coffee.

269. S-12 is an indeterminate variety of tomato.

270. Satputia is a single fruited variety.

271. Second earthing up of potato crop is done to keep the soil loose and destroy weeds.

272. Seed crop of okra is generally raised in the summer season in northern plains.

273. Seed of Brussels sprouts can be easily produced in the plains of North India.

274. Seed production of temperate types of turnip cultivars should be done in plains.

275. Seed production of tropical types of turnip cultivars should be done in the hills.

276. Seed rate of colocasia for one hectare is about 80 to 120 quintals per hectare.

277. Seed requirement of okra is generally higher in rainy season crop than in spring-summer season crop.

278. Seedlings of variety Punjab Naroya when transplanted in August produce good bulbs in December-January under North Indian conditions.

279. Seeds of Chinese cabbage can be produced under North Indian plain conditions.

280. Seeds of French bean may rot in the ground if soil temperature is lower than 15°C.

281. Seeds of okra will not germinate below 20°C.

282. Seed-to-seed or *in situ* method of seed production gives higher seed yield of carrot.

283. Sel-120 is the first root knot resistant variety of tomato.

284. Short days are beneficial for tuber production.

285. Single fruit of *palak* contains ten to thirteen seeds.

286. Single stem training of tomato is most common.

287. Snake gourd is a dioecious, perennial bush.

288. Snake gourd is grown only in subtropical climate.

289. Snap melon is not a winter season crop.

290. Snap melon is widely cultivated all over the country.

291. Soil, climate and disease-free conditions are factors influencing location of seed producing areas.

292. Some of the smooth seeded varieties are comparatively more resistant to a number of rotting organisms active at high soil temperature.

293. Some people are allergic to the pollen as well as green pods of broad bean.

294. Spears of asparagus are used for *saag* preparation.

295. *Spinacea oleracea* belongs to Chenopodiaceae family.

296. Spinach is a warm-season crop.

297. Spinach leaves provide high contents of phosphorus and magnesium.

298. Sponge gourd can be grown under waterlogging conditions.

299. Sponge gourd is dioecious.

300. Sponge gourd is generally propagated by vine cutting.

301. Sponge gourd is more commonly cultivated in Europe and the Americas.

302. Sponge gourd is self pollinated.

303. Succulency and tenderness of the spinach leaves increase under low atmospheric humidity.

304. Sugarbeet is a long day plant.

305. Summer squash is one of the most nutritive and wholesome vegetables.

306. Sunscalding in tomato may be avoided by not staking the plants.

307. Sweet chillies (*Capsicum annuum*) are cooked as vegetable.

308. Sweet gourd is mainly grown in Cochin.

309. Sweet potato belongs to the species *batatas* of the genus *Ipomoea*.

310. Sweet potato contains 16 per cent starch, 4 per cent sugar, that is 20 per cent alcohol-producing materials.

311. Sweet potato does best on slightly high acidic soil.

312. Sweet potato is a native of tropical America.

313. Sylvia is an edible podded variety of pea.

314. Tapioca is propagated by cuttings from mature plants.

315. Tapioca tubers become ready in 3-4 months.

316. Temperature is more important than day length in seed stalk development of onion.

317. The botanical name of kakrol is *Momordica cochinchinensis*.

318. The botanical name of pea is *Allium sativum*.

319. The critical stage for irrigation in onion is flowering.

320. The cucumber seed retains viability for about 5 years and good seed gives 80-90 per cent germination.

321. The cultural operations of the vegetables belonging to the same family are always similar.

322. The distance of sowing of cowpea is 20-30 cm row to row and 10 to 15 cm plant to plant.

323. The edible part in colocasia is leaves.

324. The edible portion of cabbage is 'head' which consists of thick leaves overlapping lightly on growing bud.

325. The edible portions of Brussels sprouts are leaves.

326. The fleshy root of radish is modified root and it develops from both the primary root and the hypocotyle.

327. The fruit of snake gourd are not round in shape.

328. The fruits of kartoli are not edible.

329. The leaves and tender shoots of pointed gourd are used to prepare the soup for convalescents.

330. The lima bean is a bush type plant.

331. The plant of winter squash may tolerate frost.

332. The plants of broad bean bear upright pods in the axil of the leaves along the stem.

333. The recommended seed rate for spinach is about 60 kg/ha.

334. The ripe fruits of long melon are not consumed.

335. The seed production of Brussels sprouts is done in the hills.

336. The seed rate for planting one hectare area of winter squash is about 10 to 12 kg.

337. The seed rate of onion is 10-12 kg/ha.

338. The seed rate of tomato is 400-500 g/ha.

339. The seedlings obtained from double transplanting nursery bed produce higher yield because such seedlings are strong and stout and they establish early and put on good vegetative growth.

340. The seeds of broad bean are big, therefore, they are planted individually, adopting double row system of planting.

341. The seeds of Pusa Barsati variety of cowpea have striped spots.

342. The snap melon fruits never burst.

343. The soil reaction for the successful cultivation of turnip should be around neutral to slightly alkaline.

344. The tuber production in potato totally stops at 36°C.

345. The vegetable prepared from white brinjal fruits is said to be beneficial for persons suffering from diabetes.

346. The vegetables are not blanched before filling into cans or bottles sterilized under 10 lb pressure and containing 3 per cent solution of salt and sugar each.

347. The vines of watermelon are good source of mulching which guard sandy soils during the summers from wind erosion.

348. The yield of dry chillies is 2-2.5 tonnes/ha.

349. The yield of tomato is 20-25 tonnes/ha.

350. There are more chances of getting high quality carrot seeds when seed crop is raised by root-to-seed method.

351. Thinning in carrot becomes an essential operation for maintaining proper distance.

352. Thinning in turnip is an essential operation.

353. Thinning in wax gourd is an essential operation.

354. TMV is the most serious viral disease of tomato in India.

355. Tomato can only be grown in winter season.

356. Tomato fruits have been detected to maintain nicotine when grafted on tobacco root stock.

357. Tomato requires a warm sunny weather for its proper ripening, colour, quality and high yield.

358. Tomato varieties which are suitable for processing are not suitable for vegetable purposes.

359. Though Nantes is a European variety of carrot and its seed can be produced in plain.

360. Triploid watermelons are seedless, sweet, firm and suitable for transportation.

361. Tuber production in potato is maximum at 30°C.

362. Turnip does not transplant well, hence direct sowing is adopted.

363. Turnip is found growing wild in Russia and Siberia.

364. Turnip may cross with radish, rutabaga and cabbage.

365. Under long-day condition, tuberisation in potato is affected adversely but vegetative growth but is affected favourably.

366. Underdecomposed organic matter may cause forking or deformed roots.

367. Vegetables commonly grown under forcing structures are tomato and cucumber.

368. Vegetables need sterilization as heat resistant bacteria remain unaffected by water boiling at 115.5°C or more.

369. Vines of sweet potato grow at the expense of tuber formation when soil temperature goes above 40°C.

370. Wart disease in potato is more serious in tropical climate.

371. Waterlogging improves the quality of colocasia corm.

372. Wax gourd is a monoecious annual climber.

373. When turnip is allowed to grow under shade, foliage grows at the expense of root development.

374. Whip-tail is a disease of cauliflower.

375. White skinned tubers of sweet potato generally store well better than red-skinned tubers.

376. Winter squash does not require any manures and fertilizers.

377. Winter squash is a dioecious.

378. Winter squash is *Cucurbita moschata*.

379. Yam is propagated by tubers.

380. Yam or *ratalu* is a rich source of carbohydrates.

381. Yield of fresh green chillis is 10 times higher than that of dry chilli.

382. Yield of potato raised for seed purpose is 60-80 Q/ha.

383. YVM disease is transmitted only by an insect vector known as whitefly.

# Glossary

**Acicular :** Long, narrow and cylindrical; *i.e.*, needle-shaped as the leaves of onion, etc.

**Acid foods :** Foods having pH 4.5 to 3.7 which are usually spoiled by non-spore forming aciduric, butyric anaerobes, etc. *e.g.*, products of tomato, etc.

**Alliin :** A colourless, odourless, water soluble amino acid present in the uninjured bulb of garlic which, on crushing, breaks down in presence of the enzyme alliinase to allicin, the principal ingredient of the odoriferous diallyl disulfide.

**Allogamy :** Highy cross pollination as seen in cucurbits, cole crops, radish, turnip, carrot, onion, *palak*, amaranthus, spinach, beet, etc.

**Androdioecious :** A sex form in dioecious species where staminate flowers on one plant and bisexual flowers on another plant of the same species are borne.

**Androecious :** A sex form in monoecious species where only staminate flowers are produced, giving rise to supermale plant.

**Aroids :** A group of vegetable crops under the family Araceae where edible plant parts are corm and cormels *e.g.*, *Colocasia* spp. *Amorphophallus* sp., etc.

**Asiatic carrot :** Carrot cultivars which do not require any low temperature treatment for flowering and produce seed freely in the plains of India. *e.g.*, Pusa Kesar.

**Autogamy :** High self-pollination where cross-pollination is less than 5 per cent, as seen in cowpea, cluster bean, dolichos bean, pea, tomato, fenugreek, French bean, etc.

**Autopolyploid** : A polyploid containing more than two copies of the same genome of a single species *i.e.*, autotriploid (3x), autotetraploid (4x), autohexaploid (6x), *e.g.*, potato, sweet potato.

**Base temperature** : The threshold temperature level below which plant do not develop. Each plant has its own base temperature. *e.g.*, pea (4.4 °C), French bean (10° C), asparagus (5.5 °C), spinach (2° C), pumpkin (13° C), tomato (15° C), etc.

**Beaded root** : Root possessing swellings at frequent intervals, seen in *Basella, Momordica,* etc.

**Berry** : Fleshy, superior (sometimes inferior), usually many seeded fruits, developing commonly from a syncarpous pistil (rarely from a single carpel) with axil or parietal placentation, *e.g.*, tomato, etc. With the growth of the fruit, the seeds separate from the placentae and lie free in the pulp.

**Bhasinda** : The underground stem of *Nelumbium,* a popular vegetable in North India.

**Biennial** : Plant having a two-year lifecycle, vegetative in the first season and reproductive in the second season, and this transition from vegetative to reproductive stage often requires environmental trigger such as vernalization or photoperiod. *e.g.*, cabbage, onion, carrot, etc.

**Bisexual** : Presence of both male and female parts in the same flower. *e.g.*, brinjal, tomato, etc.

**Black leaf speck of cabbage** : Small, sharply sunken brown or black specks on leaves which occur under refrigeration in transit and storage and under sharp temperature drops in the fields, also found in Chinese cabbage and cauliflower.

**Blanching (cultural)** : Exclusion of light from the edible parts of salad crops like asparagus, leek, cauliflower, etc., which makes the crop crisp, reduces acrid flavour, improves flavour and tenderness.

**Bolter** : Sporadic occurrence of abnormally big sized tuber in potato.

**Bolting** : Significant stem elongation that proceeds flowering; also includes the case of premature emergence of flower stalk.

**Bud pollination** : Artificial pollination in the bud stage; practiced in the selfing of cabbage.

**Bulb** : A specialised underground organ consisting of a short, fleshy, usually vertical stem axis bearing at its apex a growing point or a flower primordium enclosed by thick, fleshy scales, *e.g.*, onion, etc.

**Bulb crops** : Vegetable crops under the genus *Allium* which include onion, garlic, leek, shallot and chive, whose bulbs are eaten raw or cooked or they and their leaves are used to flavour other vegetables, meat, fish and sauces.

**Bulbil** : Vegetative part that is actually the modification of flower(s). It develops into plant directly without formation of seeds; seen in onion, garlic, etc.

**Bulblets** : Miniature bulbs produced around the base of the mother bulb due to development of meristem in the axil of scale leaves.

**Bulb scale** : Fleshy 'leaves' that together form the bulb.

**Bulb-to-seed method** : A method of seed production of bulb crops where the bulbs harvested during warm weather are selected, stored and again replanted in winter for seed production.

**Bulb tunic** : The dead, papery, leathery or fibrous covering that surrounds most bulbs.

**Capsanthin** : A carotenoid pigment responsible for the characteristic orange-red colouration of ripe chilli.

**Caruncle** : An outgrowth near the hilum of the seeds as seen in dolichos bean

**Celery lettuce** : Stem type cultivar of lettuce, grown for its thick stem which is eaten after peeling.

**Chasmogamy** : A built-in breeding mechanism where pollination follows opening of flowers which favours self-pollination as seen in tomato, brinjal, chilli etc.

**Chemical dormancy** : Type of seed coat dormancy in which germination inhibiting chemicals *viz.,* various phenols, coumarin and abscisic acid accumulated in the fruit as well as in the seed coverings strongly inhibit germination; found in cucurbits, tomato, etc.

**Cleistogamy** : A built-in breeding mechanism where flowers remain closed at the time of pollination which favour self-pollination, as seen in lettuce.

**Cole crops** : A group of vegetable crops which originated from wild cliff cabbage of Mediterranean region and belonging to genus Brassica (Brassicaceae) which include cabbage, cauliflower, Broccoli, Brussels sprouts, knol-khol, Chinese cabbage, etc., whose leaves, unopened flower buds, inflorescences or swollen stems are used as cooked or raw vegetables.

**Conical roots** : When the root is broad at the base and gradually tappers towards the apex like a cone as in carrot.

**Coreless carrot** : Good quality cultivars of carrot in which the core or xylem is small and deeply pigmented so that the cortex or phloem and the core is evenly coloured.

**Corm :** Bulky, short and vertical undergound modified stem in which foods are stored as in Elephant's foot (*Amorphophallus* sp.), etc.

**Cover crops :** Crops that are grown both for the protection of the soil from erosion and for soil improvement. *e.g.,* cowpea.

**Cucurbits :** A large and diverse group of vegetable crops under Cucurbitaceae, used as vegetables (pumpkin, different gourds, etc.) pickles (cucumber) and as desert fruits (muskmelon, watermelon).

**Cultigroup :** An intraspecific category below subspecies which includes cultivated types such as *Vigna unguiculata* cultigroup sesquipedalis (vegetable cowpea).

**Cultivar :** An assemblage of cultivated plants which is clearly distinguished by any character and which, when reproduced, sexually or asexually, retain the distinguishing characters.

**Curd :** Edible part of cauliflower which is actually the repeatedly branched prefloral fleshy apical meristem.

**Curd size index :** A curd character of cauliflower which is the equatorial x polar diameter of the curd.

**Cytoplasmic and genic male sterility :** Sterility of pollen grains which is governed by the interaction between sterile cytoplasm and recessive gene, seen in onion, beet, carrot, etc.

**Day-neutral plant :** Plant in which flowering is not influenced by day length. *e.g.,* tomato, cucumber, okra, asparagus, capsicum, snap bean, etc.

**Decompound leaf** : When the leaf is more than thrice pinnate as in carrot, etc.

**Degreening** : The process of decomposing the green pigment in fruits by applying ethylene (1000-2000 ppm) or similar metabolic inducers to give a fruit its characteristic colour as preferred by consumers, generally followed in citrus fruits but also practiced in banana, mango, tomato, etc.

**Dehaulming** : Removal of the top portion (haulm) of potato in the seed crop to avoid the infestation of virus carrying insect vectors.

**Dehiscent fruit** : Fruit whose pericarp bursts to liberate the seeds at maturity as seen in okra, etc.

**Dioecious** : Plant species in which unisexual flower, staminate or pistillate, is borne on separate plants, as in pointed gourd, etc.

**Earthing up** : The process of putting the soil just near the base of stems of certain crops like potato, cassava, banana, etc. to provide support and to prevent root exposure.

**Epicalyx** : A series of small sepal-like bracts forming an outer calyx beneath the true calyx, as in okra, etc.

**Epigeous germination** : A pattern of germination where the hypocotyl elongates and raises the cotyledons above the ground, as seen in tomato, beans, gourds, etc.

**Epigny** : The phenomenon in which the thalamus completely encloses the ovary and gets fused with it, and bears the sepals, petals and stamens on the top of the ovary, as seen in cucumber, etc.

**European carrot :** Carrot cultivars which are biennial in nature and require low temperature (4.8 - 10°C) treatment for certain periods for flowering, hence do not produce seeds in plains of India, *e.g.*, Nantes, Chanteny, Imperator, etc.

**Fasciculated roots :** Swollen tuberous roots which are developed in a cluster or fascicle at the base of the stem as in asparagas, etc.

**Floating garden :** A type of vegetable garden found in the lakes of Kashmir valley where vegetables are grown on a floating base prepared with some grass (*Typha*), compost and other organic matters.

**Foliaceous stipules :** A large paired leafy outgrowth as seen in pea, etc.

**Full slip :** A harvesting index of muskmelon or cantaloups for local market when the fruit can easily be removed (slip) with a slight pressure from the stem leaving a clean stem cavity.

**Fusiform root :** When the root is swollen in the middle and gradually tappers towards the apex and the base, being more or less spindle-shaped in appearance as in radish.

**Gamopetalous :** Petals united together partially or wholly as seen in tomato, brinjal, etc.

**Gamosepalous :** Sepals united together partially or wholly as seen in brinjal, chilli, etc.

**Garden for vegetable processing :** A type of vegetable farming where vegetables are produced with a sole objective of supplying them to the processing factories.

**Genic male sterility :** Sterility of pollen grains which is governed by a single recessive gene as found in squash, pumpkin, muskmelon, Brussels sprouts, cabbage, cauliflower, lettuce, sprouting broccoli, etc.

**Gourd :** Generally it refers to the fruit of cucurbits. Actually this epithet refers to the fruit character : hard and tough rind upon complete maturation as in bottle gourd, pumpkin or summer squash, even though the term gourd is applied to other fleshy fruits like bitter gourd or snake gourd whose skin do not become tough when ripe.

**Green pepper :** Tender, semi-mature green pepper (*Piper nigram*) spike which is used commercially in pickles.

**Growth crack of sweet potato :** Longitudinal or transverse splits and fissures due to irregular on interrupted growth.

**Guar gum :** The mucilaginous seed flour of guar (*Cyamopsis tetragonolobus*) is valued as guar gum (Galactomannan) used in textile, paper, cosmetic and oil industries throughout the world. It is also a useful absorbent for explosives.

**Hakuran :** An artificial amphidiploid of cabbage, as Chinese cabbage produced through embryo culture technique, which is a good leafy vegetable.

**Half-hardy vegetable :** Vegetable crops which can thrive well in cool weather condition but cannot tolerate frost. *e.g.*, beet, carrot, cauliflower, lettuce, spinach, etc.

**Head :** Edible portion of cabbage, Chinese cabbage and head lettuce which is a structurally distinct, compact leafy portion made up of numerous overlapping leaves covering the terminal bud.

**Head shape index :** A head character of cabbage, Chinese cabbage, calculated by dividing mean head length (cm) with mean head width (cm).

**Head-to-seed method :** A method of seed production practiced in cabbage where the selected plants with fully matured heads are lifted prior to snowfall, stored and again replanted at the onset of spring for seed production.

**Hermaphrodite :** A sex form where only bisexual flowers are produced as in tomato, brinjal, chilli, etc.

**Hub crop :** Crop which has the greatest comparative advantage over other crops in a sequential cropping system *e.g.* vegetable crops.

**Hypogeous germination :** A pattern of germination where the lengthening of the hypocotyl does not raise the cotyledons above the ground and only the epicotyl emerges, as seen in pea, etc.

**Hypogyny :** The phenomenon in which the ovary occupies the highest position on the thalamus while the stamens, petals and sepals are separately and successively inserted below the ovary, as seen in brinjal, etc.

**Internal browning of tomato :** Gray-brownish discolouration of internal tissues in green fruit which extend to the surface and form lesion that remain greenish or yellow in ripe fruit; caused by water imbalance and high temperature and/or nutrient imbalance.

**Inulin :** A storage polysaccharide ($C_6H_{10}O_5$) found in the foots of Jerusalem artichoke, dahlia and a few other plants.

**Long-day plant :** A plant which requires a day longer than its critical day length for flowering. *e.g.,* lettuce, radish, onion, cabbage, carrot, spinach, beet, etc.

**Lycopene :** Red pigment found in ripe tomato which is a straight chain derivative of carotene with no vitamin activity. Its chemical composition ($C_{40}H_{56}$) is same as that of carotene.

**Market gardening :** Vegetable farming for supply of vegetables to the consumers in the local market; one of the most intensive types of vegetable farming.

**Monoecious :** The condition in which staminate and pistillate flowers are borne separately on the same plant, as in cucurbits (in the separate inflorescence but produced sequentially).

**Multivitamin green :** Chekurmanis (*Sauropus androgymus*), a perennial leafy vegetable crop is called so because of the availability of various vitamins like A, B, C, D, F and K from this vegetable crop.

**Napiform root :** Root which when swollen become spherical at the upper part and sharply tapering at the lower part, as in beet, turnip, etc.

**Nursery bed :** A prepared area where seed is sown or into which transplants or cuttings are planted.

**Oleoresin :** A natural combination of resinous substances and essential oils present in the fruits of certain crop plants like chilli.

**Parietal placentation :** When placentae bearing the ovules develop on the inner wall of the one-chambered ovary corresponding to the confluent margins of carpels as seen in radish, etc.

**Pepo :** Fleshy, many seeded fruit which develops from an inferior, one-celled or spuriously three-celled, syncarpous pistil with parietal placentation, *e.g.,* cucumber, melons, squash, gourds, etc.

**Perianth :** When the calyx and corolla do not differ much in shape and colour, they together are said to form the perianth as seen in onion, garlic, etc.

**Peripheral embryo :** Embryo which encloses endosperm or perisperm tissue as seen in *Amaranthus,* etc.

**Photodormancy :** A type of physiological dormancy of seed where germination of seed is sensitive to light i.e., seeds of some plants require light to germinate whereas others require darkness. *e.g.,* lettuce seed require light and *Allium, Amaranthus,* etc. seed require darkness for germination.

**Phytoalexin :** A phenolic substance having antifungal principle, synthesised by plants in response to parasite invasion or infection by certain fungi, e.g., pisatin, phaseolin, trifolirhizin, orchinol and isocumarin from pea and bean pods, and carrot root, respectively.

**Pie plant :** Rhubarb (*Rheum raponticum*), one of the oldest cultivated vegetable crops, is commonly known as "pie plant".

**Pinnatipartite :** When the incision of leaf margin is more than half way down towards the mid-rib, as in radish.

**Planting ratio :** The male and female plants when planted in a certain proportion to ensure proper pollination and fertilization *e.g.*, 10 : 1 (female : male) ratio in pointed gourd.

**Poi :** The pressure cooked taro (*Colocasia esculenta*) corms after being passed through strainer are allowed to ferment which gives an acidic product called 'poi'.

**Polypetalous :** When the petals remain free from each other as seen in cabbage, radish, etc.

**Positional sterility :** A type of male sterility, also called functional sterility, where pollens are functional but anthers fail to dehise, found in some mutants of tomato.

**Pot herbs :** A group of vegetable crops whose foliage and sometimes immature stem are used as cooked vegetables; also called leafy vegetables or green. *e.g.*, *palak*, amaranthus, spinach, basella, etc.

**Pricking :** A method of raising secondary nursery for the crops having very small seeds; in the case of high density sowing, it helps to develop a thinner stand in the nursery. In cole crops and lettuce, pricking is done when first pair of true leaves develop.

**Quercetin :** A pigment which imparts coloration to the outer skin on onion bulb.

**Raceme :** Inflorescence where the main axis is elongated and bears laterally a number of flowers which are all stalked, the lower flowers having longer stalks than the upper ones, as in radish, etc.

**Recurrent apomixis :** A type of apomixes where embryo sac develops from egg mother cell or from some adjoining cell without complete meiosis. The embryo develops directly from the diploid egg nucleus without fertilization, seen in onion, etc.

**Root crops :** A group of vegetable crops whose swollen tap roots, and in some cases hypocotyl along with tap root such as carrot, beet, radish, turnip, rutabaga etc., are cooked or eaten raw.

**Root forking :** Branching of tap roots in the root crops, particularly in radish and carrot, due to the presence of impediment, undecomposed organic matter or plant refuse in the soil.

**Root-to-seed method :** A method of seed production in root crops where the fully matured roots are harvested, selected and after giving proper root and shoot cuts, they are replanted for seed production.

**Root tuber :** The fleshy root of a herbaceous perennial plant with buds or eyes in the upper regions. *e.g.,* sweet potato.

**Sagittate :** Arrow shaped with the basal lobes directed downwards. *e.g.,* leaves of some aroids.

**Salad crops :** Green leafy vegetables which are usually consumed raw with oil, vinegar and various other condiments, *e.g.,* lettuce, endive, celery, chicory, parsley, etc.

**Scooping :** Removal of central portion of the curd for easier initiation of flower stalk in cauliflower.

**Seedless watermelon :** It refers to autotriploid (3x) watermelon which is both male and female sterile due to unequal chromosomal distribution in meiosis resulting in seedless condition in the fruit.

**Seed potato :** Potato tubers used for planting.

**Seed-to-seed method :** A method of seed production where the plants are allowed to produce seed in its original place of growing.

**Self-incompatibility :** Failure of fertilization even though both male and female parts of the bisexual flower are fully functional as seen in cabbage, cauliflower, radish, etc.

**Shoot apex culture :** A tissue culture procedure for eliminating virus or other pathogens from plant parts where excision and aseptic culture of the small segment of the terminal growing points is done because the terminal growing point of a plant is often free from virus and other pathogens even if the rest of the plant is infected; practiced in potato, sweet potato, cassava, etc.

**Simla mirch :** Bell-shaped, non-pungent, mild and thick fleshed.

**Spadix :** A spike with a fleshy axis which is enclosed by one or more large, often brightly coloured, bracts called spathes, as in aroids, etc.

**Spear :** The shoot which is the edible part of asparagas.

**Species :** A category of taxonomic classification lower than a genus or subgenus and above that of a subspecies of variety.

**Spike :** The inflorescence in which the main axis is elongated and the lower flower opens earlier than the upper ones as in raceme, but the flowers are sessile, as seen in amaranthus, etc.

**Sporophytic self incompatibility :** A category of self incompatibility where the incompatibility reaction of the pollen is controlled by the genotype of the pollen-producing plants (sporophyte), *i.e.,* the action is pre-meiotic. In this incompatibility system there are smaller number of alleles involved that may display either dominance or may have individual action. Seen in *Brassica oleracea, Raphanus* sp., etc.

**Steckling :** Matured root of carrot, radish, turnip, etc. to be replanted after over-wintering for seed production in root-to-seed method.

**Stolon :** Slender, underground lateral stems arising from buds on the underground portion of the stem which enlarge at its tip to produce the tuber, as in potato.

**Stump method :** A method of seed production in cabbage where the head after full maturity is cut off just below the base, keeping the stem with outer whorl of leaves intact.

**Supermarket on a stalk :** The winged bean (*Psophocarpus tetragonolobus*) plant is described so, as six different foods are supplied by this plant : leaves like spinach, succulent shoots resembling large thin asparagus, fried flowers for making a sweet garnish, tender pod as vegetable, the seed and the underground tubers are exceptionally rich in proteins.

**Sweet pepper :** It is the chilli of commercial value (*Capsicum annuum*).

**Tabasco pepper :** Small fruited, very pungent chilli (*Capsicum frutescens*).

**Tenderometer :** An instrument by which toughness of the seed coat and firmness of pulp is determined and is mostly used to determine the seed quality of pea where high value of tenderometer indicates low quality.

**Tender vegetable :** Vegetable crops which cannot withstand frost and some of them even do not thrive in cool weather, *e.g.*, brinjal, okra, chilli, cucurbits, sweet potato, cowpea, tomato, cassava, beans, etc.

**Trifoliate :** A compound leaf with three separate leaflets as in French bean, cowpea, etc.

**Tuber :** A special kind of swollen modified stem structure that functions as an underground storage organ as in potato, Jerusalem artichoke, etc.

**Tubercle :** Small aerial tuber produced in the leaf axils as seen in Yam (*Dioscorea alata*).

**Tunic :** The papery or fibrous coats covering bulbs and corms.

**Tunicated bulbs :** Type of bulb where the outer leaves are usually thin, membranous and dry and completely ensheath the inner portion and the central axis of the bulb like a tunic as the bulb of *Allium*.

**Undeveloped embryo :** Partially developed torpedo-shaped embryos that may attain a size upto one half that of the seed cavity at maturity as seen in carrot, etc.

**Utilization index :** Ability of root and tuber crops for better accumulation of photosynthates in the storage organ which can be judged by root : shoot ratio.

**Vacuum cooling :** A technique of cooling vegetable (leafy vegetables, asparagus, Brussels sprout, etc.) having a high surface to volume ratio, rapidly and uniformly by boiling off some of their water at 1°C and at low pressure (5 mm mercury) into a sealed container. The produce is cooled by evaporation of water from the tissue surface and is more rapid than hydrocooling.

**Vegetable :** An edible plant or plant part eaten cooked or raw as a main part of a meal, side dish, or appetizer.

**Vegetable forcing :** A specialized type of vegetable farming where vegetables are grown out of their normal season. Vegetable forcing requires some special structures like glasshouse, hot bed, cold frame, etc.

**Vegetative apomixis :** A type of apomixis where flowers are replaced in the plants by bulbils or buds which fall to the ground like seeds, found in *Allium cepa var. Viviparum, Poa bulbosa,* etc.

**Vexiliary :** Of the five petals, when the posterior one is the largest and almost covers the two lateral petals as in pea, dolichos bean, etc.

**Waxing :** A short term storage technique of fresh fruits and vegetables under ambient conditions by applying wax emulsion containing paraffin wax, triethanol and aleic acid which provide a thin, discontinuous layer on the fruit surface and thus curtails the respiration and transpiration resulting increase in shelf-life. It also helps to keep away the microbes when fungicides like, Benlate 50 are used in wax emulsion.

**Waxy blister of tomato :** A disorder where white to cream coloured irregular blisters, 3 to 6 mm in diameter and often more than 3 mm high occur which become light to dark brown, depressed and crack as the fruits ripen.

**White heart :** Whiteness at the central portion of watermelon instead of uniform development of pink colour from centre to rind, indicating poor quality.

**Yearling :** One-year-old bulblets which have been formed on the scale.

**Zoning of beet :** Under unfavourable conditions, particularly in hot weather, beetroot show alternate white and coloured circles when sliced, called zoning.

# APPENDIX-1

## Nutritive Value of Various Vegetables Per 100 g of Edible Portion

| Name of the vegetable | Moisture content (%) | Carbohydrate (g) | Protein (g) | Fat (g) | Vitamin A (I.U.) | Vitamin B (mg) Thiamine | Vitamin B (mg) Riboflavin | Vitamin C (mg) | Minerals Total (g) | Ca (mg) | P (mg) | Fe (mg) | K (mg) |
|---|---|---|---|---|---|---|---|---|---|---|---|---|---|
| 1 | 2 | 3 | 4 | 5 | 6 | 7 | 8 | 9 | 10 | 11 | 12 | 13 | 14 |
| Amaranthus | 85.7 | 6.3 | 4.0 | 0.5 | 9,200 | .03 | 0.10 | 99 | 2.7 | 397 | 83 | 25.5 | 341 |
| Ash gourd | 96.5 | 1.9 | 0.4 | 0.1 | 0 | 0.06 | 0.01 | 1 | 0.3 | 30 | 20 | 0.8 | — |
| Beetroot | 87.7 | 8.8 | 1.7 | 0.1 | 0 | 0.04 | 0.09 | 88 | 0.8 | 200 | 55 | 1.0 | 43 |
| Bitter gourd | 92.4 | 4.2 | 1.6 | 0.2 | 210 | 0.07 | 0.09 | 88 | 0.8 | 20 | 70 | 1.8 | 152 |
| Bottle gourd | 96.1 | 2.5 | 0.2 | 0.1 | 0 | 0.03 | 0.01 | 6 | 0.5 | 20 | 10 | 0.7 | 87 |
| Brinjal | 92.7 | 4.0 | 1.4 | 0.3 | 124 | 0.04 | 0.11 | 12 | 0.3 | 18 | 47 | 0.9 | 2 |
| Cabbage | 91.9 | 4.6 | 1.8 | 0.1 | 2,000 | 0.06 | 0.03 | 124 | 0.6 | 39 | 44 | 0.8 | 114 |
| Carrot | 86.0 | 10.6 | 0.9 | 0.2 | 3,150 | 0.04 | 0.02 | 3 | 1.1 | 80 | 30 | 2.2 | 108 |
| Cauliflower | 90.8 | 4.0 | 2.6 | 0.4 | 51 | 0.04 | 0.10 | 56 | 1.9 | 33 | 57 | 1.5 | 113 |
| Chilli (green) | 85.7 | 3.0 | 2.9 | 0.6 | 292 | 0.19 | 0.39 | 111 | 1.0 | 30 | 80 | 1.2 | 217 |
| Cluster bean | 81.0 | 10.8 | 3.2 | 0.4 | 316 | 0.09 | 0.09 | 47 | 1.4 | — | — | — | — |
| Coccinia | 93.5 | 3.1 | 1.2 | 0.1 | 260 | — | — | 15 | 0.5 | — | — | — | — |
| Cowpea | 84.6 | 8.0 | 4.3 | 0.2 | 941 | 0.07 | 0.09 | 13 | 0.9 | 80 | 74 | 2.5 | — |
| Cucumber | 96.3 | 2.7 | 0.4 | 0.1 | 0 | 0.03 | 0.01 | 7 | 0.3 | 10 | 25 | 1.5 | 50 |
| Dolichos bean | 86.1 | 6.7 | 3.8 | 0.7 | 312 | 0.1 | 0.06 | 9 | 0.9 | 210 | 68 | 1.7 | 74 |
| French bean | 91.4 | 4.5 | 1.7 | 0.1 | 221 | 0.08 | 0.06 | 11 | 0.5 | 50 | 28 | 1.7 | 129 |
| Garlic (dry) | 62.0 | 29.80 | 6.3 | 0.1 | 0 | 0.06 | 0.03 | 13 | 1.0 | 30 | 310 | 1.3 | — |

| Name of the vegetable | Moisture content (%) | Carbohydrate (g) | Protein (g) | Fat (g) | Vitamin A (I.U.) | Vitamin B (mg) Thiamine | Vitamin B (mg) Riboflavin | Vitamin C (mg) | Minerals Total (g) | Ca (mg) | P (mg) | Fe (mg) | K (mg) |
|---|---|---|---|---|---|---|---|---|---|---|---|---|---|
| 1 | 2 | 3 | 4 | 5 | 6 | 7 | 8 | 9 | 10 | 11 | 12 | 13 | 14 |
| Knol-khol | 92.7 | 3.8 | 1.1 | 0.2 | 36 | 0.05 | 0.09 | 85 | 0.7 | 20 | 35 | 0.4 | 37 |
| Lady's finger | 89.6 | 6.4 | 1.9 | 0.2 | 88 | 0.07 | 0.10 | 13 | 0.7 | 66 | 56 | 1.5 | 103 |
| Lettuce | 93.4 | 2.5 | 2.1 | 0.3 | 1,650 | 0.09 | 0.13 | 10 | 1.2 | 50 | 28 | 2.4 | 33 |
| Muskmelon | 94.0 | 5.0 | 1.0 | 0 | 3,420 | 0.04 | 0.05 | 33 | – | 0.017 | 0.016 | – | – |
| Onion | 86.8 | 11.0 | 1.2 | – | 0 | 0.08 | 0.01 | 11 | 0.4 | 180 | 50 | 0.7 | – |
| Pea | 72.0 | 15.8 | 7.2 | 0.1 | 139 | 0.25 | 0.01 | 9 | 0.8 | 20 | 139 | 1.5 | 79 |
| Pointed gourd | 92.0 | 2.2 | 2.0 | 0.3 | 255 | 0.05 | 0.06 | 29 | 0.5 | 30 | 40 | 1.7 | 83 |
| Pumpkin | 92.6 | 4.6 | 1.4 | 0.1 | 84 | 0.06 | 0.04 | 2 | 0.6 | .10 | 30 | 0.7 | 83 |
| Radish (white) | 94.4 | 3.4 | 0.7 | 0.1 | 5 | 0.06 | 0.02 | 15 | 0.6 | 50 | 22 | 0.4 | 138 |
| Round melon | 93.5 | 3.4 | 1.4 | 0.2 | 23 | 0.08 | 0.04 | 18 | 0.5 | 25 | 24 | 0.9 | 24 |
| Snake gourd | 94.6 | 3.3 | 0.5 | 0.3 | 160 | 0.04 | 0.06 | 0 | 0.5 | 50 | 20 | 1.1 | 34 |
| Snap melon | 95.7 | 3.0 | 0.3 | 0.1 | 265 | – | – | 10 | 0.4 | – | – | – | – |
| Spinach | 92.1 | 2.9 | 2.0 | 0.7 | 9,300 | 0.03 | 0.07 | 28 | 1.7 | 73 | 21 | 10.9 | 206 |
| Spinach beet | 90.7 | – | 2.3 | – | 25,000 | 0.13 | 0.28 | 40.9 | – | 2.1 | 0.7 | – | – |
| Sponge gourd | 94.3 | 4.4 | 0.8 | 0.2 | 330 | – | – | 8 | – | – | – | – | – |
| Squash (summer) | 94.0 | 2.4 | 1.0 | 0.1 | 80 | 0.05 | 0.03 | 19 | 0.6 | 10 | 30 | 0.7 | 139 |
| Tomato (green) | 93.1 | 3.6 | 1.9 | 0.1 | 320 | 0.07 | 0.01 | 31 | 0.6 | 20 | 36 | 1.8 | 114 |
| Turnip | 91.6 | 6.2 | 0.5 | 0.2 | 0 | 0.04 | 0.04 | 43 | 0.6 | 30 | 40 | 0.4 | – |
| Watermelon | 95.8 | 3.3 | 0.2 | 0.2 | 590 | 0.05 | 0.05 | 6 | – | 0.07 | 0.07 | 7.9 | – |

## Approximate life of vegetable seeds stored under cool conditions

| Vegetable | Year | Vegetable | Year |
| --- | --- | --- | --- |
| Asparagus | 3 | Lettuce | 5 |
| Bean | 3 | Muskmelon | 5 |
| Beet | 3 | Mustard | 4 |
| Cabbage | 5 | Okra | 2 |
| Carrot | 3 | Onion | 2 |
| Cauliflower | 5 | Pea | 1-2 |
| Celery | 5 | Pepper | 3 |
| Cucumber | 5 | Pumpkin | 4 |
| Egg plant | 5 | Radish | 5 |
| Endive | 5 | Spinach | 5 |
| Kale | 5 | Swiss chard | 4 |
| Kholrabi | 5 | Tomato | 4 |
| Leek | 3 | Turnip | 5 |
|  |  | Watermelon | 5 |

## Approximate time required for germination

| Vegetable crops | Days |
|---|---|
| Bean | 7-14 |
| Beet | 10-16 |
| Cole crops | 5-10 |
| Cucumber | 7-14 |
| Leek | 10-15 |
| Onion | 10-15 |
| Lettuce | 6-10 |
| Tomato | 5-10 |
| Radish | 4-7 |
| Pea | 7-14 |
| Melons | 4-8 |
| Parsley | 14-30 |
| Spinach | 7-10 |
| Carrot | 12-15 |
| Brinjal | 7-12 |

## List of Pesticides Restricted or Banned in India

| | |
|---|---|
| **a.** | **Pesticides restricted for use** |

1. Aluminium phosphide
2. Captafol
3. Carbaryl
4. Dieldrin
5. Ethylene dibromide (EDB)
6. Methyl bromide
7. Sodium cyanide
8. Lindane
9. Methyl parathion

| | |
|---|---|
| **b.** | **Pesticides banned for use in agriculture in India** |

1. Dibromochloropropane
2. Endrin (DBCP)
3. Pentachloronitrolbenzene (PCNB)
4. Pentachlorophenol (PCP)
5. Toxaphene
6. Ethyl parathion
7. Chlordane
8. Heptachlor
9. Aldrin
10. Paraquat-di-methyl sulphate
11. Nitrofen

| **Pesticides banned for use in agriculture in India** |
|---|
| 12. Nicotene sulphate |
| 13. Phenyl mercury acetate |
| 14. Tetradifern |
| 15. Calcium cyanide |
| 16. Copper Acotoarsenite |
| 17. Ethyl mercury chloride |
| 18. Menazon |
| 19. Sodium methane arsonate |
| 20. BHC (HCH) |
| 21. Phenyl mercury acetate (PMA) |
| 22. Nichotine sulphate |
| 23. DDT |
| 24. Chlorobenzilate |

## Statewise area and production of vegetables (including potato and onion) for 1990-91

| State/Union Territory | Area (ha) | Production (tonnes) |
|---|---|---|
| Andhra Pradesh | 155186 | 1452583 |
| Arunachal Pradesh | 17166 | 79947 |
| Assam | 222430 | 2132273 |
| Bihar | 843308 | 8643080 |
| Goa | 7350 | 60200 |
| Gujarat | 114600 | 1667900 |
| Haryana | 60800 | 877000 |
| Himachal Pradesh | 38680 | 476000 |
| Jammu & Kashmir | 180300 | 745000 |
| Karnataka | 351060 | 3673235 |
| Kerala | 202091 | 3229069 |
| Madhya Pradesh | 176360 | 2221000 |
| Maharashtra | 241080 | 4171340 |
| Manipur | 11819 | 50280 |
| Meghalaya | 25881 | 219224 |
| Mizoram | 5975 | 31753 |
| Nagaland | 8235 | 66860 |
| Orissa | 710288 | 7275000 |
| Punjab | 84500 | 1450008 |

| State/Union Territory | Area (ha) | Production (tonnes) |
|---|---|---|
| Rajasthan | 62868 | 307033 |
| Sikkim | 7586 | 46119 |
| Tamil Nadu | 889333 | 3796897 |
| Tripura | 30350 | 30680 |
| Uttaranchal | 57065 | 617600 |
| UP (plains) | 1202000 | 17615840 |
| West Bengal | 456000 | 4680000 |
| Andman & Nicobar | 3384 | 13200 |
| Chandigarh | 273 | 11070 |
| Daman and Diu | 33 | 330 |
| Delhi | 55010 | 627816 |
| Dadra and Nagar Haveli | 1520 | 13560 |
| Lakshdweep | 379 | 348 |
| Pondicherry | 2363 | 22337 |
| **Total** | **6225273** | **66580752** |

# APPENDEX-6

The current varietal status of important vegetable crops grown in India

| Vegetable crops | Improved cultivars |
|---|---|
| Tomato | Arka Saurabh, Arka Vikas, Pusa Uphar, Hissar Anmol, Hissar Arun, Hissar Lalit, La-Bonita, Pant Bahar, Punjab Chhuhara, Avinash-2, Punjab Kesari, Pusa |
| | Early Dwarf, Pusa Ruby, Pusa Sheetal, Roma, Sel 120, Hissar Lalima, Krishna, Matri, Naveen, Pusa 120, Pusa Divya, Pusa Gaurav. Pusa Sadabahar, Rajni, Rashmi, Ratna, Ruali. |
| Brinjal | Pusa Purple Long, Pusa Purple Cluster, Jamuni Gola, Arka Kusumakar, Arka Sheel, Arka Shirish, Manjiri Gota, Bhagyamati, Mysore Green, Annamalai, Pant Samrat, Pusa Kranti, Pusa Bhairav, Pusa Anupam, Pusa Uttam, Pusa Upkar, Pusa Bindu, Pb. Neelam, Pb. Barsati, Pb. Sadabahar, Hissar Jamuni, Pant Rituraj, Vaishali, Pragati Arka |
| | Navneet, Arka Keshav, Arka Neelkantha, Arka Nidhi, Azad Kranti, Hissar Shymal, Junagarh Long, Pb. Bahar, Pb. Chamkila, White Cluster. |

| Vegetable crops | Improved cultivars |
| --- | --- |
| Chilli | Bhagya Laxmi, Andhra Jyoti, Pusa Jwala, Sindhur, Punjab Lal, Bhaskar, Masalwadi Selection, Jwalamukhi, Jwala Sakshi, Pusa Sadabahar, Arka Lohit, Arka Abir, CH1, Ujjwala, Arpana Jawahar 218, x 235, ankeshwar 32. |
| Capsicum | California Wonder, Arka Mohini, Arka Gaurau, Arka Basant, Pusa Deepti, Bharat, Indra, Sun 1090, Green Gold, Larie, Indira, Hira, Pusa Suyog. |
| Cauliflower | Early Kunwari, Pusa Early Synthetic Pant Gobhi-3, Pusa Deepali, Improved Japanese, Pusa Hybrid 2, Pusa Sharad, Pusa Synthetic, Pusa Shubhra, Pusa Himjyoti, Pusa Giant 35, Pusa Snowball-1, Pusa Snowball K1, Ooty 1 Pusa Katki, Dania Snowball 16. |
| Cabbage | Golden Acre, Pusa Mukta Pride of India, Copenhagen Market, Pusa Drum Head, September, Late Large Drum Head, Pusa Ageti, BSS-50 (Bajrang), BSS 32 |
| | Ganesh Gol Nath Laxmi 401, Bahar, Pragati, Unnati, Kalyani, Kranti, Hari Rani Gol, Hero, Mitra, Aditya, Yamuna, Ganga, Kaveri, Quisto, Margan, Meenaxi, Kuwaxi, Vishesh Uttam, Autum Queen, Green Ball, |

| Vegetable crops | Improved cultivars |
| --- | --- |
| | Green Boy, Green Express, Stone Head, Green Challenger, H44, H113, H64, Sel 507-4, Sel 528 and Sel 507-22. Drum Head Savoy, Pusa Mukta, Red Cabbage. |
| Brussels sprouts | Improved Long Island, Early Morn, Dwarf Improved, Frontier Zwerg, Kvik, Hilds Ideal, Rubine, Jade Cross, Pearl Crystal, Doreman, Predora, Rovoka Sonara, Ladosa, Oliver, Valiant, Rogor, Rider, Boxer, Richard, Royal Marvel, Asgard, Kundry, Rasmunda, Stephen. |
| Sprouting broccoli | Pusa Broccoli Kt. Sel 1, Palam Samridhi, Punjab Broccoli 1, Packman, Pirate, Brigadier, Capoven, Zues, Hi Cream, Mars, Green Lofi, Tendan, Land Star, Green Mail, Green Dome, Shigmeri. |
| Knol-khol | White Vienna, Large Green, King of North, Purple Vienna. |
| Radish | Pusa Desi, Pusa Reshmi, Pusa Chetki, Japanese White, Punjab Safed, Arka Nishant, Chinese Pink,. White Icicle, Pusa Himani, Rapid Red White Tipped, Scarlet Globe, Scarlet Long. Janupuri Giant (Newari), Kalyani White, Nadauni. |
| Carrot | Pusa Kesar, Pusa Meghali Sel No. 29, Sel. No. 233, Nantes Half Long, Early Nantes, Chantaney, Danvers, |

| Vegetable crops | Improved cultivars |
| --- | --- |
| | Zero, Ooty-1, Pusa Yamdagini, Imperator. |
| **Beet root** | Detroit Dark Red, Crimson Globe, Crosby Egyptian, Early Wonder. |
| **Turnip** | Pusa Chandrima, Pusa Swarnima Pusa Kanchan, Pusa Sweti, Pusa Top White Globe, Golden Ball, Early Milan Red Top, Snow Ball, Punjab Safed-4 Purple Top White Globe. |
| **Onion** | Pusa Red, Pusa Ratnar, Pusa Madhavi, Punjab Selection, N53, Baswant 780, Arka Niketan, Arka Kalyan, Arka Bindu, Agrifound Light Red, Agrifound Dark Red, Agrifound Rose, Udaipur 101, Udaipur 103, Hissar 11, Kalyanpur Red Round, Punjab Red Round, Pusa White Flat, Pusa White Round, N 2-4-1, Early Grano, Brown Spanish, Udipur 102, Agrifound Red, Arka Pragati. |
| **Garlic** | Agrifound White, Yamuna Safed, G-282, Agrifound Parvati, Godavari, Sweta. |
| **Pea** | Asauji, Arkel, Jawahar Matar 3, Jawahar Matar 4, Bonneville, Arka Ajit, Jawahar Matar 1, Jawahar matar 2, UN 56 (6) Harbhajan, Lincon, Pant Uphar, P 88. |

| Vegetable crops | Improved cultivars |
| --- | --- |
| French bean | Kentucky Wonder, Contender, Pusa Parvati, Arka Komal. TKD1, KKL1, YED 1, Bountiful Jampa, Lakshmi, Pant Anupam, Premier. |
| Cowpea | Pusa Phalguni, Pusa Barsati, Pusa Dofasali, Arka Garima, Yard Long Bean, Sel 263, Arka Suman, Narendra Lobia, Pusa Komal, Pusa Rituraj, Philippines Early. |
| Cluster bean | Pusa Mausami, Pusa Sadabahar. |
| Indian bean | Pusa Early Prolific, Arka Jay, Arka Vijay, Konkan Bhushan, Rajni, Deepali. |
| Lima bean | King of Garden, Karolina Butter, Challenger, Florida Butter, Hopi, Wilbur, Handerson Bush, Burpee Bush, Ford Hook 242, Baby Potato, Baby Ford Hook. KKL 1, Putter Beans. |
| Broad bean | Aquadule Claudin, Imperial White Long Pod, Masterpiece Green Long Green, Imperial Green Long Pod, Red Epicure, Imperial White Windsor, Giant Four Seeded, Green Windsor, Imperial Green Windsor. |
| Winged bean | IIHR Selections, 20, 60, 71, WBC 2. |
|  | Unnat, Azad Kranti, Punjab Padmini Varsha, Vijay Adhunik, Panchali, Nath Sobha, Red Bhindi, Supriya, Varsha. |

| Vegetable crops | Improved cultivars |
|---|---|
| **Muskmelon** | Pusa Sharbati, Pusa Madhuras, Pusa Rasraj, Hara Madhu, Punjab Sunheri, Punjab Hybrid ($F_1$), Punjab Rasila, Durgapur Madhu, Arka Rajhans, Arka Jeet, Hissar Madhur, Hissar Saras, Pusa Sharbati. |
| **Cucumber** | Japanese Long Green, Straight Eight, Pusa Sanyog, Poinsett, Himangi, Phule Shubbangi, Solan Hybrid, Poonakhira, Sheetal, Arka Jyoti. |
| **Watermelon** | Asahi Yamato, Sugar Baby, Arka Jyoti, Arka Bedana, Arka Manik, Improved-Shipper, Durgapur Meetha, Durgapur Kesar, Asahi Yamato, New Hampshire, Midget, Pusa Bedna, Special No.1. |
| **Bottle gourd** | Pusa Summer Prolific Long, Pusa Summer Prolific Round, Pusa Meghdoot, Pusa Manjari, Pusa Naveen, Pusa Komal, Arka Bahar, Kalyanpur Long Green, Samrat, Pusa Hybrid 3, Kalyanpur Hari Lambi. NDBG 1, PBOG, Phule BTG 1, Punjab Long, Punjab Komal, Punjab Round Rajendra Chamatkar, Pusa Hybrid 1. |
| **Bitter gourd** | Pusa Do Mausami, Pusa Vishesh, Coimbatore Green, Coimbatore Long Round, Priya, Preethi, |

| Vegetable crops | Improved cultivars |
|---|---|
| | Priyanka, Arka Harit, Harkani, Phule Green, Konkan Tara. |
| Summer squash | Punjab Chhappan Kaddu, Patty PAN, Early Yellow Prolific, Australian Green, Pusa Alankar. |
| Snake gourd | TA 19, Konkan Sweta, APAU Sweta. |
| Pumpkin | Arka Chandan, Ambili, Pusa Vishwas, Pusa Vikas, Pusa Hybrid 1. |
| Pointed gourd | Chhota Hilli, Dandali, Hilli, Shankolia, Swarna Alaukik, Swarna Rekha, CHES Elite Line. |
| Ash gourd | Mudliar, APAU Shakti. |
| Chow-chow | Round White, Long White, Pointed Green, Broad Green, Oval Green, Creamy Green. |
| Ridge gourd | Pusa Nasdar, Satputia, Konkan Harita, Punjab Sadabahar, PKMI, IIHR 8. |
| Sponge gourd | Pusa Chikni, Phule Prajakta, Pusa Supriya, Pusa Sneha. |
| Round melon | Arka Tinda, Tinda Ludhiana, Tinda Tonk, Tamil Nadu Selection. |
| Potato | Kufri Sinduri, Kufri Chandramukhi, Kufri Jyoti, Kufri Badshah, Kufri Bahar, Kufri Lalima, Kufri Swarna, Kufri Jawahar, Kufri Ashoka, Kufri Sutlei, Kufri Pukhraj, Kufri |

| Vegetable crops | Improved cultivars |
| --- | --- |
|  | Chipsona. Kufri 'Giriraj, Kufri, Muthu, Kufri, Lauvkar, Kufri Deva, Kufri Megha. |
| Sweet potato | Kalmegh, Sree Bhadra, Varsha, Sree Vardhini, Samrat, Kiran, VL Sakarkand, Gouri, Sankar, Sree Nandini, Bhuban Sankar, Rajendra Sakarkand. |
| Indian spinach | All Green, Pusa Palak, Pusa Jyoti, Pusa Harit, Jobner Green, Banerjee Giant |
| Spinach | Virginia Savoy, Early Smooth Leaf, Banarsi, Khara Lucknow, Khara Palak. |
| Lettuce | Great Lakes, Slowbolt, Chinese Yellow, Imperial 859, White Boston, Dark Green. |
| Amaranthus | Chhoti Chaulai, Badi Chaulai, Pusa Kiran, Pusa Kirit, Pusa Lal Chauli. |
| Fenugreek | Rajendra Kranti, Lam Sel1, RMT1, Pusa Early Bunching, Kasuri HM 57. |
| Cassava | Sree Vishakam, Sree Sahya, Sree Prakash, Sree Harsha, Sree Jaya, Sree Vijaya, Nidhi. |
| Yams | Sree Keerthi, Sree Roopa, Sree Shilpa, Sree Latha, Sree Kala, Sree Subhra, Sree Priya, Sree Dhanya. Sree Kanchan. |

| Vegetable crops | Improved cultivars |
| --- | --- |
| Asparagus | Martha Washington, May Washington, Reading Giant, Palmetto, Colossal, Argentenil, Berr's Mammoth, Mammoth White, Jersey Queen, Jersey Giant, New Jersey, Improved Perfection, Limbrass 22, Lucullas 1813. |
| Globe Artichoke | Green Globe, Purple Globe, Violet de Provence, Catanese, Spinosa Sarda, Bianco Tarantino, Precoce-di-Jesi, Bull, Tudella Brindisino, Romonesco. |
| Rhubarb | Victoria, McDonald, Ruby Valentine, Sunrise, Strawberry, Cherry Red. |
| Celery | Fork Hook Emperor, Standard Beared, Weight Grove Giant |
| Colocasia | Satmukhi, Sree Rashmi, Sree Pallavi, White Gauriya, Kaka Kachu, Panchamukhi, Saharshamukhi, Kadma, Muktakeshi, Nadia Local, Ahina Local, Telia, Jhankhri, White Gauriya. |
| Elephant's foot | Gajendra, Santragachi. |

# APPENDEX-7

## Distinguished characters of improved cultivar

| Vegetable variety/year of identification | Characteristics |
| --- | --- |
| **Brinjal (long)** | |
| 'Pusa Purple Long' (1975) | Susceptible to bacterial wilt, fruits long (20-25 cm) purple, glossy, suitable for ratooning. |
| 'Pusa Purple Cluster' (1975) | Moderately resistant to bacterial wilt, non-prickly, fruits borne in clusters of 4-9 fruits, 10-12 cm long and deep purple |
| 'PH 4' (1975) | Fruits long to medium, dark purple, flesh light-green. |
| 'Pusa Kranti' (1975) | Fruits uniform thick, oblong, 15-20 cm long, dark purple. |
| 'Pant Samrat' (1981) | Resistant to Phomopsis blight, tolerant to bacterial wilt, less infestation by shoot borers, fruit borers and Jassids. |
| 'Azad Kranti' (1983) | Non-prickly, fruits uniform thick, oblong, 15-20 cm long, dark purple with shining green calyx and less seeded. |
| **Brinjal (round)** | |
| 'Arka Navneet' (1981) | Fruit oval, deep purple, flesh soft white and a few seeded. |
| 'Pant Rituraj' (1985) | Field resistance to bacterial wilt, fruits round, less seeded, good flavour and keeping quality. |
| 'T3' (1975) | Fruits small, light purple, moderately resistant to little leaf and bacterial blight. |

| Vegetable variety/year of identification | Characteristics |
|---|---|
| 'Jamuni Gole Baingan' (1975) | Non-prickly, fruits shining purple, early, first harvest 654 days after planting. |
| **Brinjal (green)** | |
| 'Arka Kusumkar' (1981) | Fruits small, borne in clusters, good texture, skin light green. |
| 'Pusa Komal' (1981) | Plant bushy, 55-60 days duration, pods light green, non-fibrous, resistant to bacterial blight. |
| 'Bhagyalakshmi' (1975) | Fruit length 8.2 cm, width 0.7 cm, crop duration 118 days. |
| 'K 2' (1985) | Fruits pendant, length 6.1 cm, girth 4 cm, bright scarlet red fruits, tolerant to thrips. |
| **Cabbage** | |
| 'Selection 8' (1985) | Heads not perfectly round, slightly flat, moderately resistant to black rot, early. |
| **Cauliflower (group I)** | |
| 'Early Kunwari' (1975) | Very early, mid-September to mid-October maturity, curds hemispherical with even surface. |
| **Cauliflower (group II)** | |
| 'Pusa Deepali' (1975) | Early, late-October maturity, curds compact, self blanching, white, medium in size. |
| 'Pant Shubra' (1981) | Maturity 120 days, curd compact, slightly conical, non-ricy. |
| **Cauliflower (group III)** | |
| 'Improved Japanese' (1975) | Late-November to mid-December maturity, curds compact, white and large. |

| Vegetable variety/year of identification | Characteristics |
|---|---|
| 'Pusa Shubra' (1985) | Duration 125-130 days, resistant to black rot, average curd weight 700-800 g, highly tolerant to ricyness, curds compact and white. |
| **Cauliflower (group IV)** | |
| 'Pusa Snowball' (1975) | Late variety, suitable for cool season, optimum temperature for curd initiation and development 10-16°C, curds compact, medium snow white. |
| 'Pusa Snowball 2' (1975) | Late variety, curds ready by end of January (11-15°C) |
| 'Snowball 16' (1975) | Self blanched, curds snow white compact, maturity 90 days after planting. |
| 'Pusa Snowball' (K 1) (1983) | Self blanched, curds snow white compact, maturity 90-95 days after planting, resistant to black rot. |
| **French bean** | |
| 'VL Boni 1' (1985) | Dwarf, stem and leaves green, pods long, round, fleshy, stringless pole type, ready for harvest 45-60 days after sowing. |
| **Muskmelon** | |
| 'Hara Madhu' (1975) | Fruits large, round, slightly tapering at stalk-end, average fruit weight 1 kg, flesh green, crisp, very sweet. |
| 'Pusa Sharbati' (1975) | Fruit medium round to oval, with green stripes on outer skin, flesh thick orange with cantaloupe flavour, less juicy. |
| 'Pusa Madhuras' (1975) | Fruit weight 1 kg, flesh salmon orange, juicy, rich in flavour, very sweet (12-14 per cent TSS) |

| Vegetable variety/year of identification | Characteristics |
|---|---|
| 'Arka Rajhans' (1975) | Fruit weight 1.25-2.0 kg, flesh thick, white, firm texture, sweet, moderately resistant to powdery mildew, excellent keeping quality. |
| 'Punjab Hybrid' (1985) | Fruits light yellow, light green sutures, rind netted, flesh orange, juicy, sweet, suitable for distant transportation, moderately resistant to powdery mildew. |
| 'Arka Jeet' (1975) | Fruits small, flat in shape, round, orange yellow, flesh white, very sweet, medium soft texture, excellent flavour, high Vitamin C content. |
| 'Durgapur Madhu' (1975) | Fruits medium oblong, light green, flesh light green, juicy, very sweet. |
| **Okra** | |
| 'Sel.2' (1985) | Suitable for summer and *kharif,* fruits green, long, 5-edged and tender, resistant to yellow-vein mosaic. |
| **Onion (red)** | |
| 'Punjab Sel.' (1975) | Bulb globular, quite firm, good keeping quality, good for dehydration, tolerant to purple blotch and thrips. |
| 'Pusa Ratnar' (1975) | Bulbs bronze deep red, obovate to flat globular, less pungent, maturity 125 days after planting. |
| 'Pusa Red' (1975) | A short to indeterminate day length type, bulbs medium in size, bronze in colour, flat to globular, less pungent, good keeping quality, maturity 125-140 days after planting. |

| Vegetable variety/year of identification | Characteristics |
|---|---|
| 'N 2-4-1' (1985) | Bulb colour light red, globe shaped, pungent, TSS 11.3 per cent, good keeping quality, tolerant to Alternaria blight and thrips. |
| **Onion (white)** | |
| 'Pusa White Round' (1975) | Suitable for dehydration, TSS 12-13 per cent. |
| 'Pusa White Flat' (1975) | Bulbs medium to large, TSS 12-14 per cent, suitable for dehydration, good keeping quality. |
| 'S 48' (1975) | Suitable for *rabi,* bulb flatish round, bulb weight 80 g, suitable for dehydration, good keeping quality. |
| 'N 257-9-1-' (1985) | Globe shaped bulbs with light red colour, good keeping quality, suitable for *rabi* |
| **Peas** | |
| 'Arkel' (1975) | Pods dark green, 8.5 cm long, 7-8 seeds/pod, green seeded, first picking 60-65 days after sowing, suitable time-October. |
| 'Bonneville' (1975) | Pods borne in doubles, pod light green, 8 cm long, 5-7 seeds/pod, seeds green and sweet, first picking 80-85 days after sowing. |
| 'Jawahar Matar 1' (1975) | Seed green, wrinkled, 8-9 seeds/pod. |
| 'Jawahar Matar 4' (1975) | Pods green, seeds wrinkled, suitable for early sowing. |
| 'Early December' (1977) | Pod length 7-8 cm, 5 seeds/pod, bushy plant habit. |
| 'Pant Uphar' (1985) | Pod length 7-8 cm, susceptible to powdery mildew, first picking 75-80 days after sowing. |
| 'P 88' (1985) | First flower 72 days after sowing, 5-7 seeds/pod, mature seeds bold and wrinkled, susceptible to powdery mildew. |

| Vegetable variety/year of identification | Characteristics |
|---|---|
| **Tomato** | |
| 'Pusa Ruby' (1975) | Fruit small to medium sized, indeterminate plant type, suitable for year round planting, 25 to 30 fruits/plant. |
| 'HS 101' (1975) | Determinate plant type, suitable for *rabi*, round fruits, small to medium sized, fruits in cluster of 2-3. |
| 'S 12' (1975) | Semi-determinate, potato leaves, resistant to nematode, suitable for summer and winter. |
| 'T 1' (1975) | Indeterminate, fruit round, 4-5 fruits/truss, non-cracking, red on ripening, susceptible to tobacco mosaic virus. |
| 'Sweet 72' (1975) | Large fruited, uniform maturity, fruit flatish round, green stem end, slightly furrowed. |
| 'Pusa Early Dwarf' (1975) | Semi-determinate, fruit roundish, medium large, uniform red, ribbed suitable for rainy season. |
| 'Sioux' (1977) | Indeterminate, fruits round, smooth, medium to large, maturity 60-70 days after planting. |
| 'Pusa Gaurav' (1983) | Determinate, fruits oblong, two locules/fruit, uniform ripening, maturity 80-85 days after transplanting, excellent for processing and long transportation. |
| Punjab Chhuhara' (1981) | Determinate, fruits medium-sized, pear-shaped, firm, fleshy, usually bilocular, less seedy, less sour, most suitable for long transportation. |

| Vegetable variety/year of identification | Characteristics |
|---|---|
| 'KS 2' (1985) | Fruit flatish round, slightly furrowed, 4-5 locules/fruit, moderate yielder. |
| 'Pant Bahar' (1985) | Indeterminate, fruits ripen 78 days after transplanting, resistant to *Verticillium* and *Fusarium* wilts, good storage and processing qualities. |
| **Watermelon** | |
| 'Durgapura Madhu' (1975) | Elongate, fruit yellowish coloured, thin skin, light green with very attractive flavoured flesh, TSS 13-15 per cent, average fruit weight 0.4-0.7 kg, picking 90-95 days after sowing. |
| 'Sugar Baby' (1975) | Fruit small to medium, round, skin dark green, flesh deep red, fine texture, very sweet, TSS 10-12 per cent |
| 'Arka Jyoti ($F_1$)' (1985) | $F_1$ hybrid between 'Gomoon Sweet' and Indian variety, fruits round, skin light green with dark green strips, flesh deep pink, excellent texture, very sweet with excellent flavour. |

# KEY TO MULTIPLE CHOICE

| | | |
|---|---|---|
| 1. | (a) | Pusa Meghdoot |
| 2. | (d) | Pusa Harit |
| 3. | (b) | 2 or 3 |
| 4. | (c) | 13,300 |
| 5. | (d) | Japanese |
| 6. | (a) | Pole type |
| 7. | (a) | Chayote |
| 8. | (b) | Chow-chow |
| 9. | (c) | 33.3% |
| 10. | (c) | Mattar Ageta-6 |
| 11. | (b) | Knol-khol |
| 12. | (c) | 1972 |
| 13. | (b) | 5-8 |
| 14. | (d) | 50,000 |
| 15. | (d) | 300 g |
| 16. | (a) | Cruciferae |
| 17. | (d) | All of the above |
| 18. | (c) | 15 to 20 |
| 19. | (b) | Amaranthaceae |
| 20. | (b) | Cross-pollinated |
| 21. | (a) | $C_3$ |
| 22. | (a) | Vitamin A |
| 23. | (b) | Watermelon |
| 24. | (c) | Carrot |
| 25. | (a) | Greenhouses |
| 26. | (c) | Kufri Chandramukhi |
| 27. | (b) | Mithi Phali |
| 28. | (a) | Pusa Sanyog |
| 29. | (d) | 400-500 m |
| 30. | (d) | 400 m |
| 31. | (d) | Bhindi |
| 32. | (a) | Muskmelon |
| 33. | (b) | Watermelon |
| 34. | (b) | Powdery mildew |
| 35. | (b) | Winter squash |
| 36. | (a) | Early season |
| 37. | (b) | Dioecious |

| | | |
|---|---|---|
| 38. | (d) | All of the above |
| 39. | (c) | Perennial |
| 40. | (c) | Three years |
| 41. | (c) | Virus |
| 42. | (a) | 21°C to 23°C |
| 43. | (b) | Taiwan |
| 44. | (b) | Tricolor |
| 45. | (a) | X-rays |
| 46. | (b) | Cross-pollinated |
| 47. | (a) | 2 to 6 seeds |
| 48. | (c) | 12 months |
| 49. | (b) | 15-25°C |
| 50. | (b) | 4.5°C |
| 51. | (c) | Pusa Sawani |
| 52. | (c) | Solasodine |
| 53. | (d) | Potato |
| 54. | (b) | 0°C |
| 55. | (c) | Fungus |
| 56. | (b) | Bacterial disease |
| 57. | (b) | Seed |
| 58. | (c) | Non-parasitic cause |
| 59. | (a) | Calcium |
| 60. | (a) | Coast of Malabar |
| 61. | (a) | Africa |
| 62. | (c) | Melongena |
| 63. | (c) | India and Africa |
| 64. | (a) | 60 x 30 - 45 cm |
| 65. | (b) | Gamma rays |
| 66. | (b) | Maharashtra |
| 67. | (b) | Faba bean |
| 68. | (b) | Acidic soil |
| 69. | (a) | B |
| 70. | (c) | Turnip |
| 71. | (c) | Unfavourable temperature and water supply |
| 72. | (b) | Kufri Alankar |

| | | |
|---|---|---|
| 73. | (a) | Sudden heavy irrigation after a dry spell |
| 74. | (b) | 60 x 45 cm |
| 75. | (b) | Capitata |
| 76. | (c) | 0°C at 90-95% R.H. |
| 77. | (b) | N, P |
| 78. | (d) | Self-incompatibility |
| 79. | (a) | Green sprouting |
| 80. | (b) | Sweet pepper |
| 81. | (c) | Umbelliferae |
| 82. | (c) | K |
| 83. | (a) | Insects |
| 84. | (a) | Carotene |
| 85. | (b) | Cross-pollinated |
| 86. | (b) | Below 10°C |
| 87. | (a) | Above 25°C |
| 88. | (a) | Higher temperature |
| 89. | (d) | 3.9 |
| 90. | (c) | Virus |
| 91. | (a) | 0°C with 85-90 per cent R.H. |
| 92. | (a) | Mediterranean coast |
| 93. | (c) | 15-20°C |
| 94. | (b) | Umbelliferae |
| 95. | (c) | 30 days |
| 96. | (b) | Lima bean |
| 97. | (d) | Entire fruit |
| 98. | (b) | Nemagon |
| 99. | (d) | Spinach |
| 100. | (a) | Spring |
| 101. | (a) | Drying and canning |
| 102. | (b) | 2000 m |
| 103. | (b) | PAU (Ludhiana) |
| 104. | (c) | Cyamopsis |
| 105. | (c) | Mannogalacton |
| 106. | (c) | 45 x 20 cm |
| 107. | (a) | Self-pollinated |
| 108. | (a) | Aspargine |
| 109. | (c) | White |
| 110. | (b) | Mid |
| 111. | (a) | French bean |
| 112. | (a) | Zn |
| 113. | (c) | Central Africa |
| 114. | (b) | Cowpea 263 |
| 115. | (b) | A period of drought followed by sudden watering |
| 116. | (c) | Wind |
| 117. | (b) | Aphids |
| 118. | (c) | 5.5 and 6.7 |
| 119. | (a) | *Cucurbita pepo* |
| 120. | (a) | Peru |
| 121. | (d) | Esculentum |
| 122. | (b) | January to February |
| 123. | (c) | 80°C and 30% R. H. |
| 124. | (c) | 13°C to 21°C |
| 125. | (b) | Whiptail |
| 126. | (b) | Pithiness |
| 127. | (a) | 3-4 |
| 128. | (b) | 50°C |
| 129. | (c) | Varanasi |
| 130. | (c) | .16 |
| 131. | (b) | Fungus |
| 132. | (a) | Mid-September to mid-November |
| 133. | (b) | 40 days after sowing |
| 134. | (c) | Mesocarp, endocarp, placentae |
| 135. | (b) | Mesocarp |
| 136. | (a) | Pericarp and placentae |
| 137. | (c) | Flower bud |
| 138. | (b) | Second |
| 139. | (b) | Third |
| 140. | (a) | A and B |
| 141. | (c) | Hollow-heart |
| 142. | (c) | Pusa Alankar |
| 143. | (b) | Broad bean |

| | | |
|---|---|---|
| 144. | (a) | 25 - 30 days |
| 145. | (b) | Karperenko |
| 146. | (a) | 50-60 days |
| 147. | (b) | 45 - 50 days |
| 148. | (c) | Late |
| 149. | (b) | Stem |
| 150. | (d) | GA₃ |
| 151. | (d) | All of the above |
| 152. | (c) | Hermaphrodite |
| 153. | (b) | 116 g |
| 154. | (d) | Red ripe stage |
| 155. | (d) | PNR-7 |
| 156. | (c) | 50°C |
| 157. | (c) | 2500 ppm |
| 158. | (a) | Thimet 10G |
| 159. | (a) | Four days before transplanting |
| 160. | (c) | Green mature stage |
| 161. | (b) | Mature green stage |
| 162. | (a) | Potato-Onion-Green manure |
| 163. | (b) | Sandy loam |
| 164. | (a) | October-November |
| 165. | (a) | 15-20 to 1 |
| 166. | (c) | 45 x 30 |
| 167. | (c) | 15 x 7.5 cm |
| 168. | (c) | 3 ppm |
| 169. | (c) | 45 x 7.5 cm |
| 170. | (b) | 25-50 ppm |
| 171. | (b) | 10°C |
| 172. | (c) | 15-20°C |
| 173. | (b) | 10°C |
| 174. | (c) | 2000-2500 |
| 175. | (a) | 20-25 |
| 176. | (c) | Black carrots |
| 177. | (c) | Hard ripe stage |
| 178. | (a) | August-September |
| 179. | (a) | 60 x 45 cm |
| 180. | (b) | 10 x 15 cm |
| 181. | (a) | 800 m |
| 182. | (a) | Last week of August |
| 183. | (c) | 70 per cent |
| 184. | (c) | South and Central America |
| 185. | (c) | Tweed Wonder |
| 186. | (b) | 0°C and 90-95% R.H. |
| 187. | (c) | Painted bug |
| 188. | (a) | Long and medium style |
| 189. | (b) | 60-80 days |
| 190. | (c) | Cucurbitacin |
| 191. | (c) | Karathane |
| 192. | (b) | Chenopodiaceae |
| 193. | (b) | Beta |
| 194. | (c) | Below 10°C |
| 195. | (b) | Hortense |
| 196. | (d) | 1700 |
| 197. | (c) | Sativum |
| 198. | (a) | Multiple bulb |
| 199. | (a) | 10-20 |
| 200. | (b) | - 4 |
| 201. | (d) | Jamunanagar Local |
| 202. | (a) | Suckers |
| 203. | (b) | Potato |
| 204. | (a) | Crisp head lettuce |
| 205. | (b) | Decrease in nutritional quality |
| 206. | (c) | Bush type |
| 207. | (a) | Greshof and Doy (1972) |
| 208. | (a) | 4-6 |
| 209. | (b) | 3-4 weeks |
| 210. | (a) | End of December or first week of January |
| 211. | (a) | 60 x 30-45 |
| 212. | (b) | Satputia |

213. (a) I.D. Tyagi (1973)
214. (a) 75
215. (a) 5.3-6.0
216. (b) Root cuttings
217. (c) By earthing
218. (b) India
219. (a) BH-2
220. (b) CH-1
221. (b) 2 to 3 months
222. (c) Certified seed
223. (c) Boron
224. (c) October-November
225. (a) Choudhary and Singh (1971)
226. (b) Pointed gourd
227. (c) Flowers and fruits
228. (c) 700
229. (a) Methionine
230. (b) Hills
231. (b) Higher altitudes in the hills
232. (c) February-March
233. (b) June-July
234. (a) High temperature
235. (c) 2 per cent
236. (d) 4-5
237. (a) Degree days
238. (b) 20-30
239. (d) 20-25
240. (a) Potato
241. (a) Excessive watering and fertilization
242. (d) 350
243. (c) 10 quintals
244. (b) Pusa Ruby
245. (b) October-November

246. (c) India
247. (c) Cantaloupes
248. (b) PCPA
249. (d) 20-25
250. (b) Dull sound when fruit is thumped
251. (c) Tubers
252. (b) Second
253. (c) Hyacinth bean
254. (a) Self-pollinated
255. (a) Self-incompatible
256. (a) 1000 m
257. (b) 10
258. (d) Stem cutting
259. (b) Radish
260. (c) Tubers
261. (b) Tuberous root
262. (b) Dioecious
263. (b) Corniculata
264. (d) Seed
265. (b) Gongylodes
266. (b) Late blight
267. (d) Kufri Red and Kufri Kundan
268. (c) Whitefly
269. (b) Crispa
270. (b) *Karam sag*
271. (c) Branched stem and leaves
272. (a) Seed
273. (b) 12 to 15°C
274. (a) Self-pollinated crop
275. (b) 5.8 to 6.6
276. (c) 30°C
277. (b) Lunatus
278. (b) Twice
279. (b) Guatemala

| | | |
|---|---|---|
| 280. | (a) | 85-90 days |
| 281. | (a) | Mycoplasma |
| 282. | (c) | Leaf-hopper |
| 283. | (b) | Rainy season |
| 284. | (a) | North India |
| 285. | (c) | Calcium |
| 286. | (a) | 30°C |
| 287. | (b) | September-October |
| 288. | (a) | 30 x 75 cm |
| 289. | (a) | 0°C to 1°C and 90% R.H. |
| 290. | (d) | 400 m |
| 291. | (a) | Cylindrica |
| 292. | (a) | Upper portion |
| 293. | (a) | Monoecism |
| 294. | (d) | All of the above |
| 295. | (a) | Maharashtra and Karnataka |
| 296. | (b) | Carrot fly |
| 297. | (c) | Stem and root |
| 298. | (b) | Watermelon |
| 299. | (b) | 6.0 and 6.7 |
| 300. | (c) | North-West India |
| 301. | (c) | Punjab Sunehri |
| 302. | (c) | 3 |
| 303. | (c) | 15-16 |
| 304. | (c) | 10-15 |
| 305. | (a) | 6-8 |
| 306. | (b) | September-October |
| 307. | (a) | Pusa Meghdoot |
| 308. | (c) | 45 x 75 cm |
| 309. | (c) | Often-cross-pollinated |
| 310. | (d) | 20 |
| 311. | (d) | 600 |
| 312. | (d) | 850 |
| 313. | (d) | 2470 |
| 314. | (b) | 240 |

| | | |
|---|---|---|
| 315. | (a) | 700-1000 |
| 316. | (a) | 460-500 |
| 317. | (a) | 600 |
| 318. | (d) | All |
| 319. | (a) | Hissar-2 |
| 320. | (c) | 3-4 |
| 321. | (c) | Agrifound Dark Red |
| 322. | (c) | 5.5 and 6.6 |
| 323. | (b) | 6.0 - 6.8 |
| 324. | (b) | 5.8 - 6.7 |
| 325. | (c) | 5.5 and 6.5 |
| 326. | (b) | 6.0 - 7.0 |
| 327. | (b) | 20-30°C |
| 328. | (c) | 15°C and 85-90% R. H. |
| 329. | (b) | 30 |
| 330. | (c) | 30°C |
| 331. | (b) | October |
| 332. | (a) | Maharashtra and Karnataka |
| 333. | (a) | Carotene |
| 334. | (b) | Africa |
| 335. | (b) | 2.2 per cent |
| 336. | (c) | Saline-alkaline |
| 337. | (d) | All of the above |
| 338. | (d) | Sylvia |
| 339. | (c) | Punjab Chhuhara |
| 340. | (a) | Kharif |
| 341. | (d) | 70-100 kg |
| 342. | (b) | Asparagus |
| 343. | (a) | Malaysia |
| 344. | (a) | Melo |
| 345. | (b) | Pusa Rituraj |
| 346. | (c) | Petiole |
| 347. | (a) | Spears |
| 348. | (b) | 2, 4- Dichlorophenoxy acetic acid |
| 349. | (b) | April, August and January |

| | | |
|---|---|---|
| 350. | (d) | Tender |
| 351. | (c) | Basellaceae |
| 352. | (c) | Cucumber |
| 353. | (c) | Dioecious plant |
| 354. | (d) | Vine cutting |
| 355. | (a) | 13°C |
| 356. | (b) | By wind |
| 357. | (b) | Honey bees |
| 358. | (b) | Japan |
| 359. | (b) | 60 x 15-20 cm |
| 360. | (b) | Long-day |
| 361. | (b) | South America |
| 362. | (c) | C |
| 363. | (a) | Tubers |
| 364. | (b) | 20 |
| 365. | (d) | Kufri Sinduri |
| 366. | (c) | Cucurbits |
| 367. | (c) | High early yield |
| 368. | (b) | Physiological disorder |
| 369. | (b) | 6.0-6.5 |
| 370. | (c) | America |
| 371. | (c) | 110-120 days |
| 372. | (c) | Late |
| 373. | (b) | October-November |
| 374. | (a) | Sponge gourd |
| 375. | (a) | *Palak* |
| 376. | (a) | Carrot |
| 377. | (c) | Both summer and rainy season |
| 378. | (b) | A hybrid variety of muskmelon |
| 379. | (b) | Tenderometer |
| 380. | (b) | Raphanus |
| 381. | (a) | Cross-pollinated |
| 382. | (b) | June-July |
| 383. | (b) | June-July |
| 384. | (d) | Karperenko |
| 385. | (a) | Long thin pods |

| | | |
|---|---|---|
| 386. | (d) | Early and good quality sprouts |
| 387. | (a) | Thick leaf stalks |
| 388. | (a) | Shallow |
| 389. | (c) | Brinjal |
| 390. | (a) | Nucleus seed |
| 391. | (a) | Early |
| 392. | (a) | Punjab Selection |
| 393. | (c) | October-November |
| 394. | (b) | 6.0 - 7.0 |
| 395. | (c) | Perennial |
| 396. | (d) | 4-5 years |
| 397. | (a) | 1200 m |
| 398. | (b) | Cabbage |
| 399. | (a) | 8-12 |
| 400. | (c) | Trenches |
| 401. | (b) | 1.5 - 2.0 m x 45 - 60 cm |
| 402. | (d) | 50:50 |
| 403. | (a) | October-November |
| 404. | (b) | Dioecious |
| 405. | (b) | Round |
| 406. | (a) | 2-3 |
| 407. | (c) | 1500 m |
| 408. | (b) | Cauliflower |
| 409. | (b) | 15°C |
| 410. | (a) | Early spring |
| 411. | (b) | Mid-May to end of June |
| 412. | (c) | South-East Asia |
| 413. | (c) | 45 x 15 cm |
| 414. | (c) | 45 x 30 |
| 415. | (b) | 75 x 30 cm |
| 416. | (a) | Vitamin A |
| 417. | (c) | Both |
| 418. | (b) | South-West Asia |
| 419. | (c) | 65 days |
| 420. | (d) | 75 days |
| 421. | (c) | February-March |

| | | |
|---|---|---|
| 422. | (d) | 2500 ppm |
| 423. | (a) | Miniature cabbage |
| 424. | (c) | India |
| 425. | (d) | 26.6 |
| 426. | (d) | 16 |
| 427. | (d) | Vine cuttings |
| 428. | (c) | Tropical America |
| 429. | (b) | Spinach |
| 430. | (c) | Ridomil MZ |
| 431. | (c) | Pusa Himani |
| 432. | (a) | Smooth sponge gourd |
| 433. | (d) | Petioles |
| 434. | (b) | 1200 m |
| 435. | (a) | 200 m |
| 436. | (c) | January-February |
| 437. | (b) | November |
| 438. | (a) | Broad bean |
| 439. | (a) | Cluster bean |
| 440. | (b) | Broad bean |
| 441. | (a) | 6.5 |
| 442. | (c) | 6.0-7.0 |
| 443. | (a) | 60 |
| 444. | (b) | 15-20°C |
| 445. | (b) | 18-24°C |
| 446. | (c) | Evening |
| 447. | (a) | Calcium oxalate |
| 448. | (c) | Moderately tolerant |
| 449. | (a) | Peru and Mexico |
| 450. | (c) | A and C |
| 451. | (c) | Based on methods of culture |
| 452. | (d) | 80-90 |
| 453. | (c) | Orange |
| 454. | (c) | Light orange |
| 455. | (b) | *C. moschata* |
| 456. | (a) | 10-25 ppm |

| | | |
|---|---|---|
| 457. | (a) | India |
| 458. | (d) | All of the above |
| 459. | (a) | Pusa Dofasli |
| 460. | (b) | Mustard |
| 461. | (a) | 45 x 15-20 |
| 462. | (d) | Shoot |
| 463. | (c) | Swollen stem |
| 464. | (d) | Solitary and hermaphrodite |
| 465. | (c) | Pale yellow |
| 466. | (d) | Deep yellow |
| 467. | (b) | Morning |
| 468. | (a) | 1/3rd |
| 469. | (a) | Lycopersicon |
| 470. | (d) | Rutin |
| 471. | (c) | Andhra Pradesh |
| 472. | (b) | 5.2 to 6.4 |
| 473. | (c) | Thrip |
| 474. | (a) | 0.5 km |
| 475. | (d) | Excellent |
| 476. | (c) | Andhra Pradesh |
| 477. | (b) | 30 cm |
| 478. | (d) | 0.5% |
| 479. | (b) | 20°C |
| 480. | (a) | Kitchen garden |
| 481. | (c) | Cucumber |
| 482. | (b) | Aphid |
| 483. | (a) | Amaranth |
| 484. | (a) | Detriot Dark Red |
| 485. | (a) | 6.0-7.5 |
| 486. | (c) | Andhra Pradesh |
| 487. | (c) | Green coloured |
| 488. | (a) | Lettuce |
| 489. | (c) | Sugar Baby |
| 490. | (a) | Yellow vein mosaic |
| 491. | (a) | Common bean mosaic |
| 492. | (c) | Leaf curl |
| 493. | (b) | Mid-February |

| | | |
|---|---|---|
| 494. | (b) | Aphid |
| 495. | (c) | Europe |
| 496. | (d) | 650 |
| 497. | (b) | 10°C |
| 498. | (c) | 22°C |
| 499. | (d) | 27°C |
| 500. | (b) | Cracking |
| 501. | (b) | Anthocyanin |
| 502. | (c) | India |
| 503. | (c) | Africa |
| 504. | (b) | To get off season yield |
| 505. | (b) | Punjab and Kashmir |
| 506. | (c) | Ethiopia |
| 507. | (d) | Capsicin |
| 508. | (b) | 115 g |
| 509. | (a) | Capsanthin |
| 510. | (b) | 8-10 |
| 511. | (b) | 2-3 |
| 512. | (a) | 8-10 |
| 513. | (c) | 15-20 |
| 514. | (c) | 4-5 kg |
| 515. | (c) | 25-30 kg |
| 516. | (b) | January-February |
| 517. | (a) | 2½ months old |
| 518. | (c) | Within seven days from transplanting |
| 519. | (b) | Cross-pollination |
| 520. | (a) | North India |
| 521. | (c) | 30 x 20 cm |
| 522. | (c) | 45 x 30 cm |
| 523. | (b) | 40-50 g |
| 524. | (d) | HCN |
| 525. | (c) | 24 |
| 526. | (c) | 5-7 |
| 527. | (a) | One day before transplanting |

| | | |
|---|---|---|
| 528. | (a) | Less |
| 529. | (c) | 14-20 days |
| 530. | (b) | 400-500 g |
| 531. | (d) | 30-40 kg |
| 532. | (c) | 2.5-3.5 |
| 533. | (b) | 400-500 g |
| 533. | (c) | EMS |
| 534. | (b) | Gamma rays |
| 535. | (b) | Gamma rays |
| 536. | (d) | Orange to yellow |
| 537. | (a) | White |
| 538. | (b) | Pusa Bedana |
| 539. | (a) | Mid-August to beginning of September |
| 540. | (c) | 20°C |
| 541. | (b) | 30°C |
| 542. | (c) | Cruciferae |
| 543. | (a) | Root vegetable |
| 544. | (d) | All of the above |
| 545. | (a) | Floating garden |
| 546. | (c) | Six |
| 547. | (c) | 4 |
| 548. | (d) | Arka Bahar |
| 549. | (c) | Punjab No.8 |
| 550. | (b) | Pusa Chikni |
| 551. | (b) | Arka Suryamukhi |
| 552. | (b) | Punjab Barsati |
| 553. | (a) | Okra |
| 554. | (a) | Retourt |
| 555. | (c) | Vitamins |
| 556. | (a) | Blanching |
| 557. | (a) | Tomato |
| 558. | (a) | 60 x 30 cm |
| 559. | (c) | Aphids |
| 560. | (a) | White flies |
| 561. | (a) | Zinc |

| | | |
|---|---|---|
| 562. | (c) | Lanatus |
| 563. | (d) | All |
| 564. | (b) | 5.0 |
| 565. | (b) | 21°C |
| 566. | (c) | Hardy |
| 567. | (a) | Annual |
| '568. | (a) | 1 : 1 |
| 569. | (d) | Orange yellow |
| 570. | (b) | Shimla |
| 571. | (a) | Ovary, corolla and stigmatic lobes |
| 572. | (d) | Tomato |
| 573. | (c) | Onion |
| 574. | (c) | Knol-khol |
| 575. | (a) | Pusa Ketaki |
| 576. | (b) | Brinjal |
| 577. | (b) | Okra |
| 578. | (d) | 1000 m |
| 579. | (b) | Brinjal |
| 580. | (a) | Pride of India |
| 581. | (b) | Arka Jyoti |

| | | |
|---|---|---|
| 582. | (b) | M.H. |
| 583. | (a) | 150 g |
| 584. | (b) | Sweet potato-Stolon |
| 585. | (b) | Bitter gourd |
| 586. | (d) | Short-duration legume vegetables |
| 587. | (c) | Purple blotch |
| 588. | (d) | Agrifound Light Red |
| 589. | (b) | Sandy loam |
| 590. | (c) | Vegetable garden for processing |
| 591. | (a) | Spinach |
| 592. | (d) | Potato |
| 593. | (c) | Bitter gourd |
| 594. | (a) | Pea |
| 595. | (d) | Japanese |
| 596. | (d) | Deficiency of molybdenum |
| 597. | (a) | Early |
| 598. | (d) | All |
| 599. | (b) | Turnip |
| 600. | (c) | Vitamin A |

# KEY TO FILL IN THE BLANKS

1. High, long day
2. Blossom-end rot, sunscald
3. Pusa Sanyog
4. Home/kitchen garden
5. Wrinkle
6. Red, white
7. Black leg, black rot
8. Market
9. Brown heart
10. Pusa Jyoti
11. Arkel
12. Africa
13. Temperature, day length
14. India
15. Khol-rabi
16. Beet leaf
17. Black mould
18. Calabreeze
19. Leaf curl
20. Leek, stem and leaves
21. *Chyote*
22. Four
23. Punjab-8
24. Punjab Chhuhara
25. Chow-chow
26. Pruning
27. Pusa Harit
28. *Amaranthus tricolour, A. blitum*
29. Immature, mature, *petha*
30. Low, cool
31. High temperature, dry wind
32. Arka Tinda
33. Defoliation
34. Pusa Bedana
35. Co 1, Co 2
36. Sweet-potato weevil (*Cylas formicarius*)
37. Arka Suryamukhi
38. Parthenocarpically
39. Wilted
40. European
41. *Myzus persicae*
42. Red-skinned
43. Long melon
44. Pusa Savani
45. boron, molybdenum
46. 1.5
47. Satputia
48. maleic hydrazide
49. high, long
50. persistant
51. 60-80
52. 115, 70
53. broad bean
54. coleworts
55. Spermatophyta
56. vegetable
57. A&B
58. $C_4$
59. Colocasia blight
60. A.P. Maharashtra, Mysore, T.N.
61. Kentucy Wonder
62. Mithi Phali
63. pumpkin
64. flower buds
65. medicinal
66. spears
67. dioecious
68. 3, 10-15
69. 30ºC
70. winter, summer
71. several
72. Chenopodiaceae

73. 0ºC, 90%
74. 80ºC, 30%
75. 15-25
76. 2.2, 75
77. wormicidal
78. topical
79. bower
80. Dithane Z-78
81. heart rot
82. *Cuccumis melo, utilissimus*
83. *Citrullus vulgaris, fistulosus*
84. *caudatus*
85. *Momordica charantia*
86. frost
87. Africa
88. 60 t/ha
89. long warm season
90. perennial, annual
91. India
92. South America
93. flower parts
94. Brachium
95. Belgium
96. 250 sqm
97. cross
98. Potassium
99. 6.0-6.8
100. Cruciferae
101. A
102. annual, biennial
103. Iron
104. molybdenum
105. acidity
106. 0, 85-90
107. hills
108. 5.5-6.7
109. perennial, annual
110. A, C
111. 45-60 x 20-30
112. stem cutting
113. Japan
114. warm
115. puffy
116. boron
117. acidic soil
118. 18-24ºC
119. 1.5-2.5, 60-90
120. cultivated
121. aphids
122. vine crop
123. yard long bean, catjung, southern pea
124. upper
125. whiptail
126. 285
127. thiourea
128. South America
129. tendril
130. 8-10
131. four
132. mid May, end of June
133. mid-September, mid-November
134. *esculentum, pimpinelifolium*
135. appetizing
136. *Amorphophallus compannulatus*
137. Gujrat
138. 100-150
139. blanching
140. protein
141. east
142. powdery
143. nitrogen, atmosphere
144. mineral, protein, Vit A, Vit C
145. fibrous mass
146. 15-25 cm
147. 1%, 1-2

148. male sterile
149. degree days, heat units
150. 25 x 35 cm, 100
151. tropical
152. 75-90, 60-70
153. slow, suberized or cutinized
154. South and Central America
155. South America
156. CH-1
157. young, tender
158. very tender
159. tender
160. hyrids
161. amaryllidaceae
162. cool
163. cloves
164. Mediterranean
165. 3-4
166. watermelon, musk melon
167. above 90
168. maleness, femaleness
169. suckers/off shoots
170. round head, ball head
171. more
172. Vegetable forcing
173. 10-13, 85-90
174. cauliflower
175. bursting/splitting
176. Lima bean
177. humidity
178. seed-to-seed
179. charcol rot
180. Pusa Alankar
181. Sun-scalding
182. heat and drought
183. April-May
184. head-to-seed
185. rainfed
186. netting and russeting
187. typha
188. boats
-189. April-May
190. 17th century
191. high
192. *kharif*
193. finger millet
194. June-July
195. wind
196. *Camote, Kumara*
197. root-to-seed
198. mid-October, end November
199. July-September, February-March
200. June-July
201. Africa
202. South Asia
203. 1822
204. stem cutting
205. oblong tubers
206. field
207. illuminant
208. tuberous root
209. gongylodes
210. mid-January, April
211. perishable
212. bigger
213. *porrum*
214. bulb
215. *compositeae*
216. 12-15⁰C
217. seeds
218. 30⁰C
219. November
220. tropical, sub-tropical
221. bitter
222. March-April
223. Maharashtra,
224. cool

225. 20-22°C
226. damping off
227. cucurbits, peas
228. Pusa Chetaki
229. monoecious, dioecious
230. day neutral
231. cucurbitacin
232. β-carotene, Vit A
233. north-west
234. December to March, April to May
235. Sarda Melon
236. gymnospermae
237. stakes, *Jhala*, thatch
238. September to mid-November
239. allicin
240. Maharashtra
241. amaryllidaceae
242. B
243. 750-1000 mm
244. seed, bulb
245. 15 x 7.5 cm
246. August, December- January
247. iodine
248. long, medium
249. 6.0-6.8
250. 5.8, 6.5
251. 30°C
252. cool, wet, cloudy
253. receptacle
254. India
255. Lignosus
256. hilly
257. anthrocyanin
258. Dec-Jan
259. coastal
260. perennial
261. dioecious, separate
262. Assam

263. heart & brain
264. stem, root
265. cuttings
266. nitrogen
267. Solanaceae
268. Nitrogen and Phosphorus$_s$
269. Incas
270. tuber, true seeds
271. short day
272. black heart
273. Spanish
274. scab
275. Maleic hydrazide
276. solid head
277. isothiocyanates
278. red
279. Tetra-2, Pusa Rassal
280. March, August
281. spring, rainy
282. fenugreek
283. mid-December, late-February
284. Norin
285. summer, rains
286. vegetable
287. F$_1$ hybrid
288. Mediterranean
289. cooler
290. thick leaf stalks
291. Siberia
292. muffled dull/dead
293. blossom end
294. less
295. less
296. sandy loam, organic matter
297. 45 x 45
298. onion
299. *Gnorimoschema operculelia*
300. rainy season
301. tropical, temperate

302. 45 x 30 cm
303. himalayas, 1200
304. higher altitudes
305. 20°C
306. monoecious, dioecious
307. tuber production
308. 2-3
309. archipelago
310. cluster
311. 75 x 30
312. Truck
313. cool
314. annual, biennial
315. India
316. flower initiation
317. miniature cabbage
318. 45-50
319. India
320. February, March
321. vegetable-marrow
322. American
323. 20
324. 30 cm, 45 cm
325. Tropical America
326. drought
327. acid
328. Colchicine
329. flowers, fruits
330. one year
331. *Cucurbita moschata*
332. *Citrullus vulgaris*
333. cabbage, Knol-khol
334. quercetin
335. seed-to-seed
336. bulb formation, enlargement
337. potato, water melon
338. Ethiopia
339. buttoning
340. evening, morning

341. cooling
342. good
343. powdery
344. Vit. A, iron
345. TMV, CMV, curly top
346. purple blotch
347. ordinary temperature
348. 6.0, 7.5
349. Y.V.M.V.
350. bacterial blight
351. red pumpkin beetle
352. Common bean mosaic
353. caulis, floris
354. Cantaloupe
355. Tien Shan
356. Peru
357. Africa
358. China, India
359. Ethiopia
360. Abyssinia
361. 1943
362. allyl-propyl-disulphide
363. dialyl disulphide
364. pods, vegetable
365. lycopine
366. 21°C
367. 125
368. cloves/bulblets
369. 1250
370. 1875-2500
371. 5-7
372. 7-10
373. normal flowers
374. stolon ends
375. floating garden
376. 40-50, 40-50
377. 45
378. 30 cm, 20 cm
379. tomato, cucumber

380. troquer, to barter
381. 250-300
382. 24⁰C, 18⁰C
383. 4
384. 18
385. indeterminate, determinate
386. 3-4
387. two
388. 1-15
389. fully ripe
390. 500-800
391. Portugese
392. 6.0, 7.0
393. trenches
394. December and early January
395. 20⁰C
396. calcium
397. 3-4
398. 20-30, 10-15
399. 20-25, 75-90

400. 2.2
401. farming
402. Calcium
403. leafy
404. *lanatus*
405. 1629
406. 5
407. *petha*
408. Chinese
409. Bonavist/Hyacinth bean
410. Mediterranean
411. yellow tinge
412. full slip stage
413. rectangular
414. European type
415. Knol-khol
416. north India
417. Mediterranean region
418. faba bean

# KEY TO TRUE OR FALSE

| # | | # | | # | | # | | # | |
|---|---|---|---|---|---|---|---|---|---|
| 1. | F | 39. | T | 77. | T | 115. | F | 153. | T |
| 2. | F | 40. | T | 78. | T | 116. | F | 154. | T |
| 3. | F | 41. | T | 79. | F | 117. | T | 155. | T |
| 4. | T | 42. | T | 80. | F | 118. | T | 156. | F |
| 5. | F | 43. | F | 81. | F | 119. | F | 157. | F |
| 6. | T | 44. | T | 82. | T | 120. | F | 158. | T |
| 7. | T | 45. | T | 83. | T | 121. | T | 159. | T |
| 8. | T | 46. | F | 84. | F | 122. | F | 160. | F |
| 9. | T | 47. | T | 85. | F | 123. | T | 161. | T |
| 10. | T | 48. | F | 86. | F | 124. | F | 162. | T |
| 11. | F | 49. | F | 87. | T | 125. | F | 163. | T |
| 12. | T | 50. | T | 88. | F | 126. | F | 164. | T |
| 13. | F | 51. | F | 89. | T | 127. | T | 165. | F |
| 14. | F | 52. | F | 90. | T | 128. | T | 166. | F |
| 15. | T | 53. | T | 91. | F | 129. | F | 167. | F |
| 16. | T | 54. | T | 92. | T | 130. | T | 168. | T |
| 17. | F | 55. | T | 93. | F | 131. | T | 169. | F |
| 18. | F | 56. | F | 94. | T | 132. | T | 170. | T |
| 19. | T | 57. | T | 95. | F | 133. | F | 171. | T |
| 20. | T | 58. | T | 96. | T | 134. | T | 172. | T |
| 21. | F | 59. | F | 97. | F | 135. | F | 173. | T |
| 22. | F | 60. | T | 98. | T | 136. | F | 174. | F |
| 23. | F | 61. | F | 99. | T | 137. | F | 175. | T |
| 24. | T | 62. | T | 100. | T | 138. | T | 176. | F |
| 25. | T | 63. | T | 101. | F | 139. | F | 177. | F |
| 26. | T | 64. | T | 102. | F | 140. | F | 178. | T |
| 27. | T | 65. | T | 103. | T | 141. | F | 179. | F |
| 28. | F | 66. | T | 104. | F | 142. | F | 180. | T |
| 29. | T | 67. | F | 105. | F | 143. | T | 181. | F |
| 30. | T | 68. | T | 106. | T | 144. | T | 182. | T |
| 31. | T | 69. | F | 107. | F | 145. | T | 183. | T |
| 32. | T | 70. | F | 108. | F | 146. | F | 184. | T |
| 33. | T | 71. | F | 109. | T | 147. | T | 185. | T |
| 34. | F | 72. | F | 110. | F | 148. | F | 186. | F |
| 35. | F | 73. | F | 111. | F | 149. | T | 187. | F |
| 36. | T | 74. | F | 112. | F | 150. | T | 188. | F |
| 37. | F | 75. | T | 113. | F | 151. | F | 189. | T |
| 38. | T | 76. | F | 114. | T | 152. | F | 190. | F |

| | | | | | | | | | |
|---|---|---|---|---|---|---|---|---|---|
| 191. | F | 230. | F | 269. | T | 308. | F | 347. | T |
| 192. | F | 231. | T | 270. | F | 309. | T | 348. | T |
| 193. | T | 232. | T | 271. | T | 310. | T | 349. | T |
| 194. | T | 233. | T | 272. | F | 311. | F | 350. | T |
| 195. | T | 234. | T | 273. | F | 312. | T | 351. | T |
| 196. | F | 235. | T | 274. | F | 313. | T | 352. | T |
| 197. | F | 236. | F | 275. | F | 314. | T | 353. | T |
| 198. | T | 237. | F | 276. | F | 315. | F | 354. | F |
| 199. | T | 238. | T | 277. | F | 316. | T | 355. | F |
| 200. | T | 239. | F | 278. | F | 317. | T | 356. | T |
| 201. | F | 240. | T | 279. | T | 318. | F | 357. | T |
| 202. | T | 241. | T | 280. | T | 319. | F | 358. | F |
| 203. | F | 242. | F | 281. | T | 320. | T | 359. | F |
| 204. | F | 243. | T | 282. | T | 321. | F | 360. | T |
| 205. | T | 244. | T | 283. | T | 322. | T | 361. | F |
| 206. | F | 245. | T | 284. | T | 323. | F | 362. | T |
| 207. | T | 246. | T | 285. | F | 324. | T | 363. | T |
| 208. | F | 247. | T | 286. | T | 325. | F | 364. | T |
| 209. | T | 248. | F | 287. | F | 326. | T | 365. | T |
| 210. | F | 249. | F | 288. | F | 327. | T | 366. | T |
| 211. | T | 250. | T | 289. | T | 328. | F | 367. | T |
| 212. | F | 251. | F | 290. | F | 329. | T | 368. | F |
| 213. | T | 252. | T | 291. | T | 330. | F | 369. | F |
| 214. | T | 253. | T | 292. | T | 331. | T | 370. | F |
| 215. | T | 254. | F | 293. | T | 332. | T | 371. | T |
| 216. | T | 255. | T | 294. | F | 333. | F | 372. | T |
| 217. | F | 256. | F | 295. | T | 334. | T | 373. | T |
| 218. | T | 257. | T | 296. | F | 335. | T | 374. | T |
| 219. | F | 258. | T | 297. | F | 336. | F | 375. | F |
| 220. | T | 259. | T | 298. | F | 337. | T | 376. | F |
| 221. | F | 260. | T | 299. | F | 338. | T | 377. | F |
| 222. | F | 261. | T | 300. | F | 339. | T | 378. | F |
| 223. | T | 262. | T | 301. | T | 340. | T | 379. | T |
| 224. | F | 263. | F | 302. | F | 341. | T | 380. | T |
| 225. | F | 264. | F | 303. | F | 342. | F | 381. | F |
| 226. | T | 265. | T | 304. | T | 343. | T | 382. | F |
| 227. | T | 266. | T | 305. | F | 344. | T | 383. | T |
| 228. | T | 267. | F | 306. | T | 345. | T | | |
| 229. | F | 268. | T | 307. | T | 346. | F | | |

# INDEX